Fundamentals of Fire Protection for the Safety Professional

Fundamentals of Fire Protection for the Safety Professional

Third Edition

DON PHILPOTT

Lanham • Boulder • New York • London

Published by Bernan Press
An imprint of The Rowman & Littlefield Publishing Group, Inc.
4501 Forbes Boulevard, Suite 200, Lanham, Maryland 20706
www.rowman.com

86-90 Paul Street, London EC2A 4NE

Copyright © 2022 by The Rowman & Littlefield Publishing Group, Inc.

All rights reserved. No part of this book may be reproduced in any form or by any electronic or mechanical means, including information storage and retrieval systems, without written permission from the publisher, except by a reviewer who may quote passages in a review. The Rowman & Littlefield Publishing Group, Inc., does not claim copyright on U.S. government information.

British Library Cataloguing in Publication Information Available

Library of Congress Cataloging-in-Publication Data

Names: Philpott, Don, author.
Title: Fundamentals of fire protection for the safety professional / Don Philpott.
Description: Third edition. | Lanham : Bernan Press, [2022] | Includes bibliographical references and index.
Identifiers: LCCN 2022024227 (print) | LCCN 2022024228 (ebook) |
  ISBN 9781641434751 (paperback) | ISBN 9781641434768 (epub)
Subjects: LCSH: Fire protection engineering.
Classification: LCC TH9146 .F47 2022 (print) | LCC TH9146 (ebook) |
  DDC 628.9/22—dc23/eng/20220613
LC record available at https://lccn.loc.gov/2022024227
LC ebook record available at https://lccn.loc.gov/2022024228

∞™ The paper used in this publication meets the minimum requirements of American National Standard for Information Sciences—Permanence of Paper for Printed Library Materials, ANSI/NISO Z39.48-1992.

Contents

1	Introduction to Industrial Fire Protection	1
2	Chemistry and Physics of Fire	19
3	Common and Special Hazards	45
4	Mechanical and Chemical Explosions	83
5	Building Construction	103
6	Life Safety in Buildings	129
7	Hazardous Processes	149
8	Alarm and Detection Systems	171
9	Fire Extinguishment	189
10	Fire Program Management	223
11	Creating a Comprehensive Emergency and Evacuation Plan	263
Appendix: Building Design Guidelines for Emergency Mitigation		313
Glossary		335
Solutions to Chapter Questions		375
Index		395

1

Introduction to Industrial Fire Protection

"Safety science" is a twenty-first-century term for everything that goes into the prevention of accidents, illnesses, fires, explosions, and other events that harm people, property, or the environment (ASSE and BCSP 2007, 3). Fire is a hazard that can strike any workplace. Of all the events that can strike, fire losses can be one of the greatest threats to an industrial organization in terms of financial losses, loss of life, loss of property, and property damage. Therefore, industrial fire protection and prevention are crucial components of any safety professional's job, be it serving as a loss control consultant or a safety manager. Fires can strike any type of workplace at any time, resulting in property damage, injuries, and deaths. The adverse financial effects can be felt by an organization long after the fire is extinguished.

FIRE PROTECTION AND FIRE PREVENTION
Industrial fire protection and prevention involves recognizing those situations that may result in an unwanted fire, evaluating the potential for an unwanted event, and developing control measures that can be used to eliminate or reduce those fire risks to an acceptable level. As is the case with any safety control measure, these controls can range from engineering strategies to administrative strategies or a combination of the two. Included in fire protection and prevention is emergency response. Emergency response involves organizing, training, and coordinating skilled employees about

emergencies such as fires, accidents, or other disasters (ASSE and BCSP 2007, 6).

It is important to distinguish between fire prevention and fire protection. Each term is unique, and the responsibilities of the safety professional for each aspect differ. Fire prevention is the elimination of the possibility of a fire being started. To start, every hostile fire requires an initial heat source, an initial fuel source, and something to bring them together (NFPA 2008, 3–11).

Prevention can occur through successful action on the heat source, the fuel source, or the behavior that brings them together (NFPA 2008, 3–12). Examples of programs that can be instituted in the workplace to prevent fires include housekeeping programs and inspection programs. Housekeeping can eliminate unwanted fuel sources and ignition sources. Inspection programs can effectively identify fire ignition and fuel hazards, then take appropriate steps to eliminate them.

Safety professionals also recognize that prevention will never be 100 percent successful. Therefore, it is necessary to plan and design to mitigate damages when fire occurs (NFPA 2008, 3–15). This process is referred to as fire protection. Fire protection strategies are the activities designed to minimize the extent of the fire. Fire protection includes reducing fire hazards by inspection, layout of facilities and processes, and design of fire detection and suppression systems (ASSE and BCSP 2007, 5).

It is important, therefore, to include in workplace fire safety planning considerations for fire suppression or extinguishment and evacuation of persons in the event of a fire emergency (www.osha.gov). Fire extinguishment systems include sprinkler systems, rated fire doors and walls, portable fire extinguishers, and standpipe hose systems. Evacuation of persons includes means of egress, detection-and-notification systems, and emergency planning and preparedness.

Fire protection requires the development of an integrated system of balanced protection that uses many different design features and systems to reinforce one another and to cover for one another in case of the failure of anyone (NFPA 2008, 3–15). The National Fire Protection Association (NFPA) describes fire protection as a series of six opportunities to intervene against a hostile fire, arrayed along a timeline of potential growth in fire severity (NFPA 2008, 3–16).

1. Prevent the fire entirely.
2. Slow the initial growth of the fire.
3. Detect fire early, permitting effective intervention before the fire becomes too severe.
4. Provide ability for automatic or manual suppression.
5. Provide ability to confine the fire in a space.
6. Move the occupants to a safe location.

Fire protection includes the use of active systems such as automatic detection systems and passive fire protection systems that stop fire and smoke (NFPA 2008, 3–16). As one can see, the activities designed to prevent fires are different from those activities geared toward minimizing the extent of the fire once it has occurred. Thus, an effective fire safety program requires both prevention and protection aspects. Throughout this book, fire prevention and fire protection strategies will be presented for a variety of common fire hazards found in the workplace.

WILDFIRES

Year 2020 was the worst year on record for wildfires with scores of fatalities, thousands of structures destroyed, and millions of acres burned across fifteen western states. Wildfires do not discriminate, and business and industrial structures are just as vulnerable as residential buildings. All fire protection and prevention planning must, therefore, consider the risks from wildfires and how to mitigate them.

According to the experts, the threat of wildfires will continue to grow with the flames fanned by global warming. August 2020 was the warmest on record for California, according to the National Oceanic and Atmospheric Administration (NOAH) and each of the past six years were at least 1 to 2 degrees Celsius (1.8 to 3.6 degrees Fahrenheit) warmer than the historical average.

By the beginning of October 2020, California had seen six of its twenty largest wildfires in history and two of its most deadly ones. More than 7,500 structures had been damaged or destroyed, and nearly 5,000 square miles (3.7 million acres) had been burned, according to Cal Fire.

Oregon and Washington states were also hit hard. Nearly 1,000 square miles had burned in Washington state. In Oregon, 500,000 people faced evacuation as fires raged and more than 1,500 square miles had burned, according to Governor Kate Brown.

The Threat

Wildfires can break out at any moment, putting businesses and homes at risk. A single spark from a lightning strike, a fallen electrical pole, a discarded cigarette, or a campfire can quickly erupt into a massive wildfire that can cause irreparable damage. Even worse, over the past three years, many of the most damaging fires were deliberately lit.

Fighting wildfires requires massive amounts of resources. Reuters reported that the deadly Camp Fire in northern California ended up as the single most expensive natural disaster in the United States in 2018, costing $16.5 billion.

Wildfires are also different from other natural disasters. Wildfires have a unique profile compared to other natural disasters such as hurricanes and tornadoes. They come with little or no warning and they can last for weeks or even months. Wildfires can also be started by humans, either accidentally or maliciously, at any time or anywhere.

Because of the aforementioned factors, it can be a challenge to take preemptive steps other than taking general precautions like clearing brush and trimming trees near structures. However, a fast-moving wildfire fanned by high winds can easily jump fire breaks and engulf a structure in seconds. In these cases, the only option is to evacuate and ensure that everyone gets out safely. That is why you must have an evacuation plan as part of your emergency planning procedures (For more information on evacuation planning, see chapter 11).

Planning

If you live in an area that could be vulnerable to wildfires, it is important to have an effective early warning system in place. Wildfires can spread rapidly—up to 15 miles an hour—therefore early detection is critical to effect a successful evacuation. Early warning will also tell you in which direction the wildfire is moving so that you can take the safest evacuation route. A threat monitoring system is the most effective way to detect wildfires as quickly as possible. You can set up the system to warn you automatically—via SMS, email, or push notification—whenever a wildfire emerges that could impact your people or business (by forecasting the spread of the fire and cross-referencing your business and employees' locations).

Impacts

Wildfires can impact your property, equipment, assets, and people. Even if your structure is not affected by the wildfire you may have to evacuate, and business and productivity will be lost. Following the wildfire, you may not have immediate access to critical supplies to restart production as roads in the area may have been damaged and remain closed. If you are in the direct line of the wildfire, do you have plans to protect or evacuate critical equipment?

People, of course, must always be your top priority. If your building is in the line of the wildfire, you must evacuate your staff. Even if your structure is not at risk, your employees are likely to live nearby and their homes may be threatened, and they and their families may be ordered to evacuate, significantly reducing your workforce.

If you are forced to close for quite a while, you may lose business as your customers switch to other suppliers.

There are other consequences following major wildfires. Staff may not come in to work as they look after their families or repair their homes. They may even move out of the area altogether. Transportation routes—both road and rail—can also be affected causing problems with receiving supplies and shipping orders. Utilities are also usually a casualty of wildfires, especially power plants affected by downed lines. Unless you have your own generator, you probably cannot start operations again until the power is restored.

Many businesses do not have the liquidity to withstand significant shutdowns and increased expenses as a result of a major wildfire, so the consequences can be fatal.

Mitigation

There is little you can do to stop a raging wildfire but there are steps that you can take to mitigate its impact. As discussed earlier, an early warning system is essential so that you can act on the threat as soon as possible. It will also keep you informed about the latest developments and, most importantly, help you keep your employees protected by a safe and orderly evacuation.

You should secure your supply chain and have plans in place with vendors and suppliers to ensure continuation of supplies as soon as you are back up and running. Have alternate sources available in case your suppliers themselves are impacted and are unable to deliver. Finally, communicate regularly with your employees, suppliers, and customers so that everyone is kept informed.

IMPORTANCE OF FIRE SAFETY

Protecting the workplace from fires is a major job responsibility of safety managers. Not only do they have to ensure that the property is adequately protected to prevent catastrophic financial losses to the organization, but there is also the moral obligation to protect the workers and members of the community from the devastating effects that a fire can have upon the entire community. Over the years, there have been numerous examples of the effects that industrial fires have had upon both the workers involved and the communities in which they occurred. Some examples of the largest and most devastating industrial fires in the United States include the Triangle Shirtwaist Company fire, the Imperial Foods processing plant fire, the Phillips Petroleum chemical plant explosion and fire, and the Crescent City, Illinois, fire.

HISTORICAL MAJOR FIRE LOSSES IN THE UNITED STATES

Throughout U.S. history, several major fires have resulted in significant loss of life and property damage. Table 1.1 summarizes these events (NFPA 2008, 3–10).

Trends in Fires in the United States

From a historical perspective, fire death rates have steadily decreased over the years. In the United States, fire and burn fatality rates were at their peak in 1918 with 9.9 deaths per 100,000 persons which have now fallen to a low of 1.2 fatalities per 100,000 persons in 1998 (NFPA 2008, 3–6).

When examining the fire impact of industrial occupancies, the NFPA is the foremost authority on keeping track of reported fires, fire injuries, and property losses. From the deadliest manufacturing fire in U.S. history, the 1911 Triangle Shirtwaist factory fire (146 fatalities) to today, the United States has made steady progress in improving fire and life safety in the manufacturing sector (NFPA 2008, 3–26). However, there is still much work to be done as can be seen from more recent fires such as Imperial Foods fire in Hamlet, North Carolina, in 1991 that killed twenty-five workers and injured fifty-four (OSHA 2013).

INTRODUCTION TO INDUSTRIAL FIRE PROTECTION

Table 1.1. Ten Deadliest U.S. Fires and Explosions through 2020

Fires	Number of Deaths
World Trade Center, New York, September 11, 2001	2,666
S.S. Sultana steamship boiler explosion, April 27, 1865	1,547
Forest Fire, Peshtigo, WI, October 8, 1871	1,152
General Slocum excursion steamship fire, New York, June 15, 1904	1,030
Iroquois Theatre, Chicago, IL, December 30, 1903	602
Cloquet fire, Northern MN, October 12, 1918	559
Cocoanut Grove nightclub, Boston, MA, November 28, 1942	492
S.S. Grandcamp and Monsanto Chemical Company, Texas City, TX, April 16, 1947	468
Forest fire, Hinckley MN, September 1, 1894	418
Monongah Mine coal mine explosion, Monongah, WV, December 6, 1907	361

Economic Impact of Industrial Fires in the United States

The total cost of fire includes the losses that fire causes, such as human losses (e.g., lives lost, medical treatment of injuries, pain and suffering) and economic losses (e.g., property damage, business interruption), and the cost of provisions to prevent or mitigate the cost of fire, such as fire departments, insurance, and fire protection equipment and construction (Hall 2013, iii).

These losses can be further classified as direct losses and indirect losses. Direct losses also include estimated fires that occur but go unreported. Indirect loss refers to costs of temporary housing, missed work, and lost business, and it may also refer to intangible losses, such as heirlooms or pets (Hall 2013, 2). The total loss due to fires in the United States in 2014 was $328.5 billion (Zhuang 2017). Broken down by category, these losses were $53 billion in economic losses, $41.9 billion in fire department expenditures, $23.6 billion in net fire insurance expenditures, and $57.4 billion in building construction costs (Zhuang 2017).

The economic impact of fires in industry can be felt as losses in business sales and production. Employment can be impacted as workers who once worked in a facility involved in a fire may lose time at work waiting for operations to get back up and running. In some cases, they may never return to work since the business is forced to close following the fire. Thus, individuals involved and their family members suffer the loss of income. The municipality suffers losses in taxes since workers are now not bringing in an income. Loss of income means decreased spending in local businesses and, as a result, lower sales tax revenues to the local government as well. Because unemployment can increase from the loss of work or jobs in the area, increased demands upon social services are experienced. Finally, if the company ceases to exist following a major fire, unemployed workers may have no alternative but to move to another area to seek employment, resulting in population shifts in a region.

INTERNATIONAL FIRE EXPERIENCE

Worldwide, from 1979 to 2018, fire death rates per million population have consistently fallen throughout the industrialized world, with North American and Eastern European regions' fire death rates falling faster than other regions (USFA 2011, 1), while during this period, the fire death rate in the United States declined by 66 percent. Today, the United States still has one of the higher fire death rates in the industrialized world; however, its standing has greatly improved (USFA 2011, 1). Table 1.2 provides the international fire death rates for select countries (WHO 2018).

THE INDUSTRIAL FIRE EXPERIENCE IN THE UNITED STATES

A variety of agencies keep statistics on the fire experience in the United States. These agencies include the NFPA and the BLS (Bureau of Labor Statistics). The NFPA, for example, found that in 2018, public fire departments in the United States responded to 1,318,500 fires of which 499,999 were structure fires, compared with 1,098,000 in 1977. Only 22 percent of structure fires in 2018 involved nonresidential structures.

Table 1.2. 2018 International Fire Death Rates per Million Population

Country	Rate
Switzerland	1.8
Singapore	0.9
Austria	2.5
Italy	2.0
Netherlands	1.6
Australia	2.0
Spain	2.1
Germany	2.0
United Kingdom	3.2
New Zealand	3.8
France	3.8
Slovenia	3.2
Canada	5.0
Sweden	3.7
United States	8.0
Ireland	5.5
Denmark	6.2

INTRODUCTION TO INDUSTRIAL FIRE PROTECTION

The average number of structure fires, casualties, and direct property damage per year from 2011 through 2015 in industrial and manufacturing properties are listed in table 1.3 (NFPA Research, Quincy).

Leading ignition sources in industrial and manufacturing structure fires include shop tools such as torches, heat-treating equipment, and industrial furnaces and kilns, followed by heating equipment such as boilers, space heaters, and chimney and flue fires. A summary of industrial fire ignition sources appears in table 1.4.

The BLS maintains data on the occupational injuries, illnesses, and fatalities in the United States involving fires. Data from the BLS identified the following occupational injuries and fatalities due to fires and explosions in the U.S. workplace. Fires and explosions accounted for approximately 1,510 occupational injuries in 2010 with the majority occurring in the trade, transportation, and utility industries (see figure 1.1). In 2011, there were 144 occupational fatalities, with the largest number of deaths occurring in the trade, transportation, and utility industries, with 52 deaths, followed by the manufacturing industry with 30 fatalities (figure 1.2).

OCCUPATIONAL SAFETY AND HEALTH ADMINISTRATION AND FIRE SAFETY

As the federal government agency responsible for setting the national standards for worker safety and health, Occupational Safety and Health Administration (OSHA) has established standards addressing each of the three key elements of fire safety: (1) fire prevention, (2) safe evacuation from the workplace in the event of fire, and (3) protection of workers who fight fires or who work around fire suppression equipment. These issues are addressed by a variety of detailed OSHA rules applicable to general industry (all businesses except construction, shipbuilding, and longshoring) in 29 C.F.R. § 1910 (OSHA 2013). OSHA has adopted several standards dealing with fire protection and prevention in the workplace. The following list summarizes the 29 C.F.R. § 1910 standards for fire protection and prevention (OSHA 2013):

- General Industry
- 1910, Occupational Safety and Health Standards
- 1910.36, Design and construction requirements for exit routes

Table 1.3. Structure Fires in Industrial and Manufacturing Properties by Property Use, 2011–2015 Annual Averages

Property Use	Fires		Civilian Deaths		Civilian Injuries		Direct Property Damage (in Millions)	
Utility, Defense, Agriculture, Mining	**2,690**	**(35%)**	**2**	**(22%)**	**43**	**(20%)**	**$259**	**(32%)**
Agriculture	880	(11%)	0	(0%)	7	(3%)	$50	(6%)
Unclassified utility, defense, agriculture, mining	630	(8%)	0	(3%)	11	(5%)	$36	(5%)
Utility or distribution system	490	(6%)	1	(11%)	10	(4%)	$47	(6%)
Energy production plant	200	(3%)	0	(0%)	4	(2%)	$60	(8%)
Forest, timberland, or woodland	200	(3%)	0	(4%)	1	(1%)	$2	(0%)
Laboratory	140	(2%)	0	(0%)	9	(4%)	$11	(1%)
Defense, computer or communications center	90	(1%)	7	(4%)	1	(1%)	$51	(6%)
Mine or quarry	50	(1%)	0	(4%)	1	(1%)	$2	(0%)
Manufacturing or processing	**5,080**	**(65%)**	**6**	**(78%)**	**176**	**(80%)**	**$540**	**(68%)**
Total	**7,770**	**(100%)**	**8**	**(100%)**	**219**	**(100%)**	**$799**	**(100%)**

Table 1.4a. Structure Fires in Industrial Properties by Equipment Involved in Ignition, 2011–2015 Annual Averages

Equipment Involved in Ignition	Fires		Civilian Deaths		Civilian Injuries		Direct Property Damage (in Millions)	
Electrical distribution and lighting equipment	640	(24%)	2	(100%)	4	(9%)	$142	(55%)
Wiring and related equipment	360	(14%)	2	(100%)	4	(9%)	$21	(8%)
Transformers and power supplies	180	(7%)	0	(0%)	0	(0%)	$119	(46%)
Lamp, bulb or lighting	70	(2%)	0	(0%)	0	(0%)	$1	(1%)
Cord or plug	20	(1%)	0	(0%)	0	(0%)	$1	(0%)
Heating equipment	420	(16%)	0	(0%)	7	(16%)	$11	(4%)
Heat lamp	130	(5%)	0	(0%)	1	(3%)	$2	(1%)
Fixed or portable space heater	90	(4%)	0	(0%)	3	(6%)	$4	(2%)
Confined fuel burner or boiler fire	90	(3%)	0	(0%)	0	(1%)	$0	(0%)
Central heat	30	(1%)	0	(0%)	0	(0%)	$1	(1%)
Confined chimney or flue fire	30	(1%)	0	(0%)	0	(0%)	$0	(0%)
Fireplace or chimney	20	(1%)	0	(0%)	3	(6%)	$4	(1%)
Water heater	10	(1%)	0	(0%)	0	(0%)	$0	(0%)
Other known heating equipment	20	(1%)	0	(0%)	0	(0%)	$0	(0%)
No equipment involved in ignition	390	(15%)	0	(0%)	4	(10%)	$30	(12%)
Contained trash or rubbish fire	380	(14%)	0	(0%)	0	(1%)	$0	(0%)
Cooking equipment	120	(4%)	0	(0%)	0	(0%)	$0	(0%)
Confined cooking fire	97	(4%)	0	(0%)	0	(0%)	$0	(0%)
Other known cooking equipment	20	(1%)	0	(0%)	0	(0%)	$0	(0%)
Torch, burner, or soldering iron	100	(4%)	0	(0%)	12	(28%)	$14	(5%)
Unclassified equipment involved in ignition	90	(3%)	0	(0%)	1	(3%)	$6	(2%)
Fan	40	(2%)	0	(0%)	0	(0%)	$0	(0%)
Confined incinerator overload or malfunction fire	40	(1%)	0	(0%)	1	(2%)	$0	(0%)
Conveyor	40	(1%)	0	(0%)	1	(3%)	$6	(2%)
Confined commercial compactor fire	30	(1%)	0	(0%)	1	(3%)	$0	(0%)
Hay processing equipment	30	(1%)	0	(0%)	0	(0%)	$2	(1%)
Pump	30	(1%)	0	(0%)	0	(0%)	S2	(1%)
Silo loader, unloader, or screw/sweep auger	20	(1%)	0	(0%)	0	(0%)	$0	(0%)
Unclassified laboratory equipment	20	(1%)	0	(0%)	1	(2%)	$1	(1%)
Industrial furnace or kiln	20	(1%)	0	(0%)	0	(0%)	$1	(0%)
Power sander grinder, buffer, or polisher	20	(1%)	0	(0%)	0	(0%)	$2	(1%)

Table 1.4b. Structure Fires in Industrial Properties by Equipment Involved in Ignition, 2011–2015 Annual Averages (Continued)

Equipment Involved in Ignition	Fires		Civilian Deaths		Civilian Injuries		Direct Property Damage (in Millions)	
Heat-treating equipment	10	(1%)	0	(0%)	0	(0%)	$17	(6%)
Other known equipment involved in ignition	250	(9%)	0	(0%)	10	(23%)	$24	(9%)
Total	2,690	(100%)	2	(100%)	43	(100%)	$259	(100%)

Note: Non-confined fires in which the equipment involved in ignition was unknown or not reported have been allocated proportionally among fires with known equipment involved. The same approach was used with confined cooking fires. Fires in which the equipment involved in ignition was entered as none but the heat source indicated equipment involvement or the heat source was unknown were also treated as unknown and allocated proportionally among fires with known equipment involved. Non-confined fires in which the equipment was partially unclassified (i.e., unclassified kitchen or cooking equipment, unclassified heating, cooling or air conditioning equipment, etc.) were allocated proportionally among fires in that grouping (kitchen or cooking equipment; heating, cooling or air conditioning equipment, etc.). The same approach was used with confined cooking fires. The estimated of fires involving fireplace or chimney include all fires with the confined chimney or flue incident type regardless of what may have been coded as equipment involved. Likewise, the estimated of fires involving furnaces, central heat or boilers include all fires with confined fuel burner or boiler incident type. The estimates shown should be considered upper bounds. Non-cooking confined fires were not analyzed separately. Estimates of other types of equipment exclude confined fires. Sums may not equal totals due to rounding errors. Fires are rounded to the nearest ten civilian deaths and injuries to the nearest one, and direct property damage is rounded to the nearest million dollars.
Source: NFIRS and NFPA Fire Experience Survey.

FIGURE 1.1

INTRODUCTION TO INDUSTRIAL FIRE PROTECTION 13

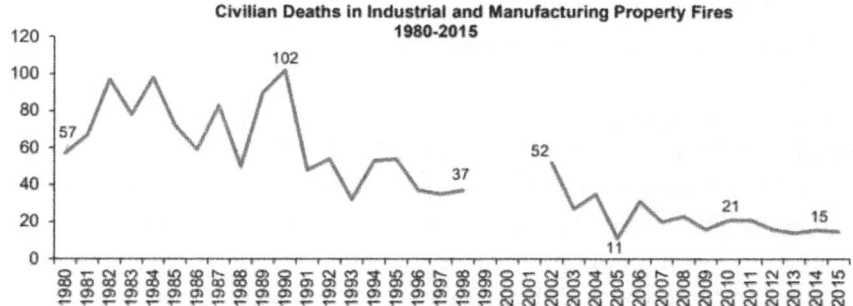

FIGURE 1.2

- 1910.37, Maintenance, safeguards, and operational features for exit routes
- 1910.38, Emergency action plans
- 1910.94, Ventilation
- 1910.103, Hydrogen
- 1910.104, Oxygen
- 1910.106, Flammable and combustible liquids
- 1910.107, Spray finishing using flammable and combustible materials
- 1910.108, Dip tanks containing flammable or combustible liquids
- 1910.109, Explosives and blasting agents
- 1910.110, Storage and handling of liquefied petroleum gases
- 1910.111, Storage and handling of anhydrous ammonia
- 1910.119, Process safety management of highly hazardous chemicals
- 1910.120, Hazardous waste operations and emergency response
- 1910.178, Powered industrial trucks
- 1910.252, General requirements
- 1910.253, Oxygen-fuel gas welding and cutting
- 1910.255, Resistance welding
- 1910.261, Pulp, paper, and paperboard mills
- 1910.263, Bakery equipment
- 1910.265, Sawmills
- 1910.266, Logging operations
- 1910.269, Electric power generation, transmission, and distribution
- 1910.272, Grain-handling facilities
- 1910.305, Wiring methods, components, and equipment for general use

- 1910.308, Special systems
- 1910.1200, Hazard communication (HAZCOM)

SOURCES OF INFORMATION
Occupational Safety and Health Administration

OSHA is responsible for the promulgation of federal legislation for safety and health in the workplace. It also has the responsibility for enforcing the standards it develops. All private employers with one or more employees are required to comply with OSHA standards applicable to their industry. Some states have state plans in which the state is responsible for developing and enforcing its safety and health standards.

National Fire Protection Association

NFPA is an international nonprofit organization established in 1896 with a mission of reducing the worldwide burden of fire and other hazards on the quality of life by providing and advocating consensus codes and standards, research, training, and education (NFPA 2013). The first set of standards developed by the NFPA can be traced back to their work in 1895 and 1896 on sprinkler system standards. At the time nine radically different sprinkler system standards were in effect within 100 miles of Boston. Seeing a need for consistency, the founding group of twenty members released sprinkler system rules "Report of Committee on Automatic Sprinkler Protection" (NFPA 1995, 97). Today, the NFPA develops, publishes, and disseminates more than 300 consensus codes and standards intended to minimize the possibility and effects of fire and other risks with membership totaling more than 50,000 individuals around the world (NFPA 201). It is also one of the largest contributors to the body of fire protection and prevention research and knowledge in the world.

United States Fire Administration

In 1973, the Commission on Fire Prevention and Control issued its report "America Burning," which documented the extent to which resources and lives were being lost to unnecessary fires in the United States (National Commission on Fire Prevention and Control 1973, iii). One of the many recommendations contained in the report was the formation of the United States Fire Administration (USFA). As an entity of the Department of Homeland

INTRODUCTION TO INDUSTRIAL FIRE PROTECTION

Security's Federal Emergency Management Agency, the mission of the USFA is to provide national leadership to foster a solid foundation for our fire and emergency services stakeholders in prevention, preparedness, and response (USFA 2013).

SUMMARY

A safety professional is a person engaged in the prevention of accidents, incidents, and events that harm people, property, or the environment (www.bcsp.org). Fire protection and prevention are just two of the many responsibilities expected of the safety professional working with industrial occupancies. The potential for a fire is always present in the workplace. The risk and the severity of a hazard in terms of losses and injuries in the workplace can vary greatly from industry to industry. These fire hazards can be properly controlled with appropriate measures.

The prevention and control of fires can be achieved through a variety of techniques, including planning a safe workplace through the selection of appropriate tools and equipment, proper work practices designed to prevent the contact of fuel sources with heat sources, the control of fires once they have started with proper fire protection and identification systems, and, finally, the protection of building occupants and the community through providing proper notification and appropriate means of egress.

History has shown us that when fire protection is not a primary concern of the organization or when appropriate measures are not taken to prevent fires and protect the workers, the results can be devastating. While there has been much improvement in the fire losses experienced in the U.S. workplace, history has also shown us that major fires and loss of life still occur and are still possible today.

Note. Building and fire codes and regulations do change and can vary from state to state and even county to county. It is incumbent on the security professional to ensure they are familiar with all the latest requirements.

CHAPTER QUESTIONS
1. Compare the fire death rates in the United States to the fire death rates in other industrialized countries.
2. Effective fire prevention requires which three components?
3. Describe the opportunities one has to intervene in a fire.

4. Differentiate between fire protection and fire prevention.
5. Describe some of the job activities a fire protection engineer may engage in.
6. Describe the trends in fire death experience in the United States since World War I.
7. What impact do fires have upon occupational fatalities and injuries in the United States?
8. What are some major sources of ignition in the workplace based on reported fires?
9. Describe some aspects of fire safety that OSHA has regulated in the workplace.
10. Describe OSHA's three key elements of fire safety.

REFERENCES

American Society of Safety Engineers (ASSE) Foundation and Board of Certified Safety Professionals (BCSP). (2007). *Career Guide to the Safety Profession*, 3rd ed. Des Plaines, IL: ASSE and BCSP.

Evarts, Ben. (2012). *Fire in U.S. Industrial and Manufacturing Facilities*. Quincy, MA: NFPA.

Hall, John R., Jr. (2013). *The Total Cost of Fire in the United States*. Quincy, MA: NFPA, Fire Analysis and Research Division, August 2013.

National Commission on Fire Prevention and Control. (1973). *America Burning*. Washington, DC: National Commission on Fire Prevention and Control.

National Fire Protection Association (NFPA). (2008). *Fire Protection Handbook*, 20th ed. Quincy, MA: NFPA.

NFPA. (2013). Overview of the NFPA. Available at http://www.nfpa.org/about-nfpa/overview (accessed November 1, 2013).

NFPA. (1995). The men who made the NFPA. NFPA Journal, May/June, 97.

Occupational Safety and Health Administration (OSHA). (2013). Available at www.osha.gov (accessed October 21, 2013).

NFPA. (2018). *Fires in Industrial and Manufacturing Properties*. Richard Campbell, NFPA Research March 2018.

U.S. Bureau of Labor Statistics. (2011). *Table R64: Number of Nonfatal Occupational Injuries and Illnesses Involving Days Away from Work by Event or Exposure Leading to Injury or Illness and Industry Division, 2010*. Washington, DC: Bureau of Labor Statistics.

U.S. Bureau of Labor Statistics. (2012). *Table A-9. Fatal Occupational Injuries by Event or Exposure for All Fatal Injuries and Major Private Industry Sector, All U.S., 2011*. Washington, DC: Bureau of Labor Statistics.

U.S. Fire Administration (USFA). (2011). *Fire Death Rate Trends: An International Perspective*. Topical Fire Report Series. Volume 12, Issue 8. Emmitsburg, MD: U.S. Fire Administration.

USFA. (2013). About the United States Fire Administration. Available at: http://www.usfa.fema.gov/about/ (accessed November 1, 2013).

Jun Zhuang, Vineet M. Payyappalli, Adam Behrendt, and Kathryn Lukasiewicz. (2017). Total Cost of Fire in the United States. Department of Industrial and Systems Engineering, University at Buffalo.

2

Chemistry and Physics of Fire

To prevent fires from occurring and to extinguish them successfully after they have started, an understanding of the chemical and physical characteristics of fire is important. The chemistry of fire involves the ways in which fires can be started and sustained at the molecular level of the fuel source. The chemical implications of the combustion process and the potential hazards of combustion by-products to workers will also be discussed. The physical aspects of fire involve its thermal properties, methods of heat transfer, and methods of extinguishment. Because fire is a chemical reaction, it is important to understand which hazardous materials pose fire hazards in the workplace and which by-products of the combustion process can often be more hazardous than the hazardous material involved in the fire. Control measures include the use of established labeling, classification systems, and handling procedures for hazardous materials. Examples of these systems include the Department of Transportation Hazardous Materials Classification Labeling System, the National Fire Protection Association (NFPA) Labeling System, and additional references on hazardous materials.

FIRE TETRAHEDRON

For many years, it was believed that the fire process could be described by what was referred to as a fire triangle. This concept basically indicated there were three elements essential to initiate and sustain fire: oxygen, heat or some other energy source, and fuel. In more recent years, we have expanded the fire triangle into what is now referred to as the fire tetrahedron (Klinoff 2012, 97). This concept contains the three elements discussed earlier; however, it adds a fourth element, which is often referred to as the chemical chain reaction. Figure 2.1 depicts the fire tetrahedron.

In terms of oxygen supply, air is the most common source of oxygen with, on average, 21 percent of air being oxygen. However, it should be noted that sources of oxygen may also include oxidizers. Oxidizers are substances that acquire electrons from a fuel in a chemical reaction and release oxygen during combustion (NFPA 2008, 17–125). Examples of common oxidizers include fluorine, chlorine, hydrogen peroxide, nitric acid, sulfuric acid, and hydrofluoric acid.

In its simplest form, the source of energy that heats the material to its ignition temperature as part of the fire tetrahedron is the ignition source or energy source. Some examples of ignition sources for fires in industrial occupancies include excessive electrical current, heating equipment, flames and sparks, and lightning. When discussing heat or energy sources, an important

FIGURE 2.1
Fire Tetrahedron

concept to understand is ignition temperature. Ignition temperature can be defined as the minimum temperature of a material required to initiate or cause self-sustained combustion of the material (Klinoff 2012, 100). Some examples of ignition temperatures of common building materials include plywood (390°C), gypsum board (565°C), carpet (300–465°C), and polystyrene (630°C) (NFPA 2008, 3–159). Ignition temperature can vary based on oxygen levels: the richer the oxygen levels, the lower the ignition temperature. The rate of heat rise, the duration of heating, and the size and shape of material also influence ignition material. A classic example illustrating how ignition material is influenced by the size and shape of material is wood.

Wood shavings ignite at a lower ignition temperature than bulk wood because of the size and shape of the material.

It should also be noted that spontaneous ignition can occur when the ignition source is slow oxidation with limited heat loss that produces a temperature rise above the ignition temperature of the material (NFPA 2008, 6–5). An example of spontaneous ignition in industrial occupancies are oil-soaked rags stored in 55-gallon waste drums.

The third element to the fire tetrahedron is fuel. Most fires involve a fuel that contains carbon and hydrogen, such as wood, paper, and flammable and combustible liquids and gases (Klinoff 2012, 99). Another potential fuel is a combustible metal such as aluminum or magnesium. For fire-extinguishment purposes, the fuels are classified as follows (NFPA 2008, 17–124):

- Class A: ordinary combustibles such as wood and paper
- Class B: flammable and combustible liquids, petroleum greases, and flammable gases
- Class C: fuels involving energized electrical equipment
- Class D: combustible metals such as aluminum, magnesium, titanium, and zirconium
- Class K: combustible cooking media

Each fuel classification also has a unique symbol. Fire extinguishers are an example of a piece of equipment that uses these fuel classification symbols. A summary of the fuel classifications and their symbols appears in figure 2.2.

The fourth and final element to the fire tetrahedron is the chemical chain reaction. This chemical chain reaction occurs within the material itself when

Symbol	Types of Materials
 **A** Ordinary Combustibles	Ordinary combustibles wood, paper
 **B** Flammable Liquids	Flammable and combustible liquids, oils, grease
 **C** Electrical Equipment	Electrical fires
 **D** Combustible Metals	Metal fires
 (cooking pan with flames)	Cooking oils

FIGURE 2.2
Classification of Fuels

the fuel is broken down by heat, producing chemically reactive free radicals which then combine with the oxidizer (Klinoff 2012, 97–98).

COMBUSTION

Combustion can be defined as an exothermic chemical reaction between some substance and oxygen. The chemical reaction between the substance and an oxidizer is referred to as oxidation (NFPA 2008, 6–3). With combustion, the energy that accompanies oxidation is commonly given off as heat and light. The speed at which oxidation occurs in cases of rusting of iron and yellowing of paper varies to that of combustion. The difference between a slow oxidation like rusting and combustion is that combustion occurs so rapidly that the heat is generated faster than it is dissipated, causing a substantial temperature rise in the substance.

The rates of combustion of gases, liquids, and solids vary depending on several factors. For solids, the rate of combustion varies primarily based on the size of the solid particles, with smaller particles having a higher rate of combustion. For flammable liquids, the rate of combustion varies based on whether the combustion occurs in a still pool, flowing current, or spray or foam. For flammable gases, the rate of combustion varies based on the extent to which the gas mixes with air prior to combustion and on the degree of motion and turbulence of the gases (NFPA 2008, 6–4).

There are four major products of combustion: heat, smoke, light, and fire gases (Schroll 2002, 19). These products of combustion are critical for fire purposes not only in terms of extinguishment but also in terms of life safety and building design. The primary loss of life in a fire is due to the toxic fire gases.

Heat affects the body in several ways, but the two major factors that determine its effect on the body include the length of exposure and the temperature. As a rule of thumb, one should not enter areas exceeding 120°F without proper personal protective equipment. Exposure to excessive heat causes an increased heart rate, dehydration, exhaustion, burns, and possibly a blockage of the upper respiratory tract.

Smoke, the second product of combustion, is the result of airborne solids and liquid particulates and fire gases that result from combustion (NFPA 2008, 18–44). Smoke is a major killer in fires, causing 50 to 75 percent of deaths. Smoke reduces visibility and irritates eyes and lungs, and in many cases, the fire gases carried in smoke are lethal. The amount and kind of fire gases will vary depending on the chemical composition of the material burning, the amount of oxygen, the temperature, and the possibility of the mixture of particles and gases producing synergistic effects (Friedman 1998, 159). The effects of these fire gases on the body are governed by the concentration of the gases present, the length of exposure time, and the physical condition of the individual exposed. In a fire situation, carbon monoxide (CO) is the gas that is produced in the greatest quantities (as high as 5–6 percent by volume), and its toxicity is based on its affinity for hemoglobin that results in formation of carboxyhemoglobin in the blood (Cote and Bugbee 2001, 56). The effects of CO are initiated at approximately 100 parts per million (ppm), where headaches will be experienced after an exposure of two to three hours; concentrations of 1,600 ppm or more are fatal in less than two hours. The current Occupational Safety and Health Administration (OSHA) permissible exposure limit for CO is 50 ppm, and National Institute for Occupational Safety and Health (NIOSH) has established an IDLH value of 1,200 ppm (USDOL 2013). Another common gas produced during a fire is carbon dioxide. It is low in toxicity and is not normally considered a significant toxin in smoke. However, carbon dioxide does increase the speed and depth of breathing, thereby increasing carboxyhemoglobin in the blood from CO. Combustion of materials containing nitrogen bonds, such as wool and silk, result in the release of hydrogen cyanide (HCN). The amount of HCN released is temperature- and material-dependent, with higher levels generated as the temperature increases (NFPA 2008, 6–13).

The toxicity of HCN is based on the fact that it inhibits cells from using oxygen (a condition called histotoxic hypoxia). Levels of HCN up to 135 ppm are lethal within thirty minutes (Cote and Bugbee 2001, 57). It should also be noted that during combustion, oxygen is consumed; therefore, oxygen-deficient atmospheres are an important consideration when considering the toxicity of smoke. The usual concentration of oxygen in air is 21 percent; levels below 17 percent cause diminished muscular control. In addition to the toxicity of smoke, it is also important to consider the effect smoke has on vision. Smoke obscures the passage of light, thereby possibly blocking the visibility of exits and impeding escape from a fire (NFPA 2008, 6–12). The development of smoke in sufficient quantities to obscure exits happens very quickly and is often the first hazard of a fire. In addition, eye irritation, which is primarily due to the concentration of irritants in the smoke, may also impact the vision of the individual trying to escape.

Unique Combustion Phenomena

Five unique combustion phenomena that may occur during a fire include explosions, deflagrations, detonations, flashovers, and back drafts. An explosion is the bursting of an enclosure due to excessive internal pressure (NFPA 2008, 18–79). In a fire situation, an example of an explosion is a boiling-liquid expanding-vapor explosion (BLEVE). To illustrate a BLEVE, consider a fire contacting the surface of a flammable liquid storage tank. The flame causes the gases in the tank to expand and open the emergency vents so that the pressure can be relieved. If the vents cannot relieve the pressure fast enough, the tank will fail and the gas will be released (NFPA 2008, 7–37–38). Sources of explosions in industry are most commonly associated with combustion explosions (44 percent) that involve fuels (Ladwig 1991, 225). Many times, the explosions are the result of improper lighting procedures or inadequate safeguards on the fuel appliances, such as low gas shutoffs. Other sources of explosions in industry include flammable liquid vapors, combustible dusts, trapped steam, gas leakage, ruptured pressure vessels, nuclear or atomic explosions, and thermal explosions due to unstable materials decomposing.

Deflagration is the burning of a gas or aerosol that is characterized by a combustion wave. The combustion wave moves through the gas and oxygen, burning until all the fuel is used. With a deflagration, the rate of travel of the combustion wave is less than the speed of sound, and no shock wave is produced. Most explosions that occur in industry are deflagrations (NFPA 2008, 18–79).

A detonation, another unique combustion phenomenon, is the burning of a gas or aerosol characterized by a shock wave. With detonation, the shock wave travels at a speed greater than the speed of sound, and the wave is characterized by extremely high pressure initiated by a very rapid release of energy (NFPA 2008, 18–79). The high pressure created from the shock wave also serves to create a heat source for igniting other combustibles.

Methods for controlling explosions include containment, suppression, venting, and isolation. Containment involves designing a container to withstand maximum pressure exerted by the explosion. Suppression stops the combustion process before most of the fuel air mixture is consumed. Suppression is like fire protection systems in that once detection of ignition occurs the extinguishing agent is applied to the protected area to extinguish the flame ball. Venting releases the energy and gases from the explosion in a controlled manner to a safe location. Isolation is a strategy to prevent an explosion at one location by passing it to another location. This can be accomplished by physical separation or using blast-resistant structures such as flame arresters or diverters, passive float valves, and actuated pinch valves (NFPA 2008, 17–154–159).

A flashover is a fire in an enclosed area that fosters the buildup of heat and when the temperature reaches the ignition temperature of the majority of combustibles in the area, there is spontaneous combustion of the combustibles in the area (Schroll 2002, 19).

The last unique combustion phenomenon is referred to as a back draft. A back draft is sometimes referred to as a smoke explosion because it is a fire in an enclosed area that consumes the oxygen supply and generates CO and heat. As the oxygen is being used up, the fire tends to smoke a lot; then, if outside air is introduced, the CO will burn rapidly with explosive force (Schroll 2002, 19).

Heat versus Temperature

It is important to make a distinction between temperature and heat. Heat is a transfer of energy between two objects due to a temperature difference, while temperature is a quantity that determines when objects are in thermal equilibrium (Giambattista, Richardson, and Richardson 2012, 506). Common units of measure for temperature are:

- Celsius (C): 0°C—water freezes; 100°C—water boils
- Fahrenheit (F): 32°F—water freezes; 212°F—water boils
- Kelvin (K): 0°K = −273.15°C or −459.67°F (also referred to as absolute zero)

Common units of heat include the following (Cote and Bugbee 2001, 48):

- Joule (J): amount of heat energy provided by 1W flowing for one second
- Calorie (cal): amount of heat required to raise the temperature of 1g of water by 1°C
- British thermal unit (Btu): amount of heat required to raise the temperature of 1 lb. of water by 1°F

When a material goes through a chemical oxidation reaction, such as combustion, it releases thermal energy. On an atomic scale, heat is the motion of atoms and molecules within the material, and as the temperature increases, the vibration of the molecules increases as well. From a fire-extinguishment perspective, another important concept related to heat is specific heat. Specific heat can be defined as the heat capacity per unit mass, and in the metric system specific heat is the amount of heat required to raise the temperature of 1 kg of a substance by 1°K (Giambattista et al., 509). The units for specific heat are cal/g·K. For fire-extinguishment purposes, it is desirable to have a material that has a high specific heat so that it will absorb the heat generated from a fire, thereby reducing the temperature to below the ignition temperature needed for combustion to occur. Water has a specific heat of 1 cal/g·K., which is much higher than many other materials, such as copper, which has a specific heat of 0.093 cal/g·K.

Heat of Combustion

The amount of heat released from a fire over a specific time (rate) is based on a material's heat of combustion. The heat of combustion can be defined as the energy released by the fire per unit mass of fuel burned (Drysdale 2011, 19). Obviously, this is important for fire purposes because the amount of heat released influences fire spread, fire extinguishment, and life safety. The unit for heat of combustion is kJ/g. Examples of heat of combustion of common materials include wood (16–19 kJ/g), flammable liquids (19.9–44.8 kJ/g), and flammable gases (10.1–12.1 kJ/g) (Friedman 1998, 41).

Heat Transfer

Heat energy flows due to a temperature difference, with the direction of the flow being from hot to cold or from a higher energy state to a lower energy state. The rate of heat transfer is often expressed as Btu per hour or joules per second. There are three major mechanisms of heat transfer: conduction, convection, and radiation. An understanding of heat transfer in these three modes allows for the design of buildings to limit heat transfer and correspondingly limit fire spread.

Heat transfer by conduction: Heat transfer by conduction occurs when the heat is transferred in the material by molecules that are vibrating and colliding with other molecules, thereby transferring their kinetic energy through the material (Friedman 1998, 62). An example of heat transfer by conduction is the simple placement of the end of a 3-ft. metal rod into a flame; the heat will transfer from one end of the rod to the other. There are four major factors that influence heat transfer by conduction: distance, temperature, cross-sectional area, and composition of material. In terms of the cross-sectional area, the larger the area, the higher the rate of heat transfer. As the distance increases, the rate of heat transfer decreases. This concept is commonly referred to as a temperature gradient ($\Delta T/d$), which is the rate of temperature change with distance (Giambattista et al. 2012, 520). In general, thermal conductivity (k) of metals decreases as temperatures increase, while the opposite is true with gases. As the temperature increases in gases, the thermal conductivity also increases. It is also worth noting that solid materials will have higher k values than gases and liquids simply because the gases have molecules that are further apart; therefore, there are fewer collisions and less heat transfer.

The fourth factor that influences heat transfer by conduction is the composition of the material. For any given material, there is a constant known as a coefficient of thermal conductivity that is directly proportional to the rate at which energy is transferred through the substance. The coefficient of thermal conductivity (k) has units of watts per meter per degree K, with higher values of k associated with good conductors of heat and smaller values with poor conductors or insulators. The insulating property of an insulator is commonly expressed as an R value, which is based on thermal resistance. The best commercial insulators, such as fiberglass, have fine particles or fibers of solid substances with a space between them filled with air. Some examples of k values for various materials include the following (Drysdale 2011, 37):

- Copper: 401 W/m°K
- Aluminum: 237 W/m°K
- Steel: 45.8 W/m°K
- Concrete: 1.4 W/m°K
- Glass: 0.76 W/m°K
- Pine: 0.14 W/m°K
- Fiber insulating board: 0.041 W/m°K
- Air: 0.026 W/m°K

To evaluate heat transfer by conduction, one can use Fourier's law of heat conduction: $I_{cd} = kA\Delta T/d$. In the preceding formula, I_{cd} represents the rate of heat flow, A is the cross-section area, d is the thickness of the material, ΔT represents the temperature difference, and k is the thermal conductivity of the material (Giambattista et al. 2012, 520).

Heat transfer by convection: Heat transfer by convection occurs because of the movement of a fluid containing the heat (NFPA 2008, 2–8). Convection occurs in fluids including both liquids and gases, and heat transfer by convection is critical to the spread of fire. Convection heating occurs in fires when the fires heats air and gases, and, in turn, the air and gases are hot enough to ignite other materials. This natural convection occurs because the heated fluid expands, resulting in a lower density than the surrounding cooler fluid, causing it to rise. The rate of heat transfer by convection is influenced by the following factors (Drysdale 2011, 52):

- Fluid properties such as density, viscosity, and thermal conductivity
- Flow parameters such as the velocity and nature of the flow
- Geometry of the surface such as cross-sectional area
- Temperature difference

Fortunately, there is one value that addresses many of these factors, that is, the convection heat transfer coefficient (h). This is a constant for a given material with the units being watts per square meter per degree C. The rate of convective heat flow can be determined using Newton's law of cooling: $I_{cu} = hAS (TS - TF)$ (Thomas 1980, 21). In this formula, Icu is the rate of heat transfer by convection, h is the convection heat transfer coefficient in $W/(m^2 \cdot °C)$, AS is the surface area in m^2 over which the fluid moves, TS is the surface temperature in °C, and TF is the fluid reference temperature in degree Celsius.

Radiant heat transfer: Heat transfer through electromagnetic waves is the basic process through which heat is transferred by radiant heat transfer. Forms of electromagnetic radiation fall into three wavelengths: infrared, visible, and ultraviolet (Giambattista et al. 2012, 526).

The electromagnetic waves associated with a fire are typically in the infrared region, with only a small fraction emitted in the visible light region. It is important to note that thermal radiation is not heat energy but was formed from heat energy and becomes heat again when it strikes an object. All matter emits thermal radiation continuously at temperatures above absolute zero (NFPA 2008, 2–10). Heat transfer via electromagnetic waves requires no medium; therefore, this type of heat transfer can travel through a vacuum because it does not need molecules to be in contact with one another like heat transfer by convection and conduction. In general, the ability of an object to emit radiation is proportional to its ability to absorb radiation; in other words, a good emitter is also a good absorber. We measure an object's ability to emit thermal radiation by a constant referred to as emissivity (e). Emissivity has no units and the values for a specific material will range from zero to one. A black body has an emissivity value of one, which is the maximum amount of thermal radiation emitted (Giambattista et al. 2012, 526). Common examples of emissivity include the following (Drysdale 2011, 62):

- Stainless steel: 0.074
- Cast iron: 0.21
- Brick: 0.75
- Plaster: 0.91
- Oak: 0.9

In addition to emissivity, two other factors that influence radiant heat transfer are temperature and distance. Radiant heat transfer increases tremendously with increase in temperature, and distance also plays a major role. Specifically, if the distance between two objects doubles, then the rate of heat transfer is reduced to one-fourth; conversely, if you have the distance between two objects, then the rate of heat transfer is increased by a factor of four.

SOURCES OF HEAT

There are numerous heat sources in the work environment capable of starting a fire or keeping it burning once it is started. The following is a description of heat sources (NFPA 1997, 1–64–67):

1. Chemical heat: Heat of combustion is the heat that is released during a substance's complete oxidation. Calorific values of fuel are expressed in joules per gram of material.
2. Spontaneous heating: Spontaneous heating is the process by which a material increases temperature without drawing heat from its surroundings. If allowed to heat up to combustion temperatures, spontaneous ignition can take place.
3. Heat of decomposition: Heat of decomposition is the heat released by the decomposition of compounds that have been formed. Acetylene is an example of a product that, once it starts to decompose, generates heat.

CHEMISTRY AND PHYSICS OF FIRE

4. Heat of solution: This is heat released when a substance is dissolved in a solution.
5. Electrical heat: Also called resistance heating, this is heat generated due to the resistance electricity encounters when traveling through a conductor.
6. Arcing: Arcing occurs when electrical energy jumps across a gap in the circuit carrying the electrical energy.
7. Sparking: This takes place when a voltage discharge is too high for a low-energy output.
8. Static electrical charge: This is an electrical charge that accumulates on the surfaces of two materials that have been brought together, then separated.
9. Lightning: This is the discharge of an electrical charge from a cloud to an opposite charge (i.e., another cloud or the ground).
10. Mechanical heat: This is the mechanical energy used to overcome the resistance to motion when two solids are rubbed together; it is also known as frictional heat.
11. Nuclear heat: This is heat energy released from the nucleus of an atom.

PHYSICS OF COMBUSTION

In addition to the chemical properties of fire, physics plays an important role in the fire behavior of materials. Fuel in a fire can be present in one of three different states of matter: solid, liquid, or gas. Combustion usually occurs when the fuel is converted to a gaseous state, because the oxidizer occurs as a gas, and it takes both oxidizer and fuel in the gaseous state for the recombination to occur (Klinoff 2012, 99).

- Solids: Factors that affect the combustion of solid fuels include the size of the particles, their moisture content, and their continuity (Klinoff 2012, 101). The smaller the solid, the less heat is needed to burn. A prime example of this is a fire involving dust, as is the case in a grain elevator fire. Because the dust particles have a greater surface area as compared to their mass, a smaller ignition source is needed to start combustion.
- Liquids: Liquids do not burn. Rather, it is the vapor of the liquid at the surface that is burning. Some characteristics of liquids that are important in terms of the fire risk of fuels are their flash point, their vapor pressure,

specific gravity, boiling point, and their vapor density. Liquids with low flash points will readily produce vapors in a concentration sufficient to support combustion in the normal work environment. The vapor pressure of the liquid provides an indication as to how readily the vapors will be released from the liquid surface. The vapor density provides an indication of whether the vapor will rise or sink when released, whereas the specific gravity provides an indication as to whether the liquid will float on water or sink. For example, liquids with specific gravities greater than 1.0 will sink in water, while liquids with specific gravities less than 1.0 will float on water (Klinoff 2012, 104–106).

- Gases: Gases can be broadly classified as either flammable or nonflammable. While oxygen itself does not burn, it serves as an oxidizer for a fire and, in some situations, lowers the ignition temperature necessary to ignite a fuel source. Some characteristics of gases that are important in terms of the fire risk of fuels are their upper and lower flammability limits and their flammable range. The flammability limits refer to the concentration of gas or vapor in air with the upper limit being the maximum concentration and the lower limit being the minimum concentration that will ignite (Klinoff 2012, 108–109). The flammable range is that proportion of gas or vapor in air between the upper and lower flammable limits, therefore the fire risk of a gas increases as the flammable range increases.

FIRE HAZARDS OF MATERIALS

Some of the more common materials found in the workplace can pose potential hazards to the occupants, should these materials become involved in a fire. The most widely used materials and the most involved in fires include wood products, textiles, masonry/stone, structural steel, gypsum, and synthetic materials. Each has its own unique characteristics and composition. Each reacts differently when involved in a fire, and each generates its own unique hazards.

Wood

For wood to burn, it must be heated to the point at which combustible gases are released from its surface. When wood burns, it releases approximately 8,000 to 12,000 Btus of energy per pound (NFPA 2008, 6–71). The physical form of wood will influence its ability to ignite and burn. A characteristic

CHEMISTRY AND PHYSICS OF FIRE

that plays a major role in the ability for wood to serve as a fuel in a fire is its ratio of surface area to its mass. Therefore, paper will flame and burn with a relatively small ignition source while heavier wood logs will require much more heat. The paper has a greater surface area exposed to oxygen source and less mass than the log; as a result, it requires less energy to continue to burn. Another characteristic of wood that influences its use as a fuel in a fire is its thermal conductivity. The thermal conductivity of a material is the measure of the rate at which absorbed heat will flow through the mass material. Wood is therefore a poor conductor of heat and a good insulator (Cote and Bugbee 2001, 65).

Moisture content also plays a significant role in the ignition and rate of combustion of wood. The moisture content of wood consists of free water (water in the cells) above the fiber saturation point (the fiber saturation point is 25–30 percent moisture content) (NFPA 2008, 6–67). The higher the water content in wood, the greater the heat required to evaporate the water from the wood.

Once the wood is heated above its autoignition temperature, ignition will occur. At this temperature, combustible gases are formed and released at the surface. If a moderate heat source is applied for long enough, it will raise the temperature of the material to its autoignition temperature. As wood is heated, it reaches a point where an exothermic reaction takes place. Air velocity around the wood can accelerate the ignition and combustion of wood. The average ignition temperature of wood is about 392°C (Quintiere 1998, 241).

Masonry/Stone

Although masonry and stone products are fire resistant, this does not mean they are not affected by fire. When exposed to a fire, the masonry and stone products can have a loss in surface material, which is commonly referred to as spalling. Spalling is also common in natural stones, especially granite, marble, and limestone. When these natural stones are self-supported, they should be avoided after exposure to a fire as collapse is a real possibility (Brannigan 2008, 53).

Metals

Structural steel is a common metal used for structural elements such as columns, beams, and walls. Structural steel today is primarily iron and carbon

(2%) and it does not burn. However, structural steel does have three negative characteristics when exposed to fire. First, structural steel conducts heat and second it elongates when heated possibly causing the structure to collapse. The last negative characteristic is that structural steel can fail at temperatures above 1,000°F. Aluminum is also a common metal used in building construction, and it is especially vulnerable in a fire as it melts at very ordinary fire temperatures (Brannigan 2008, 55).

Gypsum

Gypsum is a component in drywall or wallboard, which is a common covering for interior walls. Although the paper backing may burn, the gypsum in the wallboard is an excellent fire-retardant building material (NFPA 2008, 19–54).

Synthetic Materials

There are thousands of different types of plastics manufactured and they can be used in a variety of construction materials such as thermal and acoustical insulation, imitation wood products, ceiling and wall panels, and certain flooring materials. These plastics are typically divided into four groups based on their derivation, method of synthesis, and historical development. These four groups include plastics from natural products, plastics from condensation resins, plastics from double bond polymerization thermoplastics, and plastics from multifunctional intermediate products (NFPA 2008, 6–212–217). Some special fire hazards of plastics include their rate of burning, the smoke produced, and potentially toxic gases released during their combustion. Although plastics tend to have a higher ignition temperature than wood, some are very easily ignitable with a small flame and will burn rapidly. Some plastics, when burning, will produce large amounts of very dense sooty, black smoke. Depending on the formulation of the plastic, toxic gases such as CO, HCN, hydrogen chloride, and phosgene can be released (NFPA 2008, 6–224–225). Additional hazards of burning plastics include flaming drips where the burning plastic material may flame and drip, causing the fire to spread down and away from the source. The gases released from burning plastics can also cause corrosion of sensitive electronic equipment (NFPA 2008, 6–225).

Textiles

Fabrics are woven from various types of fibers and are common in carpets and curtains as well as furniture and clothing. The ignitability, flame spread, and overall fire hazard for textiles depend on their fiber content and construction. Textiles can be broadly classified into natural fibers and manufactured fibers. Natural fibers can be classified as cellulosic, mineral, and protein fibers. Cotton is the most important at 90 percent of all-natural fiber. Cotton is combustible, and when ignited, it produces smoke, heat, carbon dioxide, and CO (NFPA 2008, 6–75–76). Protein fibers are derived from animals and include wool and silk. As a rule, the protein fibers present less of a fire hazard because they are more difficult to ignite, burn slowly, and are easier to extinguish than cotton fibers (NFPA 2008, 6–76).

Manufactured fibers are subdivided into two groups: regenerated fibers and synthetic fibers. The most common regenerated fiber is rayon and common synthetic fibers are acrylic and nylon. Rayon, which is made from dissolved cellulose, has similar burning characteristics to cotton. Acrylics and nylon, on the other hand, can produce HCN when burned. There are also manufactured textiles made completely from inorganic materials, such as glass, metal and carbon fibers, which can be considered noncombustible (NFPA 2008, 6–77–79).

To reduce the fire potential of textiles, various fire-retardant treatments have been developed, and their use is required by life-safety and building codes. To make a textile fire retardant, chemical treatments are available; however, age, washing, and dry cleaning may reduce the effectiveness of the treatment. Also, the treatment does not make the textile flameproof; it only makes the textile resistant to ignition by a small flame. Therefore, the textile may still burn when exposed to a large flame. There are following five methods by which fire-retardant treatments work (NFPA 2008, 6–85):

1. They produce noncombustible gases that exclude oxygen.
2. Such molecules are released that interfere with the chemical chain reaction.
3. A nonvolatile char or liquid is produced which reduces oxygen to the material.
4. Finely divided particles are formed that interfere with the chain reaction.
5. The flame-retardant chemical decomposes endothermically.

Examples of textiles required to be flame retardant include aircraft textiles, carpeting, blankets, children's sleepwear, clothing, textiles in motor vehicles, tents, and curtains or draperies.

HAZARDOUS MATERIALS

The U.S. Department of Transportation (USDOT) defines a hazardous material as a substance or material capable of posing an unreasonable risk to health, safety, and property when transported in commerce (USDOT 2013a, 2). It should be noted that there is no consensus on a definition for hazardous materials, and in many cases, there is simply a listing of materials rather than a definition of the term itself, such as with the Pennsylvania Right-to-Know Law. Both the DOT and the Emergency Response Guidebook place hazardous materials into nine primary classes: explosives, gases, flammable and combustible liquids, flammable solids, oxidizers and organic peroxides, toxic and infectious substances, radioactive substances, corrosives, and miscellaneous hazardous materials (USDOT 2013c, 11). A description of each of these classes follows (USDOT 2013d, 49CFR Part 173).

- Class 1 is explosives. Explosives are any chemical compound, mixture, or device, the primary purpose of which is to function by explosion, with substantially instantaneous release of gas and heat, or which by a chemical reaction is capable to function in a similar manner. There are six divisions of explosives. Division 1.1 is explosives that detonate and present a mass explosion hazard. Division 1.2 is explosives that present a projection hazard but not a mass explosion hazard. Division 1.3 is explosives that present a fire hazard and a minor blast and/or projection hazard. Division 1.4 is explosives that present a minor explosion hazard with the effects confined to the package and no projection hazards. Division 1.5 is explosives that have a mass explosion potential, but they are very insensitive, that is, low probability of initiation under normal conditions of transport. Division 1.6 is explosives like Division 1.5, but they do not have a mass explosion potential.
- Class 2 is gases and it is divided into three divisions. Division 2.1 is flammable gases. Division 2.2 is a nonflammable nonpoisonous gas. This division would include compressed gases, liquefied gases, pressurized cryogenic gases, compressed gases in solution, asphyxiant gases, and

oxidizing gases. Division 2.3 is for all gases that are poisonous (toxic) by inhalation. These gases are known to be toxic to humans and therefore present a hazard during transportation or is presumed to be toxic to humans based on animal studies where testing has determined an LC50 of not more than 5,000 mL/m^3.
- Class 3 is flammable and combustible liquids. A flammable liquid is defined as a liquid with a flash point of 140°F or less while a combustible liquid has a flash point above 140°F but less than 200°F. Examples of flammable liquids include acetone, gasoline, and methyl alcohol; and examples of combustible liquids include fuel oils and ethylene glycols.
- Class 4 is flammable solids and it is divided into three divisions. Division 4.1 is for flammable solids and this includes desensitized explosives, self-reactive materials, and readily combustible solids. An example of a flammable solid would be magnesium or aluminum powder. Division 4.2 is spontaneously combustible materials and this division includes either pyrophoric or self-heating materials. A pyrophoric material can spontaneously ignite within 5 minutes when exposed to air without an external ignition source. Self-heating materials have a gradual reaction of the substance with oxygen and produce heat. When the rate of heat production is greater than the rate of heat loss the material will heat up and this may lead to self-ignition. Division 4.3 is water-reactive substances or sometimes referred to as "Dangerous When Wet Materials." These materials may become spontaneously flammable or give off toxic or flammable gases when in contact with water.
- Class 5 is oxidizers or organic peroxides, and it is divided into two divisions. Division 5.1 is for oxidizers which are materials that generally cause or enhance the combustion of other materials by yielding oxygen. Examples of oxidizers include ammonium nitrate fertilizer, hydrogen peroxide solution, or chlorate. Division 5.2 is for organic peroxides which are any organic compounds containing oxygen in the bivalent –O–O– structure and which may be considered a derivative of hydrogen peroxide, where one or more of the hydrogen atoms have been replaced by organic radicals.
- Class 6 is toxic or infectious substances, and it is divided into two divisions. Division 6.1 is for a poisonous (toxic) material which is any material, other than a gas, which is known to be so toxic to humans as to afford a hazard to health during transportation, or which, in the absence of adequate data on

human toxicity is presumed to be toxic to humans because it falls within any one of the following categories when tested on laboratory animals:
- A liquid or solid with an LD50 for acute oral toxicity of not more than 300 mg/kg;
- A material with an LD50 for acute dermal toxicity of not more than 1,000 mg/kg;
- A dust or mist with an LC50 for acute toxicity on inhalation of not more than 4 mg/L;
- A material with a saturated vapor concentration in air at 68°F greater than or equal to one-fifth of the LC50 for acute toxicity on inhalation of vapors and with an LC50 for acute toxicity on inhalation of vapors of not more than 5,000 mL/m3; or
- Is an irritating material, with properties like tear gas, which causes extreme irritation, especially in confined spaces.
- Division 6.2 is for infectious substances which are any material known or reasonably expected to contain a pathogen that can cause disease in humans or animals. Infectious substances are further classified into two categories:
- Category A is an infectious substance in a form capable of causing permanent disability or life-threatening or fatal disease in otherwise healthy humans or animals when exposure to it occurs.
- Category B is an infectious substance that is not in a form generally capable of causing permanent disability or life-threatening or fatal disease in otherwise healthy humans or animals when exposure to it occurs.
- Class 7 is radioactive materials. Radioactive materials are any material or combination of materials containing radionuclides where both the activity concentration and the total activity in the consignment exceed the values specified in the table in CFR 173.436 or values derived according to the instructions in CFR 173.433. Examples of radioactive materials include plutonium, cobalt, and uranium.
- Class 8 is corrosive substances. Corrosive materials are liquids or solids that cause full thickness destruction of human skin at the site of contact within a specified period or have severe corrosion rate on steel or aluminum. Examples of corrosive materials include hydrochloric acid, sulfuric acid, and caustic soda.
- Class 9 is miscellaneous hazardous materials. These are materials which present a hazard during transportation, but which do not meet the

definition of any other hazard class. This would include materials with anesthetic, noxious, or similar properties, which could cause extreme annoyance or discomfort, or the material meets the definition in 49 CFR 171.8 for an elevated temperature, hazardous substance, hazardous waste, or marine pollutant. Class 9 used to include five subclasses referred to as "Other Regulated Materials" (ORMs). Except for ORM Class D, the other four classes are no longer used. An ORM class D is a material such as a consumer commodity that, although otherwise subject to the DOT regulations, presents a limited hazard during transportation due to its form, quantity, and packaging. An example of an ORM class D would be cartridges used for fastening devices.

SOURCES OF INFORMATION ON HAZARDOUS MATERIALS

There is a variety of reference sources a safety professional can use when addressing the handling, storage, and shipping of hazardous materials. The first reference is the Fire Protection Guide to Hazardous Materials published by the NFPA. This reference is useful to a safety professional because it provides recommendations for labeling, shipping, storing, and extinguishing hazardous materials. The guide includes six documents. The first document is titled "Matrixes." This document provides a quick reference to chemicals based on chemical name, synonym, and Chemical Abstract Service (CAS) number (NFPA 2010, iii). The second document titled "Hazardous Chemicals Data Compilation" includes the fire hazard properties, NFPA 704 Hazard Ratings, and recommendations on storage and fire fighting for 325 chemicals. The third document titled "Fire Hazard Properties of Flammable Liquids, Gases and Volatile Solids" lists in a table format the fire hazard properties of 1,300 flammable substances. Properties include flash point, boiling point, CAS number, NFPA 704 Hazard Rating, and NFPA 30/OSHA flammable/combustible liquid classification. The fourth document titled "Compilation of Hazardous Chemical Reactions" is a guide to hazardous chemical reactions of over 3,600 mixtures of two or more chemicals that may cause fires, explosions, or detonations. The fifth document is titled "Recommended Practice for the Classification of Flammable Liquids, Gases, or Vapors and of Hazardous Locations for Electrical Installations in Chemical Process Areas." This document provides parameters used to determine the degree and extent of hazardous locations (NEC Groups) for gases, liquids, and vapors for 264 chemicals. The sixth document is titled "Standard System for the Identification of the Hazards of Materials for Emergency Response" (NFPA 2010, iii). This document provides criteria for the NFPA 704 Hazard Rating, specifically how the system determines the degree of health, flammability, and instability hazard of materials. The blue area in the diamond rating is used to identify health hazards; the red identifies fire hazards; the yellow identifies instability hazards; and the white section is used for miscellaneous categories, such as water-reactive chemicals, pressurized chemicals, or those needing specific types of personal protective equipment. The ratings that go within this labeling system range from a score of 0 to indicate no hazard to 4 which indicates a severe hazard (NFPA 2010, 704–8).

Another useful reference source for the safety professional regarding hazardous materials is the USDOT, which has the federally mandated responsibility for developing and enforcing the hazardous materials regulations included within Title 49 of the Code of Federal Regulations. These federal regulations focus on the transportation of hazardous materials. Within USDOT, the Pipeline and Hazardous Materials Safety Administration (PHMSA) was created. PHMSA's mission is to protect people and the environment from the risks of hazardous materials transportation. This is accomplished by establishing national policy, setting and enforcing standards, education, conducting research to prevent incidents, and preparing the public and first responders to reduce consequences if an incident does occur (http://www.phmsa.dot.gov/about/mission). The Hazardous Materials Table (HMT) is a particularly important reference for hazardous materials and is the backbone of the Hazardous Materials Regulations (USDOT 2013b, 1). This table is included in 49 CFR § 172.101, and a sample of this table is shown in table 2.1.

The USDOT has extremely specific regulations regarding the identification of hazardous material during transportation using placards and labeling. The requirements for all labels and placards are included in 49 CFR § 172.400 and §172.519. In addition to Title 49, the USDOT has developed a guidebook referred to as the Emergency Response Guidebook. This guidebook was originally created in 1996 in collaboration with Canada and Mexico so that one hazardous materials emergency response guidebook would be used for all North America. The primary purpose of this guidebook was to aid first aid responders in identifying the specific or generic classification of the hazardous materials and to protect themselves and the public during the response

Table 2.1. Excerpt of the Hazardous Materials Table

Symbols	Hazardous materials description and proper shipping names	Hazard Class or Division	Identification Numbers	PG	Label Codes	Special Provisions (172.102)	Packaging Exceptions	Packaging Non Bulk	Packaging Bulk	Quantity limitations Passenger Aircraft/Rail	Quantity limitations Cargo Aircraft Only	Vessel Stow-Age Location	Vessel Stow-Age Other
(2)		(3)	(4)	(5)	(6)	(7)	(8A)	(8B)	(8C)	(9A)	(9B)	(10A)	(10B)
d)	Accellerene, see p-Nitrosodimethyl aniline												
	Accumulators, electric, see Batteries, wet etc												
	Accumulators, pressurized, pneumatic or hydraulic (containing nonflammable gas) see Articles pressurized, pneumatic orhydraulic (containing nonflammable gas)												
	Acetaldehyde	3	UN1089	I	3	A3, B16, T11, TP2, TP7	None	201	243	Forbidden	30 L	E	

to the hazardous materials incident. This guidebook is divided into four sections. The first two sections are a listing of hazardous materials both numerically (using the four-digit DOT identification number) and alphabetically by chemical name.

The third section describes potential hazards that materials may display in terms of fire, explosion, and health effects upon exposure. This section also outlines suggested public-safety measures, emergency response actions, and first aid. The fourth section is an additional reference for isolation and protective action distances where applicable (USDOT, Canada, and Mexico 2012).

The Chemical Transportation Emergency Center (CHEMTREC) was established in 1971 by the Manufacturer Chemists Association, located in Washington, DC. CHEMTREC provides a clearinghouse of information on hazardous materials for emergency response to such incidents. Some of the resources available to the emergency responders from CHEMTREC include (Chemtrec 2013):

- Access to chemical product specialists and hazardous materials experts through CHEMTREC's database of over 30,000 manufacturers, shippers, carriers, public organizations, and private resources.
- An expansive electronic library of over 5 million Safety Data Sheets (SDS).
- Advice on emergency medical treatment to on-scene medical professionals treating victims of product exposure from medical experts and chemical toxicologists.

CHAPTER QUESTIONS
1. What is combustion?
2. What is specific heat?
3. With heat transfer by conduction, as the temperature increases, the thermal conductivity of metals will do what?
4. In the NFPA 704M labeling system, the blue-, red-, and yellow-colored diamonds represent what?
5. When assessing the effect of heat on the body from a fire, what are the two main factors that determine severity?
6. What are the four elements of the fire tetrahedron?

7. What is a hazardous material as defined by USDOT?
8. Class K fires involve what type of fuel?
9. What are some examples of control methods for explosions?
10. The primary purpose of this reference guidebook is to aid first aid responders in identifying the specific or generic classification of hazardous materials and to protect themselves and the public during the response to the hazardous materials incident.

REFERENCES

Brannigan, Francis L. (2008). *Building Construction for the Fire Service*. Quincy, MA: NFPA.

Chemtrec-The Right Information at the Right Time. www.chemtrec.com/about/history

Cote, Arthur, and Bugbee, Percy. (2001). *Principles of Fire Protection*. Quincy, MA: NFPA.

Drysdale, Dougal. (2011). *An Introduction to Fire Dynamics*. New York: John Wiley & Sons.

Friedman, Raymond. (1998). *Principles of Fire Protection Chemistry and Physics*. Quincy, MA: NFPA.

Giambattista, Alan, Betty Richardson, and Robert Richardson. (2012). *College Physics*. Boston, MA: McGraw-Hill.

Klinoff, Robert. (2012). *Introduction to Fire Protection*, 4th ed. Clifton Park, NY: Delmar Learning.

Ladwig, Thomas H. (1991). *Industrial Fire Prevention and Protection*. New York: Van Nostrand Reinhold.

NFPA. (1997). *Fire Protection Handbook*, 18th ed. Quincy, MA: NFPA.

NFPA. (2008). *Fire Protection Handbook*, 20th ed. Quincy, MA: NFPA.

NFPA. (2010). *Fire Protection Guide to Hazardous Materials*, 14th ed. Quincy, MA: NFPA.

NFPA. (2013). *NFPA 704 Standard System for the Identification of the Hazards of Materials for Emergency Response*. Quincy, MA: NFPA.

Quintiere, James G. (1998). *Principles of Fire Behavior*. Albany, NY: Delmar Publishers.

Schroll, Craig R. (2002). *Industrial Fire Protection Handbook*. Lancaster, PA: Technomic Publishing Co.

Thomas, Lindon C. (1980). *Fundamentals of Heat Transfer*. Englewood Cliffs, NJ: Prentice Hall.

USDOL. (2013). General Facts - Carbon Monoxide. www.osha.gov/OshDoc/data-GeneralFacts/carbonmonoxide

USDOT. (2013a). Training Module version 5.1: Introduction. www.phmsa.dot.gov

USDOT. (2013b). Training Module version 5.1: Module 1 The Hazardous Materials Table. www.phmsa.dot.gov

USDOT. (2013c). How to Use the Hazardous Materials Regulations. www.phmsa.dot.gov

USDOT. (2013d). Title 49CFR: Transportation - Part 173 Shippers - General Requirements for Shipments and Packagings. www.ecfr.gov

USDOT, Transport Canada, and Secretariat of Communications and Transportation of Mexico. (2012). *Emergency Response Guidebook*. Washington, DC: U.S. Government Printing Office.

3

Common and Special Hazards

Materials and processes commonly found in the industrial workplace pose unique fire hazards. The presence of electrical sources of ignition, in addition to the use of hazardous materials like flammable and combustible liquids and gases, creates the potential for major fire hazards. The proper installation and maintenance of electrical equipment should be part of every fire prevention program.

Some of the more common hazardous materials, from a fire standpoint, used in the workplace include flammable and combustible liquids, liquefied petroleum (LP) gases, hydrogen gas, oxygen, and acetylene. To control potential fuel and oxidizer sources, proper handling and storage requirements should be followed. Work procedures involving housekeeping and the control of ignition sources should also be considered as an integral part of the fire protection program.

ELECTRICITY AS AN IGNITION SOURCE

Electrical failures or malfunctions are a factor in approximately 16 percent of industrial structure fires (Evarts 2012, 2). Common causes for fires that start because of electrical failures include short circuits, ground faults, or other electrical failures.

In a workplace, several electrical sources can be involved in a fire. Examples of these sources include production equipment, electrical wiring, and heating equipment. Two common electric safety codes in the United States are the National Fire Protection Association (NFPA)'s National Electrical Code (NEC) and American National Standards Institute (ANSI)'s National Electrical Safety Code. The NEC provides for the practical safeguarding of persons and property from hazards arising due to the use of electricity. The NEC was first issued in 1897 by the NFPA and is updated every three years (NFPA 2008, 8–137).

Electrical fires can be the result of a variety of electrical problems at workplace. Improper use of equipment, improper installation, and improper maintenance of equipment are some of the common reasons for electrical fires. Examples of sources of electrical fires at workplace include the following:

- Misuse of electric cords: This includes running the cords under rugs, over nails, or through high traffic areas and the use of extension cords as permanent wiring.
- Poor maintenance: This entails a lack of a preventive maintenance program that should be designed to identify and correct potential problems before they occur.
- Ground failure: Failure to maintain a continuous path to ground can expose entire electrical systems to damage, and can expose the workers using unprotected equipment to electrical hazards.
- Damaged insulation: Insulation protecting current-carrying wires can become damaged over time, resulting in exposed wires. If the exposed hot and neutral wires touch, they can create a short circuit and an ignition source for fires.
- Sparking: Friction sparking is a form of mechanical heat created when two hard surfaces, at least one which is metal, impact (NFPA 1997, 1–66).
- Circuit overload: A circuit becomes overloaded when there are more appliances on the circuit than it can safely handle. When a circuit is overloaded, the wiring overheats and the fuse blows or the circuit breaker trips.
- Short circuit: A short circuit occurs when a bare hot wire touches a bare neutral wire or a bare grounded wire (or some other ground). The flow of extra current blows a fuse or trips a circuit breaker.
- Arcing: This occurs when an electric circuit that is carrying current is interrupted either intentionally or unintentionally (NFPA 1997, 1–66).

HUMAN ERROR IN ELECTRICAL FIRES

When equipment failure is the proximate cause of a fire, nearly all fires involving electrically powered equipment stem from foreseeable or avoidable human error (NFPA 2008, 8–135). Examples of these human errors include improper installation, lack of maintenance, improper use, and carelessness or oversight.

HAZARDOUS LOCATIONS AND THE NATIONAL ELECTRICAL CODE

The environment in which the electrical equipment is placed may present a fire hazard. Environments where concentrations of flammable vapors, ignitable fibers, or combustible dusts are present in sufficient concentrations could be ignited by the electrical equipment and installations in that area. To prevent the possibility of electrical equipment and wiring from igniting flammable and combustible vapors, all electrical wiring and equipment used in hazardous locations must meet proper design requirements. Sections 500.5 and 500.6 of the NEC classify locations according to the potential for the presence of hazardous materials in the atmosphere (NFPA 70 2014, 384–387). Furthermore, because the electrical equipment may create an ignition source, design specifications have been developed for the equipment to reduce the risk of fire and explosion hazard.

The NEC classifies hazardous locations into Classes I, II, and III. These major classes are further subdivided into Divisions 1 and 2. The classes differentiate between the type of material that may be present in the air and the divisions differentiate between the circumstances that may lead to the presence of hazardous concentrations of the material. Within each class, groupings are assigned. The groupings identify the hazardous material present in the environment. These classes, divisions, and groups are also used by the Occupational Safety and Health Administration (OSHA) for the purposes of classifying hazardous locations and selecting safe electrical equipment for these areas.

Class I Locations

Class I locations are those where flammable gases or vapors are or may be present in the air in quantities sufficient to produce explosive or ignitable mixtures. The most common types of Class I locations are those involved in the handling or processing of volatile flammable liquids such as gasoline, naphtha, and benzene (Earley and Sargent 2011, 728). Class I, Division 1 locations are locations where any of the following conditions exist (NFPA 70 2014, 384):

1. Ignitable concentrations of flammable gases or vapors exist under normal operating conditions.
2. Ignitable concentrations of such gases or vapors may exist frequently because of repair or maintenance operations or because of leakage.
3. Breakdown or faulty operation of equipment or processes might release ignitable concentrations of flammable gases or vapors and might also cause simultaneous failure of electric equipment.

Examples of Class I, Division 1 locations include locations where volatile flammable liquids or liquefied flammable gases are transferred from one container to another, the interiors of spray booths, areas containing open tanks of volatile flammable liquids, and all other locations where ignitable concentrations of flammable vapors or gases are likely to occur in the course of normal operations.

Class I, Division 2 locations are locations where any of the following conditions exist (NFPA 70 2014, 385):

1. Volatile flammable liquids or flammable gases are handled, processed, or used; however, the liquids, vapors, or gases will normally be confined within closed containers or closed systems from which they can escape only in case of accidental rupture or breakdown of such containers or systems or in case of abnormal operation of equipment.
2. Ignitable concentrations of gases or vapors are normally prevented by positive mechanical ventilation but might become hazardous through failure or abnormal operation of the ventilating equipment.
3. The location is adjacent to a Class I, Division 1 location; ignitable concentrations of gases or vapors might occasionally be communicated to the area unless such communication is prevented by adequate positive pressure ventilation from a source of clean air and effective safeguards against ventilation failure are provided.

This classification usually includes locations where volatile flammable liquids or flammable gases or vapors are used but would become hazardous only in case of an accident or some unusual operating condition.

COMMON AND SPECIAL HAZARDS 49

Class I locations can also be further subdivided into groups. A group is based on the specific type of hazardous material in the location. The Class I Hazardous Location Atmospheres Groups include the following (NFPA 2014, 386–387):

- Group A typical: acetylene
- Group B typical: hydrogen
- Group C typical: ethylene
- Group D typical: propane

Therefore, a Class I, Division 1, Group C location is a location in an environment in which ethylene vapors or a closely similar material are present in sufficient concentration to ignite under normal conditions.

Class II Locations

Class II locations are those that are hazardous due to the presence of combustible dust. A Class II, Division 1 location may include any of the following conditions (NFPA 70 2014, 385):

1. Combustible dust is in the air under normal operating conditions in quantities sufficient to produce explosive or ignitable mixtures.
2. Mechanical failure or abnormal operation of machinery or equipment might cause the production of such explosive or ignitable mixtures and might also provide a source of ignition through simultaneous failure of electric equipment, operation of protection devices, or from other causes.
3. Combustible dusts of an electrically conductive nature may be present in hazardous quantities.

Examples of combustible dusts that are electrically nonconductive include dusts produced in the handling and processing of grain and grain products, pulverized sugar and cocoa, oil meal from beans and seed, and dried hay. Only Group E dusts are electrically conductive for classification purposes. Dusts containing magnesium or aluminum are particularly hazardous, and the use of extreme caution is necessary to avoid ignition and explosion.

Class II, Division 2 locations are those where the following conditions exist (NFPA 70 2014, 385):

1. Combustible dust is not normally in the air in quantities sufficient to produce explosive or ignitable mixtures, and dust accumulations are normally insufficient to interfere with the normal operation of electrical equipment or other apparatuses, but combustible dust may be in suspension in the air as a result of infrequent malfunctioning of handling or processing equipment.
2. Combustible dust accumulations on, in, or in the vicinity of the electrical equipment may be sufficient to interfere with the safe dissipation of heat from electrical equipment or it may be ignitable by abnormal operation or failure of electrical equipment.

Class II Hazardous Location Atmosphere Groups include the following (NFPA 70 2014, 386-387):

- Group E: Atmospheres containing combustible metal dusts, including aluminum, magnesium, and their commercial alloys, or other combustible dusts whose particle size, abrasiveness, and conductivity present similar hazards in the use of electrical equipment.
- Group F: Atmospheres containing combustible carbonaceous dusts that have more than 8 percent total entrapped volatiles or dusts that have been sensitized by other materials so that they present an explosion hazard, including carbon black, charcoal, coal, and coke dusts.
- Group G: Atmospheres containing combustible dusts not included in Group E or F, including flour, grain, wood, plastic, and chemicals.

Class III Locations

Class III locations are those that are hazardous because of the presence of easily ignitable fibers or flyings but where such fibers or flyings are not likely to be in suspension in the air in quantities sufficient to produce ignitable mixtures (NFPA 70 2014, 385). Class III, Division 1 locations are locations where easily ignitable fibers or materials producing combustible flyings are handled, manufactured, or used. Such locations usually include some parts of rayon, cotton, and other textile mills, combustible fiber manufacturing and processing plants, clothing manufacturing plants, and woodworking plants. Easily ignitable fibers and flyings include rayon, cotton (including cotton linters and cotton waste), jute, and hemp. Class III, Division 2 locations are locations where easily ignitable fibers are stored or handled, other than in the process of manufacturing.

DEFINING HAZARDOUS LOCATIONS

In addition to classifying areas where a hazardous environment exists, areas beyond the hazardous location should also be identified as they may also be subject to potential environmental fire hazards. Therefore, the NECs and OSHA standards stipulate hazard classifications for areas adjacent to those in which a hazardous environment is present. Under 29 CFR § 1910.307, OSHA stipulates that NFPA 70: NEC should be followed when determining the type and design of equipment used in hazardous locations (USDOL 2013c, 29 CFR. § 1910.307). OSHA standards further stipulate that for dip-tank vapor areas, electrical wiring appropriate to the class and division shall be provided (USDOL 2013a, 29 C.F.R. § 1910.125).

SAFE DESIGN OF ELECTRICAL EQUIPMENT

OSHA established the requirements for electric equipment and wiring in locations that are classified. The type of equipment approved for use in hazardous locations must be selected on the basis of the properties of the flammable vapors, liquids, or gases or combustible dusts or fibers that may be present therein and the likelihood that a flammable or combustible concentration or quantity is present.

When equipment is designed specifically for a hazardous environment, it is deemed intrinsically safe. Intrinsic safety is a protection concept employed in potentially explosive atmospheres. Intrinsic safety relies on the electrical apparatus being designed such that it is unable to release sufficient energy, by either thermal or electrical means, to cause an ignition of a flammable gas. Equipment, wiring methods, and installations of equipment in hazardous (classified) locations should be intrinsically safe, approved for the hazardous (classified) location, or safe for the hazardous (classified) location. Requirements for each of these options are as follows (USDOL 2013c, 29 C.F.R. § 1910.307):

1. Equipment and associated wiring approved as intrinsically safe shall be permitted in any hazardous (classified) location for which it is approved.

2. Equipment shall be approved not only for the class of location but also for the ignitable or combustible properties of the specific gas, vapor, dust, or fiber that will be present.
3. Equipment shall be properly marked according to the applicable electrical codes.
4. Equipment that is safe for the location shall be of a type and design that the employer demonstrates will provide protection from the hazards arising due to the combustibility and flammability of vapors, liquids, gases, dusts, or fibers.
5. Equipment that has been approved for a Division 1 location may be installed in a Division 2 location of the same class and group.
6. General-purpose equipment or equipment in general-purpose enclosures may be installed in Division 2 locations if the equipment does not constitute a source of ignition under normal operating conditions.

Class I Electrical Equipment Requirements

When examining the requirements for electrical equipment used in hazardous locations involving Class I locations, one term is of importance: explosionproof. Explosionproof equipment has been designed and constructed to withstand an internal explosion without creating an external explosion or fire (Earley and Sargent 2011, 26). Explosionproof equipment is permitted for Class I, Division 1 or Division 2 locations (Earley and Sargent 2011, 722). Therefore, electrical equipment capable of creating ignition capable sparks or arcs should not be taken into Class I locations. Explosionproof electrical conduit and boxes are designed in a way that the energy available from an arc or spark from the equipment is not sufficient to ignite the flammable vapors. When an explosion occurs within the enclosure or conduit of explosionproof equipment, the hot gases are confined within the equipment and prevented from igniting the flammable vapors or gases outside (Earley and Sargent 2011, 728).

Class II Electrical Equipment Requirements

Class II locations entail potential hazards created by the presence of dusts in concentrations sufficient to ignite. Equipment designed for use in Class II locations may be classified as either dust-tight or dustproof. Dust-tight equipment is constructed so that dust will not enter the enclosing case under specified test conditions. Dustproof or dust-ignition-proof equipment is constructed or protected so that dust will not interfere with its successful operation. Dust-ignition-proof equipment for use in Class II hazardous locations, as defined in the NEC, is tested with respect to acceptability of operation in the presence of combustible dusts in air. Equipment intended for use in a Class II, Division 1 location must be dust-ignition-proof and designated for use in Class II, Division 1 locations. Class I equipment is not necessarily suitable for Class II locations. Dust-ignition-proof equipment used in Class II locations is designed to prevent the ignition of layers of dust which may increase the operating temperature, while Class I equipment does not address this concern (Earley and Sargent 2011, 754). Dust-proof ignition proof enclosures are not required to be explosion proof. Explosion proof enclosures are not necessarily dust proof; however, they are allowed in Class II locations if the equipment is dual rated for Class II locations (Earley and Sargent 2011, 755).

Class III Electrical Equipment Requirements

Class III areas include textile, clothing, mattress, cotton, batting, rayon, cottonseed, woodworking industries, and those that process similar materials. Easily ignitable fibers and flyings include rayon, cotton, sisal, jute, hemp, cocoa fibers, oakum, Spanish moss, excelsior, and similar materials. Electrical equipment in Class III, Division 1 locations should be installed to prevent heat buildup or entrance of fibers or flyings into enclosures where a spark could ignite them, causing fire. The equipment shall also meet maximum surface temperature ratings (NFPA 2008, 8–171).

For Class III, Division 2 locations, equipment such as lamps, lamp holders, fuses, and circuit breakers have enclosures similar to those found in Class III, Division 1 locations. The temperature limits found with Class III, Division 1 locations also apply to Class III, Division 2 locations (NFPA 2008, 8–171).

NATIONALLY RECOGNIZED TESTING LABORATORIES

OSHA requires that only approved electrical equipment be used in the workplace. Being approved for a purpose means that the equipment is suitable for a particular application as determined by a recognized testing laboratory, inspection agency, or other organization concerned with product evaluation as part of its labeling or listing program (USDOL DTSEM 2013). Nationally recognized testing laboratories (NRTLs) go through a rigorous evaluation procedure before they are recognized by OSHA. NRTLs must have the capability, control programs, complete independence, and reporting and complaint handling procedures to test and certify specific types of products for workplace safety. This means, in part, that an organization must have the necessary capability both as a product-safety-testing laboratory and as a product-certification body to receive OSHA recognition as an NRTL. Equipment listed or approved by these organizations is acceptable for use in the workplace. The following organizations are currently recognized by OSHA as NRTLs (USDOL DTSEM, 2013):

- Applied Research Laboratories (ARL)
- Canadian Standards Association (CSA) (also known as CSA International)
- Communication Certification Laboratory (CCL)
- Curtis-Straus LLC (CSL)
- FM Approvals LLC (FM)
- Intertek Testing Services NA (ITSNA) (formerly ETL)
- MET Laboratories (MET)
- NSF International (NSF)
- OPS Evaluation Services
- SGS North America, Inc.
- Southwest Research Institute (SWRI)
- TUV Rheinland PTL, LLC (TUVPTL)
- TUV SUD America, Inc. (TUVAM)
- TUV SUD Product Services GmbH (TUVPSG)
- TUV Rheinland of North America, Inc. (TUV)
- Underwriters Laboratories Inc. (UL)

FLAMMABLE LIQUIDS AND COMBUSTIBLE LIQUIDS

Flammable and combustible liquids pose a unique hazard in the workplace primarily because of the amount of fuel they can provide for a fire and the relatively low heat source necessary to ignite them. Flammable and combustible liquids are classified as either flammable or combustible based on their flash points and boiling points. The term flash point means the minimum temperature at which a liquid gives off vapor within a test vessel in sufficient concentration to form an ignitable mixture with air near the surface of the liquid. When a flammable or combustible liquid is involved in a fire, it is the vapor over the surface of the liquid that burns. The lower the flash point, the more readily a sufficient concentration of vapor is present at a lower temperature.

Classification Systems for Flammable and Combustible Liquids

There are many different acceptable classification systems for flammable and combustible liquids. Two more common systems used in industry are NFPA 30: Flammable and Combustible Liquids Code and the Global Harmonization System (GHS) which has been adopted by OSHA.

NFPA CLASSIFICATION SYSTEM

Under the NFPA Classification System, flammable liquids are liquids having a flash point below 100°F, except any mixture having components with flash points of 100°F or higher, the total of which makes up 99 percent or more of the total volume of the mixture. At normal room temperature, flammable liquids will produce a vapor sufficient to ignite without requiring heating of the liquid to do so. Flammable liquids are known as Class I liquids. Class I liquids are divided into three classes (NFPA 30 2012, 21):

1. Class IA includes liquids having flash points below 73°F and having a boiling point below 100°F.
2. Class IB includes liquids having flash points below 73°F and having a boiling point at or above 100°F.
3. Class IC includes liquids having flash points at or above 73°F but below 100°F.

Combustible liquids typically require some external heating to produce a sufficient concentration of vapors. Combustible liquids are liquids having a flash point at or above 100°F and are divided into two classes. Class II liquids include those liquids with flash points at or above 100°F and below 140°F (NFPA 30 2012, 21).

Class III liquids include those liquids with flash points at or above 140°F. Class III liquids are divided into two subclasses (NFPA 30 2012, 21):

1. Class IIIA liquids include those liquids with flash points at or above 140°F and below 200°F.

2. Class IIIB liquids include those liquids with flash points at or above 200°F. OSHA combustible liquid requirements do not apply to Class IIIB liquids, only to Class IIIA liquids.

GLOBALLY HARMONIZED SYSTEM

The GHS of classification and labeling of chemicals was adopted by the United Nations in 2003. The GHS includes criteria for the classification of health, physical and environmental hazards, as well as specifying what information should be included on labels of hazardous chemicals as well as safety data sheets (USDOL, 2014a). This new system affects several OSHA standards, Department of Transportation regulations, and Environmental Protection Agency regulations. Areas affected include labeling and classification systems. The classification system establishes the criteria for classifying chemicals according to several physical hazards, health hazards, and environmental hazards.

GHS CATEGORIES OF FLAMMABLE LIQUIDS

The GHS classification system for flammable liquids is of greatest importance for safety professionals. This system has been adopted by the OSHA and affects the Hazard Communication Standard requirements and the Flammable Liquids Standard requirements. Under the new GHS requirements, flammable liquids are classified by their flash point and boiling point characteristics. Table 3.1 summarizes this classification system (USDOL, 2013b).

Along with this classification system for flammable liquids, OSHA has also implemented additional handling requirements for flammable liquids

Table 3.1. Summary of GHS Classification System for Flammable Liquids

Category 1 Flammable Liquids	Category 2 Flammable Liquids	Category 3 Flammable Liquids	Category 4 Flammable Liquids
Category 1 includes liquids with flash points below 73.4°F (23°C) and a boiling point at or below 95°F (35°C)	Category 2 includes liquids with flash points below 73.4°F (23°C) and a boiling point above 95°F (35°C)	Category 3 includes liquids with flash points at or above 73.4°F (23°C) and at or below 140°F (60°C)	Category 4 includes liquids with flash points above 140°F (60°C) and at or below 199.4°F (93°C)

Source: USDOL, 2013b.

COMMON AND SPECIAL HAZARDS

that are heated. For example, under the Flammable Liquids Standards, when a Category 3 liquid with a flash point at or above 100°F is heated for use to within 30°F of its flash point, it shall be handled as a Category 3 liquid (USDOL, 2013b). In addition, when a Category 4 flammable liquid is heated for use to within 30°F of its flash point, it shall be handled as a Category 3 liquid and when liquid with a flash point greater than 199.4°F is heated for use to within 30°F of its flash point, it shall be handled as a Category 4 flammable liquid (USDOL, 2013b).

UPPER AND LOWER EXPLOSIVE LIMITS

As discussed previously, for a flammable liquid to burn, it must evolve enough vapors at its surface in a sufficient concentration. At temperatures below the flash point, the liquid cannot evaporate quickly enough to generate enough vapors. The concentration of flammable or combustible vapors in air also determines whether there will be an ignitable concentration or not. If the vapor concentration in air is too low, there will not be enough vapors to ignite. We refer to this concentration as being below the lower flammable limit (LFL) (also referred to as the lower explosive limit [LEL]). People commonly refer to the vapors as being "too lean."

On the other hand, if the vapor concentrations in air are above the upper flammable limit (UFL) (also referred to as the upper explosive limit [UEL]), the vapors will not ignite. This is commonly referred to as the vapors being "too rich." Therefore, flammable and combustible vapors must be present in concentrations in air above the LFL or LEL but below the UFL or UEL to burn.

From a safety standpoint, you will see that in many situations, ventilation requirements have been established by OSHA and the NFPA to reduce the potential for accidental ignition of flammable and combustible vapors. The purpose of the ventilation requirements is to keep the concentrations of vapors in the air well below their LFLs.

OSHA'S REQUIREMENTS FOR FLAMMABLE AND COMBUSTIBLE LIQUID STORAGE AND HANDLING

Flammable and combustible liquids can be stored in the workplace in a number of ways. Types of flammable and combustible liquid storage include portable container storage, tank storage, storage cabinets, and storage rooms. In 2012,

OSHA converted the requirements in the standards for handling and storage of flammable and combustible liquids from the NFPA system to the GHS system.

Containers and Portable Tanks

It is common in industry to store flammable and combustible liquids in containers. OSHA defines a container as any can, barrel, or drum. A barrel is a container that has a capacity of 42 gallons, while a safety can is an approved container with a maximum capacity of 5 gallons with a spring-closing lid and spout cover. Safety cans are designed to relieve internal pressure when they are subjected to heating. Drums have a capacity of up to 60 gallons and tanks have a capacity of more than 60 gallons. Barrels, safety cans, drums, and tanks used in the workplace must be approved for this type of use by a NRTL or the U.S. Department of Transportation (USDOT) (USDOL 2013b, 29 C.F.R. § 1910.106).

Flammable and combustible liquid containers may also be made of glass and plastic, provided that no more than a 1-gallon capacity container may be used for a Category 1 flammable (see table 3.2). The maximum allowable quantities allowed by OSHA appear in table 3.2 (USDOL 2013b, 29 C.F.R. § 1910.106).

Transferring Flammable and Combustible Liquids

A common practice in industry is the transferring of flammable and combustible liquids from one storage container to another. For example, a worker

Table 3.2. Maximum Allowable Quantities of Flammable and Combustible Liquids by Container Type

Container Type	Category 1	Category 2	Category 3	Category 4
Glass or approved plastic	1 qt	1 qt	1 gal	1 gal
Metal (other than DOT drums)	1 gal	5 gal	5 gal	5 gal
Safety cans	2 gal	5 gal	5 gal	5 gal
Metal drums (DOT specifications)	60 gal	60 gal	60 gal	60 gal
Approved portable tanks	660 gal	660 gal	660 gal	660 gal

Source: USDOT, 2013b.

Note: Container exemptions: (a) Medicines, beverages, foodstuffs, cosmetics, and other common consumer items, when packaged according to commonly accepted practices, shall be exempt from the requirements of 29 CFR 1910.106(d)(2)(i) and (ii).

may need to transfer a flammable liquid from a 55-gallon drum to a safety can for use on the production line. Several hazards may present during this type of task, including the potential for spilling the liquid, release of vapors into the work area, and the accidental ignition of the vapors. First, flammable liquids must always be kept in approved, covered containers when not in use. Where flammable or combustible liquids are used or handled, except in closed containers, means shall be provided to dispose promptly and safely of leakage or spills. To prevent possible ignition of the vapors, Category 1 or 2 flammable liquids, or Category 3 flammable liquids with a flash point below 100°F may be used only where there are no open flames or other sources of ignition within the possible path of vapor travel.

To prevent both the release of flammable vapors and the buildup of static electricity, flammable liquids should be transferred into vessels, containers, or portable tanks within a building only through a closed piping system, from safety cans by means of a device drawing through the top, or from a container or portable tank by gravity through an approved self-closing valve. Transferring by means of air pressure on the container or portable tank is prohibited.

Adequate precautions should be taken to prevent the ignition of flammable vapors. Sources of ignition include, but are not limited to, open flames; lightning; smoking; cutting and welding; hot surfaces; frictional heat; static, electrical, and mechanical sparks; spontaneous ignition, including heat-producing chemical reactions; and radiant heat.

Due to the extreme fire hazards of Category 1 or 2 flammable liquids, or Category 3 flammable liquids with a flash point below 100°F, they should not be dispensed into containers unless the nozzle and container are electrically interconnected through the use of a bonding wire (see figure 3.1).

An alternative to using a bonding wire is the use of a metallic floor plate while the fill stem is connected to the container (USDOL 2013b, 29 C.F.R. § 1910.106). To prevent the accumulation of flammable vapors in hazardous concentrations, transferring Category 1 or 2 flammable liquids, or Category 3 flammable liquids with a flash point below 100°F should not be done inside buildings unless adequate ventilation is provided. Where mechanical ventilation is required, it should be kept in operation while transferring takes place.

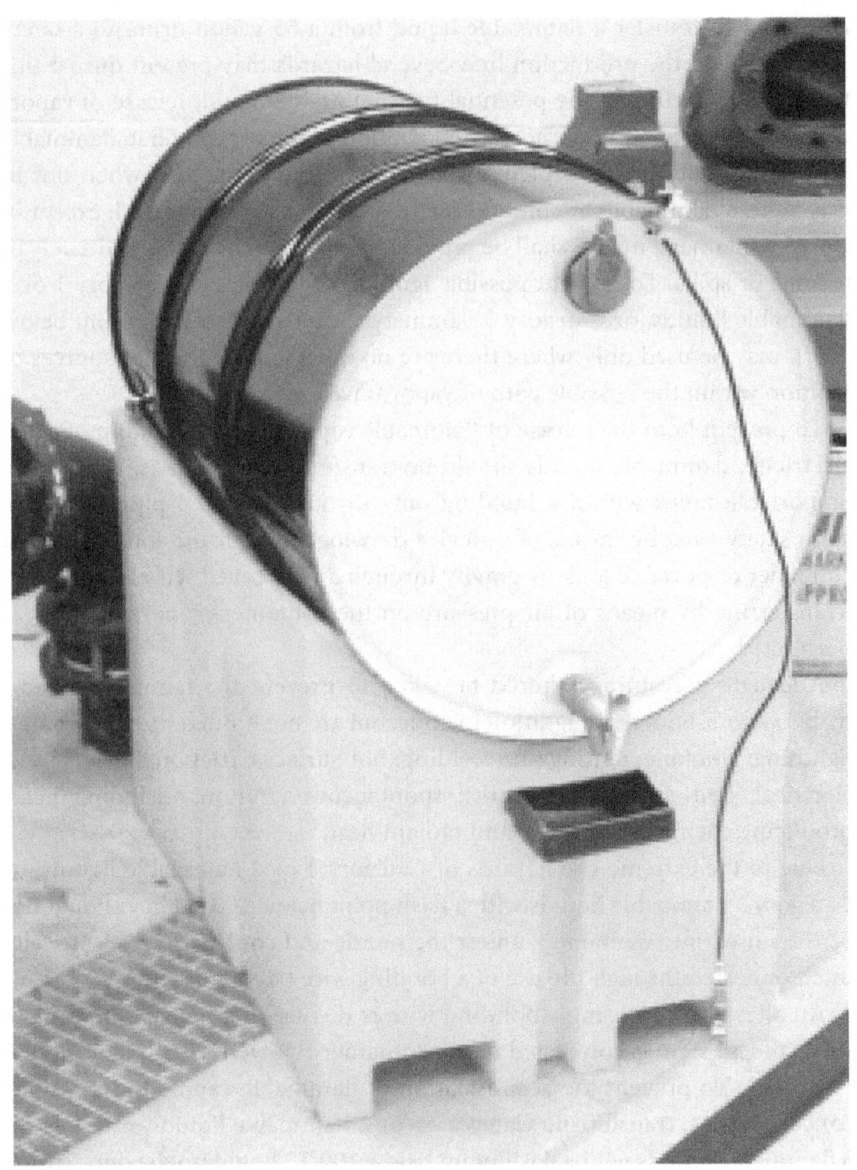

FIGURE 3.1
Typical Flammable Liquids Transferring Arrangement

Storage Cabinets

Storage cabinets are commonly used to store quantities of flammable and combustible liquids in containers, drums, and barrels. The purpose of the storage cabinet is to protect the liquids inside the cabinet and, if subjected to heat, to limit the internal temperature to not more than 325°F (USDOL 2013b, 29 C.F.R. § 1910.106). The cabinets are constructed to withstand a ten-minute fire test. Additional features of the cabinet include steel construction (wood construction is also acceptable), self-closing doors, venting to the outside where required by code, and raised at least 2 in. above the bottom of the cabinet. Storage cabinets should be approved for use for flammable liquid storage, meeting NFPA standards for the design and construction of flammable and combustible liquid storage cabinets. The cabinets should also be labeled in conspicuous lettering "Flammable—Keep Fire Away." Figure 3.2 depicts a typical flammable liquids storage cabinet.

Inside Storage Rooms

Storage rooms used in the industry specifically for the storage of flammable and combustible liquids must be professionally designed and constructed. The major design aspects of the storage room include fire protection, spill containment, ventilation, fire resistance, and proper electrical wiring and equipment. To prevent the accidental release of spilled flammable or combustible liquids from the storage room to other parts of the facility, openings such as doorways should be protected and the room should be liquid tight where the walls join the floor. Common methods for containing spills at doorways leading into the storage room involve the use of liquid-tight raised sills at least 4 in. high, ramps, or a floor area in the storage room that is at least 4 in. below the surrounding floor to hold spilled liquids within the room. An alternative method of spill control for the room can be the use of a floor drain that evacuates the spilled liquid out to a safe location, typically a storage vault located outside the facility which can be emptied. Rated self-closing fire doors should be provided to the room and if windows are provided in the room, they should be approved for such use.

Fire protection for a storage room is accomplished using fire extinguishers. OSHA requires that at least one portable fire extinguisher having a rating of not less than 12-B units be located outside of, but not more than 10 ft. from, the door

FIGURE 3.2
Approved Storage Cabinet. *Source*: Photo courtesy of Justrite Manufacturing

opening into any room used for storage. If the storage area contains Category 1, 2, or 3 liquids, at least one portable fire extinguisher having a rating of not less than 12-B units must be located not less than 10 ft. or more than 25 ft. from any storage area located outside of a storage room but inside a building. Sources of ignition from open flames and smoking are not permitted in flammable or combustible liquid storage areas (USDOL 2013b, 29 C.F.R. § 1910.106).

OSHA has also established the maximum quantity of flammable and combustible liquids to be stored in an inside storage room. Factors to consider

Table 3.3. Storage in Inside Rooms

Fire Protection Provided	Fire Resistance	Maximum Size (Floor Area)	Total Quantities Allowed (gals./sq. ft./floor area)
Yes	2 hours	500 sq ft.	10
No	2 hours	500 sq ft.	5
Yes	1 hour	150 sq ft.	4
No	1 hour	150 sq ft.	2

Source: USDOL, 2013b.

when determining maximum quantities include the fire rating construction of the room and the presence of a sprinkler system (see table 3.3). There shall be one clear aisle at least 3-ft wide in the storage room to allow for access, and containers over a 30-gallon capacity shall not be stacked one upon the other.

From a hazardous environment standpoint, flammable and combustible liquid storage rooms may be classified as either Class I, Division 1 locations or Class I, Division 2 locations, depending on the class of liquid stored and whether activities inside the storage room create flammable vapor–air mixtures. If the room is only used for storage of Category 1 or 2 flammable liquids, or Category 3 flammable liquids with a flash point below 100°F (i.e., no transferring), then the electrical wiring and equipment located in the storage room should meet Class I, Division 2 requirements. If the storage room is used for the storage of Category 3 flammable liquids with a flash point at or above 100°F and Category 4 flammable liquids, the electrical equipment should be approved for general use. If flammable vapor–air mixtures can exist under normal operations, then the electrical wiring should meet Class I, Division 1 requirements.

Ventilation is required in storage rooms to keep the flammable and combustible vapors well below their LELs. The ventilation systems should change over the entire volume of air inside the room at least six times per hour (USDOL 2013b, 29 C.F.R. § 1910.106). Where gravity ventilation is provided, the fresh air intake, as well as the exhaust outlet from the room, shall be on the exterior of the building in which the room is located.

Storage Tanks

Storage tanks are classified according to their operating pressures as low-pressure tanks, atmospheric tanks, and pressure vessels. Atmospheric tanks are designed to operate at pressures from atmospheric pressure through 0.5 psig; low-pressure tanks are designed to operate at pressures above 0.5 psig but not more than 15 psig; pressure vessels are designed to operate at pressures above 15 psig (USDOL 2013b, 29 C.F.R. § 1910.106).

Tanks used for the storage of flammable and combustible liquids must be built of steel or other approved materials that are compatible with the liquids to be stored, and their design and construction must meet approved standards. Examples of organizations that have standards for tank construction include FM Global, Underwriters' Laboratories, the American Petroleum Institute (API), and the American Society of Mechanical Engineers (ASME) Boiler and Pressure Vessels Code. The safety requirements for storage tanks for flammable and combustible liquids can be further classified based on whether they are intended for aboveground or underground use.

Outside Aboveground Tanks

Methods for controlling and preventing fires involving outside, aboveground tanks include the separation of storage tanks, diking and drainage, and venting. Separating aboveground storage tanks reduces the spread of fire from one tank to another and provides access to the tanks in the event of a fire. Several factors are examined when determining the minimum separation between two aboveground storage tanks containing flammable and combustible liquids. These factors include the diameters of the tanks, the tank capacities, and the characteristics of liquids being stored. Following are some of the criteria used to determine the minimum spacing between the aboveground storage tanks (USDOL 2013b, 29 C.F.R. § 1910.106):

- The minimum distance between any two flammable or combustible liquid storage tanks shall not be less than 3 ft.
- The distance between any two adjacent tanks shall not be less than one-sixth of the sum of their diameters.
- When the diameter of one tank is less than one-half the diameter of the adjacent tank, the distance between the two tanks shall not be less than one-half the diameter of the smaller tank.
- Where crude petroleum in conjunction with production facilities is in uncongested areas with capacities not exceeding 126,000 gallon (3,000 barrels), the distance between such tanks shall not be less than 3 ft.
- Where unstable flammable or combustible liquids are stored, the distance between such tanks shall not be less than one-half the sum of their diameters.

- The minimum separation between an LP gas container and a flammable or combustible liquid storage tank shall be 20 ft., except in the case of flammable or combustible liquid tanks operating at pressures exceeding 2.5 psig or equipped with emergency venting that will permit pressures to exceed 2.5 psig.

Venting is used on storage tanks to maintain a constant pressure inside the tank. Pressure inside the tank can change due to temperature changes and the displacement of the liquid inside by either adding or removing liquid. To control pressure changes inside the tank, air movement into and out of it must be allowed. But this can also create additional fire hazards when the flammable or combustible vapors escape through the vent piping and come in contact with a heat source. To alleviate potential fire hazards, when vent pipe outlets for tanks storing Category 1 or 2 flammable liquids or Category 3 flammable liquids with a flash point below 100°F liquids are adjacent to buildings or public ways, they must be located such that the vapors are released at a safe point outside of buildings and not less than 12 ft. above the adjacent ground level. The vent outlets must also be located so that eaves or other obstructions will not trap flammable vapors. The venting equipment must meet applicable standards. Venting also provides protection against overpressure from any pump discharging into the tank or vessel if the pump discharge pressure exceeds the design pressure of the tank or vessel.

In addition to venting, emergency relief venting is required on storage tanks to relieve excessive internal pressure caused by fires. This protection may take the form of a floating roof, a lifter roof, a weak roof-to-shell seam, or another approved pressure-relieving construction. Emergency relief venting should meet design capacities to prevent the rupture of the tank. A commercial tank-venting device must have stamped on it the opening pressure, the pressure at which the valve reaches the full open position, and the flow capacity at the latter pressure expressed in cubic ft. per hour of air at 60°F and at a pressure of 14.7 psia.

Drainage and diking are means to prevent the accidental discharge of liquid into adjacent properties or waterways or the area around an aboveground tank. OSHA requirements for diking and drainage include the following (USDOL 2013b, 29 C.F.R. § 1910.106):

1. When drainage is used, it must terminate in vacant land or other area or in an impounding basin having a capacity not smaller than that of the largest tank served.
2. When diking is used, the volume of the diked area should have a volumetric capacity no less than the greatest amount of liquid that can be released from the largest tank within the diked area.
3. If the diked area encloses more than one tank, the capacity should be calculated by deducting the volume of the tanks other than the largest tank below the height of the dike.
4. For a tank or group of tanks with fixed roofs containing crude petroleum with boilover characteristics, the volumetric capacity of the diked area shall be not less than the capacity of the largest tank served by the enclosure, assuming a full tank.
5. The capacity of the diked enclosure should be calculated by deducting the volume below the height of the dike of all tanks within the enclosure.
6. When constructing the walls for a diked area, the walls may be earth, steel, concrete, or solid masonry designed to be liquid tight and to withstand a full hydrostatic head.
7. Earthen walls 3 ft. or more in height should have a flat section at the top not less than 2 ft. wide.
8. The maximum height of the walls is an average height of 6 ft. above interior grade.
9. The diked areas shall also be kept free from loose combustible materials.

Underground Tanks

Underground storage tanks containing flammable and combustible liquids pose both fire hazards and environmental hazards. Underground storage tanks should be placed in the ground in areas that are free from the potential hazards created by extreme loading generated by buildings and vehicle traffic. The backfill material should be noncorrosive and inert material, such as clean sand, earth, or gravel, well tamped into place, ensuring that the tanks are handled carefully to prevent breaking a weld, puncturing or damaging the tank, or scraping off the protective coating of coated tanks (USDOL 2013b, 29 C.F.R. § 1910.106).

Due to the potential for release of liquid and vapors from an underground tank, the distance from any part of a tank storing Category 1 or 2 flammable liquids, or Category 3 flammable liquids with a flash point below 100°F to the nearest wall of any basement or pit shall be not less than 1 ft. and to any property line that may be built upon, not less than 3 ft. The distance from any part of a tank storing Category 3 flammable liquids with a flash point at or above 100°F or Category 4 flammable liquids to the nearest wall of any basement, pit, or property line shall be not less than 1 ft. (USDOL 2013b, 29 C.F.R. § 1910.106).

To prevent leaking and deterioration of the underground storage tanks due to corrosion, corrosion protection for the tank and its piping shall be provided by one or more of the following methods (USDOL 2013b, 29 C.F.R. § 1910.106):

- Protective coatings or wrappings
- Cathodic protection
- Corrosion-resistant construction materials

Like aboveground storage tanks, underground storage tanks should be equipped with vents to maintain adequate pressure inside the tank during filling and unfilling operations. Design considerations for the vent piping include the following (USDOL 2013b, 29 C.F.R. § 1910.106):

1. The vents must be located in a manner such that the vapors escaping from the tank through the vent cannot be accidentally ignited.
2. Vent pipes from tanks storing Category 1 or 2 flammable liquids, or Category 3 flammable liquids with a flash point below 100°F must be located so that the discharge point is outside of buildings, higher than the fill pipe opening, and not less than 12 ft. above the adjacent ground level.
3. If the vent pipe is less than 10 ft. in length or has a nominal inside diameter greater than 2 in., the outlet shall be provided with a vacuum-and-pressure-relief device, or there shall be an approved flame arrester located in the vent line at the outlet or within the approved distance from the outlet.
4. Vent pipes from tanks storing Category 3 flammable liquids with a flash point at or above 100°F or Category 4 flammable liquids shall terminate outside of the building and higher than the fill pipe opening.

Additional design considerations for underground tanks include tank supports and proper locating of tanks to protect against potential damage to the storage tanks due to floodwaters. All tanks, whether shop built or field erected, must be designed and strength tested in accordance with the applicable codes. The ASME code stamp, API monogram, or the label of the Underwriters' Laboratories on a tank shall be evidence of compliance with this strength test. All piping, valves, and fittings should be approved for use with flammable and combustible liquids. These components should be installed and maintained according to applicable codes and standards.

Tank Vehicle and Tank Car Loading and Unloading

Transferring flammable and combustible liquids from tank vehicles and tank cars poses the same types of hazards one encounters when transferring these liquids from one portable container to another. These hazards include the potential for the presence of flammable and combustible vapors and the potential to create a static electrical ignition source. To protect adjacent property, loading or unloading facilities involving Category 1 or 2 flammable liquids, or Category 3 flammable liquids with a flash point below 100°F should not be sited any closer than 25 ft. to the adjacent property or 15 ft. loading and unloading facilities for Category 3 flammable liquids with a flash point at or above 100°F or Category 4 flammable liquids (USDOL 2004b, 29 C.F.R. § 1910.106b). Equipment used in the transfer of liquids from tank cars and rail cars should be approved for use and for the category of liquid being transferred. To prevent accidental spills or overfilling, approved valves should be used that are either self-closing or automatically closing when the vehicle is full or filled to a certain level.

The transfer of liquids to or from the tank car or tank truck can create an electrical potential difference between the tank car or tank truck and the tank being transferred to. As is the case when transferring between two containers, a means of bonding the tank car or truck to the tank should be provided when Category 1 or 2 flammable liquids, or Category 3 flammable liquids with a flash point below 100°F are loaded or when Category 3 flammable liquids with a flash point at or above 100°F (37.8°C) or Category 4 flammable liquids are loaded into vehicles that may contain vapors from previous cargoes of Category 1 or 2 flammable liquids, or Category 3 flammable liquids with a flash point below 100°F (USDOL 2013b, 29 C.F.R. § 1910.106). Bonding shall be accomplished by connecting a metallic bond wire to the fill stem or to some part of the rack structure to some metallic part in electrical contact with the cargo tank of the tank vehicle. The bonding connection is made to the vehicle or tank before dome covers are raised and remains in place until filling is completed and all dome covers have been closed and secured (USDOL 2013b, 29 C.F.R. § 1910.106).

WORKPLACE PRACTICES

To protect employees and property from fire hazards involving flammable and combustible liquids in the workplace, control measures and workplace procedures should be adopted and followed. OSHA, NFPA, and various organizations require procedures to be followed in the storage and handling of flammable and combustible liquids. These regulations also stipulate requirements for housekeeping, fire protection, and the control of ignition sources.

Controlling Sources of Ignition

In locations where flammable vapors may be present, precautions should be taken to prevent ignition by eliminating or controlling sources of ignition. Sources of ignition include open flames, lightning, smoking, cutting and welding, hot surfaces, frictional heat, sparks (static, electrical, and mechanical), spontaneous ignition, chemical and physical-chemical reactions, and radiant heat.

Maintenance and Repairs

When it is necessary to do maintenance work in a flammable or combustible liquid processing area, the work shall be authorized by a responsible representative of the employer. Hot work, such as welding or cutting operations, use of spark-producing power tools, and chipping operations, shall be permitted only under supervision of an individual in responsible charge, who shall inspect the area to be sure that it is safe for the work to be done and that safe procedures will be followed for the work specified.

Housekeeping

Maintenance and operating practices shall be in accordance with established procedures, which will control leakage and prevent the accidental escape of flammable or combustible liquids. Spills shall be cleaned up promptly. Adequate aisles shall be maintained for unobstructed movement of personnel and so that fire protection equipment can adequately reach all areas of the storage room.

Combustible waste material and residues in a building or unit operating area shall be kept to a minimum, stored in covered metal receptacles, and disposed of daily.

Housekeeping can also involve the removal of dust accumulations within the plant. Careful removal of lying dust can eliminate the possibility of secondary dust explosions occurring and help to prevent some ignition sources. Installing good dust extraction systems wherever there is a particularly dusty area of the plant should do this. Dust should be removed immediately using either a high-power explosion-proof vacuum cleaner or an internal vacuuming system, which removes dust and sends it to a central filtration system.

HYDROGEN

Hydrogen is a nontoxic, colorless gas with no odor. It is flammable and may form mixtures with air that are flammable or explosive. Hydrogen may react violently if combined with oxidizers, such as air, oxygen, and halogens. Hydrogen is an asphyxiant and may displace oxygen in a workplace atmosphere.

The concentrations at which flammable or explosive mixtures form are much lower than the concentration at which asphyxiation risk is significant (Voltaic 2006, 1).

Hydrogen can be found in a variety of industries serving several useful purposes. It can be stored in containers such as cylinders or it may be part of a tank, piping, and manifold system. Hydrogen can be used in a gaseous form or stored under pressure in a liquefied form. Regardless of the state it is stored in, hydrogen poses an extreme fire hazard; therefore, the safety standards for the storage and handling of hydrogen in the workplace should be closely adhered to.

The containers, whether they are cylinders or tanks, should meet applicable design and construction requirements. For example, OSHA requires hydrogen containers used in the workplace to be designed, constructed, and tested in accordance with appropriate requirements of ASME Boiler and Pressure Vessel Code, Section VIII—Unfired Pressure Vessels, and in accordance with USDOT specifications and regulations. Construction requirements include being equipped with safety relief devices to discharge upward and unobstructed to the open air. Piping and tubing shall conform to "Industrial Gas and Air Piping," Code for Pressure Piping, ANSI B31.1-1967 with addendum B31.1-1969 and shall be suitable for hydrogen service and for the pressures and temperatures involved. Valves, gauges, regulators, and other accessories shall be suitable for hydrogen service. Each portable container and manifolded hydrogen supply unit must be legibly marked with the name "HYDROGEN." The hydrogen storage location shall be permanently placarded as follows: "HYDROGEN—FLAMMABLE GAS—NO SMOKING—NO OPEN FLAMES" or the equivalent (USDOL 2013d, 29 C.F.R. § 1910.103).

To prevent the accidental ignition of flammable gas, systems containing hydrogen should be located above ground and away from potential sources of ignition such as electric power lines, flammable liquid piping, or piping of other flammable gases. The location of a hydrogen system, as determined by the maximum total contained volume of hydrogen, shall be in the order of preference as indicated by Roman numerals in table 3.4 (USDOL 2013d, 29 C.F.R. § 1910.103).

The rooms and facilities used to house hydrogen storage and piping equipment must meet special design considerations. Design features of these facilities include proper venting, fire-resistive building construction, and approved electrical installations. The minimum distance in ft. for hydrogen systems to outdoor exposures shall be in accordance with table 3.5 (USDOL 2013d, 29 C.F.R. § 1910.103). Protective structures can be used in lieu of the distances.

Table 3.4. Maximum Capacities of Gaseous Hydrogen Systems

Nature of Location	Size of Hydrogen System		
	Less than 3,000 CF	3,000 CF to 15,000 CF	In excess of 15,000 CF
Outdoors	I	I	I
In a separate building	II	II	II
In a special room	III	III	Not permitted
Inside buildings not in a special room and exposed to other occupancies	IV	Not permitted	Not permitted

Source: USDOL, 2013d, 29 C.F.R. 1910.103.

Table 3.5. Minimum Safe Distances for Gaseous Hydrogen Systems to Exposures

Type of Outdoor Exposure		Size of Hydrogen System		
		Less than 3,000 CF	In excess of 15,000 CF	In excess of 15,000 CF
1. Building or structure	Wood frame construction[a]	10	25	50
	Heavy timber, noncombustible or ordinary construction[a]	0	10	25[b]
	Fire-resistive construction[a]	0	0	0
2. Wall openings	Not above any part of a system	10	10	10
	Above any part of a system	25	25	25
3. Flammable liquids above ground	0 to 1,000 gallons	10	25	25
	In excess of 1,000 gallons	25	50	50
4. Flammable liquids below ground: 0 to 1,000 gallons	Tank	10	10	10
	Vent or fill opening of tank	25	25	25
5. Flammable liquids below ground: in excess of 1,000 gallons	Tank	20	20	20
	Vent or fill opening of tank	25	25	25
6. Flammable gas storage, either high pressure or low pressure	0 to 15,000 CF capacity	10	25	25
	In excess of 15,000 CF capacity	25	50	50
7. Oxygen storage	12,000 CF or less[c]			
	More than 12,000 CF[d]			
8. Fast burning solids such as ordinary lumber, excelsior, or paper		50	50	50
9. Slow-burning solids such as heavy timber or coal		25	25	25
10. Open flames and other sources of ignition		25	25	25
11. Air compressor intakes or inlets to ventilating or air-conditioning equipment		50	50	50
12. Concentration of people[e]		25	50	50

[a]Refer to NFPA No. 220 Standard Types of Building Construction for definitions of various types of construction (1969 Ed.).
[b]But not less than one-half the height of adjacent side wall of the structure.
[c]Refer to NFPA No. 51, gas systems for welding and cutting (1969).
[d]Refer to NFPA No. 566, bulk oxygen systems at consumer sites (1969).
[e]In congested areas such as offices, lunchrooms, locker rooms, time-clock areas.

Liquefied Hydrogen Systems

To be liquefied, hydrogen must be compressed and cooled to $-250°C$. In this state, hydrogen still possesses many of the hazards as it does in a gaseous state.

Proper storage and handling of liquid hydrogen is key to preventing fires caused by the material. Therefore, liquefied hydrogen containers used in the workplace should be designed, constructed, and tested in accordance with appropriate requirements of the ASME Boiler and Pressure Vessel Code, Section VIII—Unfired Pressure Vessels (1968), or applicable provisions of *API Standard 620, Recommended Rules for Design and Construction of Large, Welded, Low-Pressure Storage Tanks* (2nd ed., June 1963) (USDOL 2013d, 29 C.F.R. § 1910.103). Requirements for proper storage involving liquefied hydrogen in the workplace include the following (USDOL 2013d, 29 C.F.R. § 1910.103):

1. Portable containers should be designed, constructed, and tested in accordance with USDOT specifications and regulations.
2. Containers should be legibly marked to indicate "LIQUEFIED HYDROGEN—FLAMMABLE GAS."
3. Containers should be equipped with safety relief devices as required by applicable regulations.
4. Piping, tubing, fittings, and gasket and thread sealants should be suitable for hydrogen service at the pressures and temperatures involved and shall conform to the applicable standards.
5. Valves, gauges, regulators, and other accessories should be suitable for liquefied hydrogen service and for the pressures and temperatures involved.
6. Electrical wiring and equipment located within 3 ft. of a point where connections are regularly made and disconnected must meet requirements for Class I, Group B, Division 1 locations.
7. Electrical wiring and equipment located within 25 ft. of a point where connections are regularly made and disconnected or within 25 ft. of a liquid hydrogen storage container should be in accordance with Subpart S of the OSHA standards for Class I, Group B, Division 2 locations.

The location of liquefied hydrogen storage, as determined by the maximum total quantity of liquefied hydrogen, should be in the order of preference as indicated by Roman numerals in table 3.6 (USDOL 2013d, 29 C.F.R. § 1910.103):

The minimum distance in ft. from liquefied hydrogen systems of the indicated storage capacity located outdoors, in a separate building, or in a

Table 3.6. Maximum Capacities of Liquefied Hydrogen Systems

Nature of Location	Size of Hydrogen Storage (capacity in gallons)			
	39.63 (150 liters–50)	*51–300*	*301–600*	*In excess of 600*
Outdoors	I	I	I	I
In a separate building	II	II	II	Not permitted
In a special room	III	III	Not permitted	Not permitted
Inside buildings not in a special room and exposed to other occupancies	IV	Not permitted	Not permitted	Not permitted

Source: USDOL, 2013d, 29 C.F.R. 1910.103.
Note: This table does not apply to the storage in dewars of the type generally used in laboratories for experimental purposes.

special room, to any specified exposure should be in accordance with table 3.6 (USDOL 2013d, 29 C.F.R. § 1910.103).

Portable liquefied hydrogen containers of 50 gallon or less capacity, when housed inside buildings, not located in a special room, and exposed to other occupancies, should comply with the following minimum requirements (USDOL 2013d, 29 C.F.R. § 1910.103):

Be located 20 ft. from flammable liquids and readily combustible materials such as excelsior or paper,
2. Be located 25 ft. from ordinary electrical equipment and other sources of ignition including process or analytical equipment,
3. Be located 25 ft. from concentrations of people,
4. Be located 50 ft. from intakes of ventilation and air-conditioning equipment or intakes of compressors,
Be located 50 ft. from storage of other flammable gases or storage of oxidizing gases.

Containers should be protected against damage or injury due to falling objects or work activity in the area. They should be firmly secured and stored in an upright position; welding or cutting operations and smoking should be prohibited while hydrogen is in the room. As is the case with a hydrogen gas storage location, liquefied hydrogen storage must meet requirements for the construction and location of storage buildings, venting, noncombustible construction, electrical wiring, ventilation, and explosion venting.

COMMON AND SPECIAL HAZARDS

Table 3.7. Minimum Safe Distances for Liquefied Hydrogen Systems to Exposures[a,b]

	Liquefied Hydrogen Storage (capacity in gallons)		
Type of Exposure	39.63 (150 liters) to 3,500	3,501–15,000	15,001–30,000
1. Fire-resistive building and fire walls[c]	5	5	5
2. Noncombustible building[c]	25	50	75
3. Other buildings[c]	50	75	100
4. Wall openings, air compressor intakes, inlets for air-conditioning or ventilating equipment	75	75	75
5. Flammable liquids (above ground and vent or fill openings if below ground) (see 513 and 514)	50	75	100
6. Between stationary liquefied hydrogen containers	5	5	5
7. Flammable gas storage	50	75	100
8. Liquid oxygen storage and other oxidizers (see 513 and 514)	100	100	100
9. Combustible solids	50	75	100
10. Open flames, smoking, and welding	50	50	50
11. Concentrations of people	75	75	75

[a]The distance in Nos. 2, 3, 5, 7, 9, and 12 in Table H-4 may be reduced where protective structures, such as fire walls equal to height of top of the container, to safeguard the liquefied hydrogen storage system, are located between the liquefied hydrogen storage installation and the exposure.
[b]Where protective structures are provided, ventilation and confinement of product should be considered. The 5-ft. distance in Nos. 1 and 6 facilitates maintenance and enhances ventilation.
[c]Refer to Standard Types of Building Construction, NFPA No. 220-1969 for definitions of various types of construction.

ACETYLENE

Acetylene is most often associated with its use as a fuel in welding and cutting operations. Because acetylene is highly soluble in acetone, large quantities of acetylene can be stored in small cylinders at low pressures (NSC 2009, 846). The gas may be stored in cylinders and tanks, or it may be stored in a manifold system. Acetylene consists of 92.3 percent by weight of carbon and 7.7 percent by weight of hydrogen (NSC 2009, 846). Acetylene is used as a fuel since it produces a much higher flame temperature than other fuels. The gaseous acetylene that comes from cylinders is really evolved from liquid acetone stored in an inert filler material inside the cylinder. As the tank is opened, the acetylene gas is released from the cylinder. Tipping the cylinder on its side can result in the release of liquid acetone from the cylinder.

The in-plant transfer, handling, storage, and use of acetylene in cylinders shall be in accordance with Compressed Gas Association Pamphlet G-1-1966, which is incorporated by reference as specified in 29 C.F.R. § 1910.6 (USDOL 2013e, 29 C.F.R. § 1910.102). The piped systems for the in-plant transfer and distribution of acetylene shall be designed, installed, maintained, and operated in accordance with Compressed Gas Association Pamphlet G-1.3-1959. Plants for the generation of acetylene and the charging (filling) of acetylene cylinders shall be designed, constructed, and tested in accordance with the standards prescribed in Compressed Gas Association Pamphlet G-1.4-1966 (USDOL 2013d, 29 C.F.R. § 1910.103).

When using acetylene cylinders for oxy-acetylene welding and cutting, the acetylene serves as the fuel, while the oxygen serves as the oxidizer. Hazards with acetylene in welding and cutting include the potential for fires because the fuel and oxidizer are together. Separating the fuel and oxygen cylinders when in storage by 20 ft. will reduce the potential for fires. Because the acetylene is really being formed from the liquid acetone inside the container, acetylene cylinders should always be stored in the upright position.

OXYGEN

Oxygen is a nonflammable gas, meaning that it does not burn. Oxygen is an oxidizer, serving as an oxygen source for other materials that are consumed as fuel in a fire. Introducing pure oxygen to greases and oils can result in spontaneous combustion. Therefore, equipment making up a bulk oxygen system should be cleaned to remove oil, grease, or other readily oxidizable materials before the system is put into service.

Oxygen can be stored in cylinders or in bulk. A bulk oxygen system is an assembly of equipment, comprising oxygen storage containers, pressure regulators, safety devices, vaporizers, manifolds, and interconnecting piping, with a storage capacity of more than 13,000 ft.^{3} of oxygen at normal temperature and pressure (NTP) that is connected in service or ready for service or with capacity for more than 25,000 ft.^{3} of oxygen (at NTP)

COMMON AND SPECIAL HAZARDS

including unconnected reserves on hand at the site (USDOL 2013f, 29 C.F.R. § 1910.104).

The bulk oxygen system terminates at the point where oxygen at service pressure first enters the supply line, while oxygen containers may be stationary or movable and the oxygen may be stored as gas or liquid. Bulk oxygen systems shall be designed and installed according to applicable codes. Examples of some provisions pertaining to bulk oxygen systems include the following (USDOL 2013f, 29 C.F.R. § 1910.104):

1. Bulk oxygen storage systems should be located above ground out of doors or should be installed in a building of noncombustible construction that is adequately vented and used for that purpose exclusively.
2. The location selected should be such that containers and associated equipment should not be exposed by electric power lines, flammable or combustible liquid lines, or flammable gas lines.
3. The minimum safe distance from any bulk oxygen storage container to exposures should be maintained.
4. Examples of exposures include combustible structures, fire-resistive structures, openings in walls, combustible liquid storage, and combustible gas storage.
5. High-pressure gaseous oxygen containers should comply with applicable codes, including the ASME Boiler and Pressure Vessel Code, Section VIII—Unfired Pressure Vessels (1968), and USDOT specifications and regulations.
6. Piping, tubing, and fittings should be suitable for oxygen service and for the pressures and temperatures involved, conforming with "Gas and Air Piping Systems," Code for Pressure Piping, ANSI, B31.1-1967, with addendum, B31.10a-1969.
7. Bulk oxygen storage containers, regardless of design pressure, should be equipped with safety relief devices as required by the ASME code or USDOT specifications and regulations.
8. Bulk oxygen storage containers, designed and constructed in accordance with the ASME Boiler and Pressure Vessel Code, Section VIII—Unfired Pressure Vessel (1968), should be equipped with safety relief devices meeting the provisions of the Compressed Gas Association Pamphlet Safety Relief Device Standards for Compressed Gas Storage Containers, S-1, Part 3.

Valves, gauges, regulators, and other accessories should be suitable for oxygen service. Any enclosure containing oxygen control or operating equipment should be adequately vented. The bulk oxygen storage location should be permanently placarded to indicate "OXYGEN—NO SMOKING—NO OPEN FLAMES" or an equivalent warning.

LIQUEFIED PETROLEUM GAS

LP gas serves a variety of uses in industry, for example, as a fuel for heating processes or to power equipment such as powered industrial trucks and vehicles and as a propellant in aerosol products. LP gas is extremely flammable. Control measures include using approved equipment to store and transfer LP gas, controlling ignition sources, and using proper handling procedures. Like other gases used in industry, LP gas is commonly stored in portable cylinders, or it may be stored in a manifold system as part of a fixed industrial process.

LP gas stored in USDOT containers should have approved valves, connectors, manifold valve assemblies, and regulators. Containers should be designed, constructed, and tested following, as appropriate, the Rules for Construction of Unfired Pressure Vessels of the ASME Boiler and Pressure Vessel Code (USDOL 2013g, 29 C.F.R. § 1910.110). Containers filled on a volumetric basis shall be equipped with a fixed liquid-level gauges to indicate the maximum permitted filling level.

Welding, repairs, and modifications to LP gas containers should be performed following approved methods to reduce the potential for fires and explosions. LP gas containers should be marked with a metal nameplate attached to the container and located in such a manner as to remain visible after the container is installed; information on the nameplate should include the following (USDOL 2013g, 29 C.F.R. § 1910.110):

1. The name and address of the supplier of the container or the trade name of the container,

The water capacity of the container in pounds or gallons, U.S. Standard,

3. The pressure in psig for which the container is designed,
4. The wording "This container shall not contain a product having a vapor pressure in excess of — psig at 100°F."
5. The tare weight in pounds or other identified unit of weight for containers with a water capacity of 300 lb. or less,

6. The maximum level to which the container may be filled with liquid at temperatures between 20°F and 130°F (except on containers provided with fixed maximum-level indicators or that are filled by weighing) in increments of not more than 20°F (this marking may be located on the liquid-level gauging device), and
7. The outside surface area in square ft..

Each individual container should be located with respect to the nearest important building or group of buildings in accordance with table 3.8 (USDOL 2013g, 29 C.F.R. § 1910.110).

Additional requirements for valves, fittings, piping, and relief valves established by OSHA include the following (USDOL 2013g, 29 C.F.R. § 1910.110):

1. Valves, fittings, and accessories connected directly to the container, including primary shutoff valves, should have a rated working pressure of at least 250 psig and should be of material and design suitable for LP gas service.
2. Piping, tubing, hoses, fittings, and relief valves used for LP gas systems should meet applicable requirements.
3. Piping, tubing hoses, fittings, and relief valves should be installed, tested, and maintained following standards.
4. Relief valve assemblies should be of sufficient size to provide the rate of flow required for the container on which they are installed that will discharge the gas to an acceptable location should the valve open.

Tank Car and Truck Loading or Unloading
Transferring LP gas from tank cars and trucks presents unique fire hazards.
To prevent the accidental release of the gas, the following precautions should be taken (USDOL 2013g, 29 C.F.R. § 1910.110):

1. The track of tank car siding should be relatively level.
2. A "TANK CAR CONNECTED" sign, as covered by USDOT rules, should be installed at the active end or at the ends of the siding while the tank car is connected.

Table 3.8. Minimum Distances Between Aboveground LP Gas Containers

Water Capacity per Container	Minimum Distances Containers		Between Aboveground Containers (ft)
	Underground (ft)	Aboveground (ft)	
Less than 125 gals[a]	10	None	None
125–250 gals	10	10	None
251–500 gals	10	10	3
501–2,000 gals	25[b]	25[b]	3
2,001–30,000 gals	50	50	5
30,001–70,000 gals	50	75	1/4 of sum of diameters of adjacent containers
70,001–90,000 gals	50	100	1/4 of sum of diameters of adjacent containers

Source: USDOL 2013d, 29 C.F.R. 1910.110.

[a]If the aggregate water capacity of a multi-container installation at a consumer site is 501 gallons or greater, the minimum distance should comply with the appropriate portion of this table, applying the aggregate capacity rather than the capacity per container. If more than one installation is made, each installation should be separated from another installation by at least 25 feet. Do not apply the MINIMUM DISTANCES BETWEEN ABOVE-GROUND CONTAINERS to such installations.

[b]The above distance requirements may be reduced to not less than 10 ft. for a single container of 1,200 gallons water capacity or less, providing such a container is at least 25 ft. from any other LP gas container of more than 125 gallons water capacity.

3. While cars are on sidetrack for loading or unloading, the wheels at both ends should be blocked on the rails.
4. The employer should ensure that an employee is always present while the tank car, cars, or trucks are being loaded or unloaded.
5. Electrical equipment and wiring should meet applicable codes for use in an area where LP gas vapors may be present.
6. Open flames or other sources of ignition should not be permitted in vaporizer rooms (except those housing direct-fired vaporizers), pump houses, container-charging rooms, or other similar locations.
7. Direct-fired vaporizers should not be permitted in pump houses or container-charging rooms.
8. Open flames from sources, such as cutting or welding, portable electric tools, and extension lights capable of igniting LP gas, should not be permitted within specified areas such as storage areas, tank car loading and unloading areas, and LP gas vehicle recharging areas.

CHAPTER QUESTIONS
1. Describe some of the common electrical hazards that can result in fire ignition sources.
2. Differentiate between the various NFPA classes of flammable and combustible liquids.
3. What import do the various classes and groups of flammable and combustible liquids have for the safety manager?
4. Differentiate between the various NFPA hazardous environment classes.
5. What process should be followed when transferring a Category 1 flammable liquid from a 55-gallon drum to a 5-gallon safety container?
6. What type of fire hazard does oxygen present in the workplace?
7. What fire prevention and spill control features would one expect to find on an aboveground storage tank that holds flammable liquids?
8. Describe the safety features found in a flammable liquid storage room.
9. Describe the safety features of an approved safety container.

REFERENCES

Earley, Mark W., and Jeffrey S. Sargent. (2011). *National Electrical Code 2011 Handbook.* Quincy, MA: NFPA.

Evarts, Ben. (2012). *Fire in U.S. Industrial and Manufacturing Facilities.* Quincy, MA: NFPA.

National Fire Protection Association (NFPA). (1997). *Fire Protection Handbook*, 18th ed. Quincy, MA: NFPA.

National Fire Protection Association (NFPA). (2008). *Fire Protection Handbook*, 20th ed. Quincy, MA: NFPA.

National Safety Council (NSC). (2009). *Accident Prevention Manual for Business and Industry: Engineering and Technology*, 13th ed. Itasca, IL: NSC.

NFPA. (2012). *NFPA 30: Flammable and Combustible Liquids Code, 2012 Edition.* Quincy, MA: NFPA.

NFPA. (2014). *NFPA 70: National Electrical Code, 2014 Edition.* Quincy, MA: NFPA.

U.S. Department of Labor (USDOL). (2013a). *Occupational Safety and Health Standards for General Industry, Subpart H: Hazardous Materials: Dip Tanks Containing Flammable or Combustible Liquids,* 29 C.F.R. § 1910.125. Washington, DC: USDOL.

USDOL. (2013b). *Occupational Safety and Health Standards for General Industry, Subpart H: Hazardous Materials: Flammable Liquids, 29 C.F.R. § 1910.106.* Washington, DC: USDOL.

USDOL. (2013c). *Occupational Safety and Health Standards for General Industry, Subpart S: Electrical: Hazardous Locations, 29 C.F.R. § 1910.307.* Washington, DC: USDOL.

USDOL. (2013d). *Occupational Safety and Health Standards for General Industry, Subpart H: Hazardous Materials: Hydrogen, 29 C.F.R. § 1910.103.* Washington, DC: USDOL.

USDOL. (2013e). *Occupational Safety and Health Standards for General Industry, Subpart H: Hazardous Materials, Acetylene: 29 C.F.R. § 1910.102.* Washington, DC: USDOL.

USDOL. (2013f). *Occupational Safety and Health Standards for General Industry, Subpart H: Hazardous Materials: Oxygen, 29 C.F.R. § 1910.104.* Washington, DC: USDOL.

USDOL. (2013g). *Occupational Safety and Health Standards for General Industry, Subpart H: Hazardous Materials: Storage and Handling of Liquefied Petroleum Gases, 29 C.F.R. § 1910.110.* Washington, DC: USDOL.

USDOL, Directorate of Technical Support and Emergency Management (DTSEM). (2013). *Nationally Recognized Testing Laboratories.* Washington, DC: USDOL. https://www.osha.gov/dts/otpca/nrtl/index.html (accessed October 24, 2013).

USDOL, Occupational Safety and Health Administration (OSHA). (2014a). *The Globally Harmonized System for Hazard Communication.* Washington, DC: USDOL. https://www.osha.gov/dsg/hazcom/global.html (accessed January 14, 2014).

USDOL, Occupational Safety and Health Administration (OSHA). (2014b). *A Guide to the Globally Harmonized System of Classification and Labelling of Chemicals (GHS).* USDOL. https://www.osha.gov/dsg/hazcom/ghs.html (accessed January 14, 2014).

Voltaix. (2006). *Material Safety Data Sheet for Hydrogen (H2).* North Branch, NJ: Voltaix.

4

Mechanical and Chemical Explosions

ANATOMY OF AN EXPLOSION

Explosions can result in massive property damage and fires. Explosions can be categorized into either mechanical explosions, as in the failure of a pressure vessel, or chemical explosions, as in the case of ignited dynamite. The term "explosion" is defined as a sudden, rapid release of energy that produces potentially damaging pressures (NFPA 2008, 2–93). The high-pressure gas released seeks equilibrium with the pressure of the surrounding environment. The dissipation of the energy from the shock wave into the environment is what can cause damage. The effects of the high-pressure gas on the environment depends on the following (NFPA 1997, 1–69):

1. The rate of the release,
2. The pressure at release,
3. The quantity of the gas released, and
4. The directional factors governing the release.

Explosions can be broadly classified into physical explosions and chemical explosions. In physical explosions, the explosion occurs as the sudden release of pressure due to mechanical means. An example of a physical explosion is the release of energy resulting in the failure of a pressure vessel. In physical explosions, there is no chemical change in the substances involved. In chemical explosions, however, there is some type of chemical reaction taking place in which the composition of the materials involved in the explosion changes form. These chemical changes can occur throughout the entire material, in which case they are referred to as uniform reactions. The chemical changes can also take place in such a way that there is a clearer distinction between changed material and unchanged material, in which case they are referred to as propagating reactions (NFPA 1997, 1–71). Explosions can be classified as detonations (the flame front speed is greater than the speed of sound in the explosion medium) or deflagrations (the flame front speed is less than the speed of sound). Detonations are much more destructive than deflagrations (NFPA 2008, 2–99).

EXPLOSIVES AND BLASTING AGENTS

The U.S. Department of Transportation (USDOT) defines an explosive as any substance or article, including a device, designed to function by explosion (i.e., an extremely rapid release of gas and heat) or something that, by a chemical reaction within itself, is able to function in a similar manner even if not designed to function by explosion (USDOT 2013a, 49 C.F.R. § 173.50). Explosives in Class 1 are divided into the following six divisions (USDOT 2013a, 49 C.F.R. § 173.50):

1. Division 1.1 consists of explosives that have a mass explosion hazard. A mass explosion is one that affects almost the entire load instantaneously.
2. Division 1.2 consists of explosives that have a projection hazard but not a mass explosion hazard.
3. Division 1.3 consists of explosives that have a fire hazard and either a minor blast hazard or a minor projection hazard or both, but not a mass explosion hazard.
4. Division 1.4 consists of explosives that present a minor explosion hazard. The explosive effects are largely confined to the package and no projection of fragments of appreciable size or range is to be expected. An external fire must not cause virtually instantaneous explosion of almost the entire contents of the package.
5. Division 1.5 consists of very insensitive explosives. This division comprises substances that have a mass explosion hazard but are so insensitive that there is little probability of initiation or of transition from burning to detonation under normal conditions of transport.

6. Division 1.6 consists of extremely insensitive articles that do not have a mass explosion hazard. This division comprises articles that contain only extremely insensitive detonating substances and that demonstrate a negligible probability of accidental initiation or propagation.

The Occupational Safety and Health Administration (OSHA) defines explosives further as including but not limited to dynamite, black powder, pellet powders, initiating explosives, blasting caps, electric blasting caps, safety fuses, fuse lighters, fuse igniters, squibs, cordeau detonant fuses, instantaneous fuses, igniter cord, igniters, small-arms ammunition, small-arms ammunition primers, smokeless propellants, cartridges for propellant-actuated power devices, and cartridges for industrial guns (USDOL 2013, 29 C.F.R. § 1910.109). Commercial explosives are those explosives that are intended to be used in commercial or industrial operations.

Materials such as trinitrotoluene (TNT) and dynamite are considered highly explosive materials since they detonate at supersonic speeds greater than 1,100 ft./s. Dynamite is a mixture consisting of an absorbent such as sodium nitrate and diatomaceous earth saturated with nitroglycerin (Schnepp and Gantt 1999, 67). Low explosives, or those that detonate at speeds less than 1,100 ft./s, include materials such as black powder. Black powder is one of the oldest explosives known. It is made up of potassium or sodium nitrate, sulfur, and charcoal.

A blasting agent is defined by OSHA as any material or mixture consisting of a fuel and oxidizer, intended for blasting, not otherwise classified as an explosive, and in which none of the ingredients are classified as an explosive, provided that the finished product, as mixed and packaged for use or shipment, cannot be detonated by means of a No. 8 test blasting cap when unconfined. Blasting agents are those materials used to initiate the higher-order explosives (Schnepp and Gantt 1999, 68). Some explosive materials are not easily detonated until an explosive train is established. The explosive train consists of a smaller precursor explosion used to start the second, higher-order explosion.

LABELS AND PLACARDS

USDOT has established the labeling requirements for hazardous materials, including explosives. Figure 4.1 depicts the proper format for explosive labels

FIGURE 4.1
Labels and Placards for Explosives Divisions 1.1, 1.2, and 1.3. *Source*: U.S. Department of Transportation

FIGURE 4.2
Labels and Placards for Explosives Division 1.4. *Source*: U.S. Department of Transportation

and placards for explosives in Divisions 1.1, 1.2, and 1.3 (USDOT 2013b, 49 C.F.R. § 172.522).

The "**" in the figure represents the appropriate division number and compatibility group letter. The compatibility group letter must be the same size as the division number and must be shown as a capitalized Roman letter (USDOT 2013b, 49 C.F.R. § 172.522).

FIGURE 4.3
Explosive Subsidiary Label. *Source*: U.S. Department of Transportation

An example label for explosives in Divisions 1.4 appears in figure 4.2. The *on the label represents the appropriate compatibility group. The compatibility group letter must be shown as a capitalized Roman letter.

Labels for explosives in Division 1.5 and 1.6 would have their appropriate division number depicted. An "EXPLOSIVE" subsidiary label is required for materials identified in USDOT's hazardous materials table. The division number or compatibility group letter may be displayed on the subsidiary hazard label. An example of the "EXPLOSIVE" subsidiary label appears in figure 4.3.

EXPLOSIVES HANDLING AND STORAGE PROCEDURES

Magazines used for the storage of explosives are classified as either Class I or Class II magazines. A Class I magazine is required where the quantity of explosives stored is more than 50 lb., while a Class II magazine may be used where the quantity of explosives stored is 50 lb. or less (USDOL 2013, 29 C.F.R. § 1910.109). Magazines used in the workplace must meet design and construction requirements. For example, magazines for the storage of certain types of explosives must be bullet-resistant, weather-resistant, fire-resistant, and ventilated sufficiently to protect the explosive in the specific locality. Safety requirements for heating sources for the magazines and ventilation requirements have also been established. In addition, the property on which Class I magazines are located and property where Class II magazines are located outside of buildings should be posted with signs reading "EXPLOSIVES—KEEP OFF" (USDOL 2013, 29 C.F.R. § 1910.109). Magazines must be located a safe distance from other magazines and locating magazines in certain types of occupancies is prohibited. The storage requirements do not apply to explosives materials such as stocks of small-arms ammunition, propellant-actuated power cartridges, small-arms ammunition primers in quantities of less than 750,000; smokeless propellants in quantities less than 750 lb.; explosive-actuated power devices in quantities less than 50 lb.; fuse lighters and fuse igniters; and safety fuses other than cordeau detonant fuses (USDOL 2013, 29 C.F.R. § 1910.109). Blasting caps, electric blasting caps, detonating primers, and primed cartridges should not be stored in the same magazine with other explosives.

Sources of ignition, such as smoking, matches, open flames, spark-producing devices, and firearms (except firearms carried by guards) are not permitted inside of or within 50 ft. of magazines. The land surrounding a magazine should be kept clear of all combustible materials for a distance of at least 25 ft., and combustible materials should not be stored within 50 ft. of magazines. Magazines should always be under the custody of a competent person and this person should be responsible for enforcing all safety precautions.

TRANSPORTING EXPLOSIVES

The transportation of explosives shall only be performed by competent employees using approved vehicles and procedures. Employees shall be prohibited from carrying potential sources of ignition while in or near a motor vehicle transporting explosives. No spark-producing metal, spark-producing metal tools, oils, matches, firearms, electric storage batteries, flammable substances, acids, oxidizing materials, or corrosive compounds shall be carried in the body of any motor truck or vehicle transporting explosives, unless the loading of such dangerous articles and the explosives comply with USDOT regulations (USDOL 2013, 29 C.F.R. § 1910.109).

Employees should be required to inform the fire and police departments of when the loading or unloading of explosives will be performed. In the event of breakdown or collision, the local fire and police departments shall be promptly notified to help contain such emergencies. Explosives shall be transferred from the disabled vehicle to another only when proper and qualified supervision is provided (USDOL 2013, 29 C.F.R. § 1910.109).

Vehicles used for transporting explosives should be in good mechanical conditions. Means should be provided to protect the explosives from sparks and moisture. The vehicles should be properly placarded according to USDOT regulations. Each motor vehicle used for transporting explosives should be equipped with a minimum of two extinguishers, each having a rating of at least 10-BC. The fire extinguisher should be readily accessible to the driver and maintained in proper working condition. OSHA also requires that the motor vehicle used for transporting explosives be given the following inspection to determine that it is in proper condition for the safe transportation of explosives (USDOL 2013, 29 C.F.R. § 1910.109):

1. Fire extinguishers should be filled and in working order.
2. All electrical wiring should be completely protected and securely fastened to prevent short-circuiting.
3. Chassis, motor, pan, and underside of body should be reasonably clean and free of excess oil and grease.
4. Fuel tank and feed line should be secure and have no leaks.
5. Brakes, lights, horn, windshield wipers, and steering apparatuses should function properly.
6. Tires shall be checked for proper inflation and defects.
7. The vehicle should be in proper condition in every other respect and acceptable for handling explosives.

The transportation of the explosives creates hazards for both the driver and the people in the area. A competent driver should always be with the vehicle. The driver or other attendant is not to leave the vehicle unattended for any reason. A vehicle is "attended" only when the driver or other attendant is physically on or in the vehicle or has the vehicle within his field of vision and can reach it quickly and without any kind of interference; "attended" also means that the driver or attendant is awake, alert, and not engaged in other duties or activities that may divert their attention from the vehicle, except for necessary communication with public officers, representatives of the carrier shipper, or the consignee or except for necessary absence from the vehicle to obtain food or to provide for his physical comfort (USDOL 2013, 29 C.F.R.

§1910.109). The vehicle should only be parked in designated and secured areas. A designated and secured area is an area that is securely fenced or walled in, with all gates or entrances locked and where parking of the vehicle is otherwise permissible, or at a magazine site established solely for the purpose of storing explosives.

USE OF EXPLOSIVES AND BLASTING AGENTS

The use of explosives and blasting agents is an extremely dangerous activity.

OSHA has promulgated standards for the protection of workers engaged in this activity and the people and property that may be in the blasting area. Some of the safety requirements associated with blasting include the following (USDOL 2013, 29 C.F.R. § 1910.109):

1. Blasting should only be performed by qualified individuals.
2. Procedures for loading explosives into blast holes, initiating the explosive charges, and dealing with misfires should adhere to applicable safety standards.
3. Sources of ignition, such as matches, open light, or other fire or flame, should be prohibited at the blasting site to control the potential hazards associated with explosives.
4. Because accidental discharge of electric blasting caps can occur from current induced by radar, radio transmitters, lightning, adjacent power lines, dust storms, or other sources of extraneous electricity, all blasting operations should be suspended and all persons removed from the blasting area during the approach and progress of an electrical storm.
5. Warning signs should be posted against the use of mobile radio transmitters on all roads within 350 ft. of the blasting operations.
6. Precautions should be taken to prevent collateral damage to buildings and structures near the blasting site.
7. Persons authorized to prepare explosive charges or conduct blasting operations shall use every reasonable precaution, including but not limited to warning signals, flags, barricades, or woven wire mats, to ensure the safety of the public and workmen.
8. Whenever blasting is being conducted in the vicinity of gas, electric, water, fire alarm, telephone, telegraph, and steam utilities, the blaster should notify the appropriate representatives of such utilities at least 24

hours in advance of blasting, specifying the location and intended time of such blasting. Verbal notice should be followed with written notice.
9. Blasting operations should be conducted during daylight hours.

OXIDIZING AGENTS

As per the general definition of an oxidizing agent it is a chemical substance in which one of the elements tends to gain electrons. NFPA 430: Code for the Storage of Liquid and Solid Oxidizers (2004) has classified oxidizing materials according to their ability to cause spontaneous combustion and how much they can increase the burning rate:

- Class 1 oxidizer: It is an oxidizer that does not moderately increase the burning rate of combustible materials with which it comes in contact.
- Class 2 oxidizer: It is an oxidizer that causes a moderate increase in the burning rate of combustible materials with which it comes in contact.
- Class 3 oxidizer: It is an oxidizer that causes a severe increase in the burning rate of combustible materials with which it comes in contact.
- Class 4 oxidizer: It is an oxidizer that can undergo an explosive reaction due to contamination or exposure to thermal or physical shock and that causes a severe increase in the burning rate of combustible materials with which it comes in contact.

The following common materials are oxidizing agents by this definition:

- Nitrates: sodium nitrate ($NaNO_3$)
- Nitrites: sodium nitrite ($NaNO_2$)
- Chlorates: potassium chlorate ($KClO_3$)
- Chlorites: sodium chlorite ($NaClO_2$)
- Dichromates: sodium dichromate ($Na_2Cr_2O_7$)
- Hypochlorites: sodium hypochlorite ($NaClO$)
- Perchlorates: perchloric acid ($HClO_4$)
- Permanganates: potassium permanganate ($KMnO_4$)
- Persulfates: sodium persulfate ($Na_2O_8S_2$)

Oxidizing agents are usually themselves not combustible but can provide oxygen to accelerate the burning of combustible materials (NFPA 2008, 7–45).

A primary fire prevention method for oxidizing agents is to ensure that they are not stored with flammable and combustible materials. Although the oxidizing agents themselves are not combustible, they will provide oxygen for the other combustible materials to burn.

Ammonium Nitrate

OSHA regulates the handling and storage of ammonium nitrate in the workplace due to the hazards it poses as an oxidizing agent and an explosion hazard. Ammonium nitrate can be in the form of crystals, flakes, grains, or pills, including fertilizer grade, dynamite grade, nitrous oxide grade, technical grade, and other mixtures containing 60 percent or more ammonium nitrate (USDOL 2013, 29 C.F.R. § 1910.109). Ammonium nitrate can be highly reactive with other types of materials. As a precaution, it should be stored in a separate building or separated by approved fire walls of not less than 1-hour fire-resistance rating from organic chemicals, acids, other corrosive materials, materials that may require blasting during processing or handling, compressed flammable gases, flammable and combustible materials, or other contaminating substances, including but not limited to animal fats, baled cotton, baled rags, baled scrap paper, bleaching powder, burlap or cotton bags, caustic soda, coal, coke, charcoal, cork, camphor, excelsior, fibers of any kind, fish oils, fish meal, foam rubber, hay, lubricating oil, linseed oil (or other oxidizable or drying oil), naphthalene, oakum, oiled clothing, oiled paper, oiled textiles, paint, straw, sawdust, wood shavings, or vegetable oils (USDOL 2013, 29 C.F.R. § 1910.109).

Additional materials for which precautions should be taken to avoid mixing with ammonium nitrate include the following:

1. Flammable liquids such as gasoline, kerosene, solvents, and light fuel oils
2. Sulfur and finely divided metals
3. Explosives and blasting agents

Storage Bins for Ammonium Nitrate

Bins used for the storage of ammonium nitrate should meet applicable OSHA standards to minimize the potential for fires and explosions due to situations such as contamination of the ammonium nitrate and the accidental storage of ammonium nitrate with incompatible materials. Guidelines to follow when using bins to store ammonium nitrate include the following (USDOL 2013, 29 C.F.R. § 1910.109):

1. Due to the corrosive and reactive properties of ammonium nitrate and to avoid contamination, galvanized iron, copper, lead, and zinc should not be used in a bin construction unless suitably protected.
2. Aluminum bins and wooden bins protected against impregnation by ammonium nitrate are permissible.
3. The partitions dividing the ammonium nitrate storage from other products that would contaminate it shall be of tight construction.
4. The ammonium nitrate storage bins or piles shall be clearly identified by signs reading "AMMONIUM NITRATE" with letters at least 2 in. high.
5. Piles or bins shall be so sized and arranged that all material in the pile is moved out periodically in order to minimize possible caking of the stored ammonium nitrate.
6. In no case the ammonium nitrate should be piled higher at any point than 36 in. below the roof or supporting and spreader beams overhead.

STORAGE OF AMMONIUM NITRATE

OSHA standards for storing ammonium nitrate apply to facilities storing quantities of 1,000 lb. or more. Buildings used for the storage of ammonium nitrate must meet the following design standards (USDOL 2013, 29 C.F.R. §1910.109):

1. Provisions should be made to prevent unauthorized personnel from entering the ammonium nitrate storage area.
2. Storage buildings should have no basements, unless they are open on at least one side.
3. Storage buildings should not be over one story in height.
4. Storage buildings should have adequate ventilation or be of a construction that will be self-ventilating in the event of fire.
5. The wall on the exposed side of a storage building within 50 ft. of a combustible building, forest, pile of combustible materials, or similar exposure hazards shall be of fire-resistive construction. In lieu of the fire-resistive wall, other suitable means of exposure protection, such as a freestanding wall, may be used.

6. The roof coverings shall meet applicable standards.
7. All flooring in storage and handling areas should be of noncombustible material or should be protected against impregnation by ammonium nitrate and should be without open drains, traps, tunnels, pits, or pockets into which any molten ammonium nitrate could flow and be confined in the event of fire.
8. Buildings and structures shall be dry and free from water seepage through the roof, walls, and floors.
9. Not more than 2,500 tons of bagged ammonium nitrate should be stored in a building or structure not equipped with an automatic sprinkler system.

Storage of Ammonium Nitrate in Bags, Drums, or Other Containers

To minimize the potential for fires and explosions involving ammonium nitrate materials stored in bags and drums, OSHA has adopted the following safety standards (USDOL 2013, 29 C.F.R. § 1910.109):

1. Bags and containers used for ammonium nitrate must comply with specifications and standards required for use in interstate commerce.
2. Bags of ammonium nitrate should not be stored within 30 in. of storage building walls and partitions.
3. The height of piles should not exceed 20 ft. The width of piles should not exceed 20 ft., and the length should not exceed 50 ft.; however, where the building is of noncombustible construction or is protected by automatic sprinklers, the length of piles should not be limited. In no case should the ammonium nitrate be stacked closer than 36 in. below the roof or supporting and spreader beams overhead.
4. Aisles should be provided to separate piles by a clear space not less than 3 ft. in width. At least one service or main aisle in the storage area shall be not less than 4 ft. in width.

BOILING-LIQUID EXPANDING-VAPOR EXPLOSIONS

The phenomenon known as a boiling-liquid expanding-vapor explosion (BLEVE) is the result of a liquid within a container reaching a temperature well above its boiling point at atmospheric temperature, causing the vessel to rupture into two or more pieces (Duval 1998, 3). A BLEVE can occur when

MECHANICAL AND CHEMICAL EXPLOSIONS

FIGURE 4.4
A BLEVE Fire

fire impinges on the liquefied petroleum (LP) tank shell at a point or points above the liquid level of the contents of the LP tank (see figure 4.4). This impingement causes the metal to weaken and fail from the internal pressure (Duval 1998, 3). In a typical BLEVE, a tank of LP gas is engulfed in a fire. As the fire heats the tank, the fluid inside rises in temperature and pressure. The temperature of the liquid inside is now above its normal boiling point and is superheated with respect to the new pressure outside the tank. As the pressure is released from the tank, either through relief devices or tank failure, the liquid inside begins to boil. This boiling can tear the tank apart, hurling pieces hundreds of yards. If the fluid is flammable, it can ignite, forming a fireball. A vapor explosion results from the rapid and intense heat transfer that may follow contact between a hot liquid and a colder, more volatile one (Berthoud 2000).

DERAILMENT OF TOLEDO, PEORIA, AND WESTERN RAILROAD COMPANY'S TRAIN NO. 20, CRESCENT CITY, ILLINOIS

One of the best-known railroad incidents occurred in Crescent City, Illinois in 1970, when ten tank cars carrying more than 34,000 gallons (128,700 L) derailed (U.S. Fire Administration [USFA] 1973, 23). Train No. 20, an east-bound freight train of the Toledo, Peoria, and Western Railroad Company, consisting of a four-unit diesel-electric locomotive and 109 cars, derailed the twentieth to the thirty-fourth cars, inclusive, at the west switch of the siding in Crescent City at about 6:30 a.m. on June 21, 1970 (NTSB 1972, 3). During the derailment, one of the tank cars was punctured and the leaking propane was immediately ignited, engulfing the other tank cars in the fire.

Included in the fifteen derailed cars were nine tank cars loaded with LP gas. Three BLEVEs resulted, generating enough force to blow people, railroad ballast, ties, and track into the street, destroying most of the business district and several homes (USFA 1973, 23). The National Transportation Safety Board determined that the probable cause of this accident was the breaking of the L-4 journal of CB&Q 182544, the twentieth car, due to excessive overheating, which permitted the truck side to drop to the track and derail the leading wheels of the car; however, the cause of the overheating could not be determined. The cause of the initial fire was the puncturing of one tank during the derailment, the jumbling of the derailed cars, and the large volume of propane released which immediately ignited and subjected the other tanks to impingement of fires. Despite the efforts of 250 firefighters and 58 pieces of apparatus, 64 people were injured, 24 living quarters were destroyed, and 90 percent of the business district was wiped out.

Dust Explosions

A dust explosion can occur when particulate solid material is suspended in air and a sufficiently energetic ignition source is present (Amyotte and Eckhoff 2010, 15). For a dust explosion to occur, certain elements must be present (NFPA 2008, 9–96–97). There must be a combustible dust suspended in air in a sufficient concentration. The second most important element is ignition of the suspended dust cloud. Potential ignition sources include electrical arcs, welding and cutting equipment, and frictional heat from mechanical equipment. The third element required for a dust explosion is oxygen. The oxygen concentration required for a dust explosion is the same as that found in the atmosphere (NFPA 2008, 9–98). The fourth element is confinement of the burning dust. If there is no confinement, the explosion pressures are minimal. As confinement increases, the explosion pressures also increase.

In chapter 3, electrical installations were shown to serve as an ignition source for dust explosions. Preventive measures for dust explosions include meeting the requirements for Class II, Division 1 and Division 2 hazardous locations, preventive maintenance on equipment in dust-prone areas of the facility, engineering controls to eliminate or reduce the development of dust in the workplace, and housekeeping measures to control dust buildup in and around equipment. Locations typically classified as Class II locations include grain elevators and grain-handling facilities.

CYLINDER FAILURES

The storage of gases in cylinders may pose two types of hazards: the combustion of the gas inside the cylinder or the combustion of the gas when released from the cylinder. A less frequent, but still significant, hazard of gas stored in containers is the danger of container failure due to overpressure resulting from the combustion of the gas while inside the container (NFPA 1997, 4–75). As a result, a combustion gas explosion occurs when the container is not strong enough to withstand the pressure generated by the combustion of the material. The most common situation resulting in a combustion explosion is the release of the flammable gas from the piping or container.

BOILERS AND UNFIRED PRESSURE VESSELS

A pressure vessel is a vessel in which pressure is obtained from an external source or by the application of heat from an indirect or direct source. The vessels may contain gases, vapors, and liquids at various pressures and temperatures. Pressure vessels can be classified as either fired or unfired pressure vessels. Fired pressure vessels use an external heat source to heat their contents, while unfired pressure vessels have no external heat source. Hazards associated with fired pressure vessels include fire hazards involving the fuel source, rupture or failure of the vessel, and explosion hazards. Hazards associated with unfired pressure vessels include the rupture or failure of the vessel and explosion hazards. Boilers can be further classified according to their heat source and operating temperatures. The following are the major classifications of boilers used in industry:

- Electric boilers: a power boiler, heating boiler, or high- or low-temperature water boiler in which the source of heat is electricity
- High-temperature water boiler: a water boiler intended for operations at pressures more than 160 psig or temperatures more than 250°F

- Hot water heating boiler: a boiler in which no steam is generated, from which hot water is circulated for heating purposes and then returned to the boiler, and which operates at a pressure not exceeding 160 psig or a temperature of 250°F at the boiler outlet
- Process steam generator: a vessel or system of vessels comprising one or more drums and one or more heat-exchange surfaces as used in waste heat or heat recovery type steam boilers.
- Unfired steam boiler: a vessel or system of vessels intended for operation at a pressure in excess of 15 psig for the purpose of producing and controlling an output of thermal energy
- Water heater supply boiler: a closed vessel in which water is heated by combustion of fuels, electricity, or any other source and withdrawn for use external to the system at a pressure not exceeding 160 psig and which should include all controls and devices necessary to prevent water temperatures from exceeding 210°F

Proper construction, maintenance, and testing of the pressure vessels are ways to prevent some of the accidents associated with these types of equipment. At present, there is no one specific standard for pressure vessels; however, some OSHA standards require that a pressure vessel be built in accordance with the industry codes and standards. Sections of the ASME Boiler and Pressure Vessel Code include material specifications, specifications for heating boilers, nondestructive testing of boilers, and recommended rules for care and operation and for inspection procedures after installation and repairs. The National Board of Boiler and Pressure Vessel Inspectors comprises the chief inspectors of jurisdictions. This board has developed its national inspection code, the National Board Inspection Code (2013), commonly referred to as the "NB code."

To ensure adequate operation of the boiler and to reduce the likelihood of fires and explosions, building owners should follow a prescribed boiler inspection program. The inspections should be conducted by a qualified person on a regular basis and should include both internal and external inspections. The external inspection should involve as complete an examination as can be reasonably made of the external surfaces and safety devices while the boiler or pressure vessel is in operation. The internal inspection should involve as complete an examination as can be reasonably made of the internal surfaces of the boiler or pressure vessel while it is shut down and while manhole plates and handhole plates are removed as required by the inspector.

Safety appliances found on a boiler include but are not limited to the following NSC (2009, 206–208):

1. Rupture disk device: a non-reclosing pressure-relief device actuated by inlet static pressure and designed to function by the bursting of a pressure-containing disk
2. Safety relief valve: an automatic pressure-relieving device actuated by a static pressure upstream of the valve, which opens further with the increase in pressure over the opening pressure
3. Temperature limit control: a control that ensures boiler is operating within acceptable temperature ranges
4. Low water cutoffs: an indicator of when water levels in the boiler have gotten to a low level that shuts down heat source to the boiler
5. Flame supervisory unit (igniter): a control that checks to ensure the gas is burning and prevents the accumulation of gas in a room
6. High- and low-gas-pressure switches: switches that monitor gas pressure going into boiler
7. Trial for ignition limiting timer (fifteen seconds): a timer that checks to ensure that the gas has been ignited and prevents the accumulation of gas in a room

Boiler Maintenance

Most of the boiler failures are due to inadequate maintenance. Inspections of pressure vessels have shown that there is a considerable number of cracked and damaged vessels in workplaces. Cracked and damaged vessels can result in leakage or rupture failures. Potential health and safety hazards of leaking vessels include poisonings, suffocations, fires, and explosion hazards. Rupture failures can be much more catastrophic and can cause considerable damage to life and property. Safe design, installation, operation, and maintenance of pressure vessels in accordance with the appropriate codes and standards are essential to worker safety and health.

Boiler tests should be conducted according to manufacturers' recommendations and applicable codes. Blow-down piping and valves are to remove sludge and other impurities in the boiler water that could build up and seriously impede the efficiency and safety of the boiler (NSC 2009, 192). Tests and inspections should be recorded in a boiler log. When alterations, retrofits, or repairs are necessary on a boiler, they should be made so that the object is at least as safe as the original construction. Alterations, retrofits, and repairs should be done as though they were new construction and should comply with the applicable code or codes. Repairs or alterations by welding should be approved beforehand by an authorized inspector. All welding repairs or alterations must be in accordance with the standards for repairs and alterations to boilers and pressure vessels by welding of the NB code. All welding should be done by a qualified organization.

The design, construction, and maintenance of the boiler room itself play a role in the prevention of boiler fires and explosions. Boiler rooms should be constructed according to building codes ensuring that walls and doors meet applicable fire ratings. Fire protection should be provided. All objects in the boiler room should be located so that adequate space is provided for the proper operation and inspection, and the necessary maintenance and repair of the boiler and its controls. An adequate number of exits should be provided. The exits should be clear of obstructions. Ventilation should be provided for the room to permit satisfactory combustion of fuel and ventilation if necessary under normal operations. The minimum ventilation for coal, gas, or oil burners in rooms containing objects is based on the BTUs per hour, required air, and louvered area.

CHAPTER QUESTIONS
1. Describe the mechanics of a BLEVE.
2. What is an oxidizer?
3. What are four ways one can prevent boiler explosions?
4. What is an explosion?
5. Describe some of the safety precautions one must take when transporting explosives.
6. Define the various classes of oxidizers.
7. Describe some of the safety precautions one must take when handling ammonium nitrate.
8. Describe some of the more common safety devices found on boilers.
9. Describe some of the safety precautions one must take when using explosives.

REFERENCES

Amyotte, Paul R., and Rolf K. Eckhoff (2010). Dust Explosion Causation, Prevention and Mitigation: An Overview. *Journal of Chemical Health and Safety*, 17(1): 15–28.

Berthoud, Georges. (2000). Vapor Explosions. *Annual Review of Fluid Mechanics*, 32: 573–611.

Duval, Robert. (1998). *NFPA Alert Bulletin No: 98-1. LP-Gas BLEVES Result in Fire Fighter Fatalities.* Quincy, MA: NFPA.

National Board of Boiler and Pressure Vessel Inspectors. (2013). *National Board Inspection Code, 2013 Edition.* Columbus, OH: National Board of Boiler and Pressure Vessel Inspectors.

National Fire Protection Association (NFPA). (1997). *Fire Protection Handbook*, 18th ed. Quincy, MA: NFPA.

National Fire Protection Association (NFPA). (2008). *Fire Protection Handbook*, 20th ed. Quincy, MA: NFPA.

National Safety Council (NSC). (2009). *Accident Prevention Manual for Business and Industry: Engineering and Technology*, 13th ed. Itasca, IL: NSC.

National Transportation Safety Board (NTSB). (1972). Derailment of Toledo, Peoria and Western Railroad Company's Train No. 20 with Resultant Fire and Tank Car Ruptures, Crescent City, Illinois, June 21, 1970. NTSB Report Number RAR-72-02, adopted on March 29, 1972.

NFPA. (2004). *NFPA 430: Code for the Storage of Liquid and Solid Oxidizers.* Quincy, MA: NFPA.

Schnepp, Rob, and Paul Gantt. (1999). *Hazardous Materials: Regulations, Response, and Site Operations.* New York: Delmar Publishers.

U.S. Department of Labor (USDOL). (2013). *Occupational Safety and Health Standards for General Industry, Subpart H: Hazardous Materials, Explosives and Blasting Agents, 29 C.F.R. § 1910.109.* Washington, DC: USDOL.

U.S. Department of Transportation (USDOT). (2013a). *Hazardous Materials Regulations: Definitions, 49 C.F.R. § 173.50.* Washington, DC: USDOT.

USDOT. (2013b). *Hazardous Materials Regulations: Explosives Placards, 49 C.F.R. § 172.522.* Washington, DC: USDOT.

U.S. Fire Administration. (1973). *America Burning.* Washington, DC: National Commission on Fire Prevention and Control.

5

Building Construction

The primary reason for studying building construction is that fire protection begins during the preplanning phase of any new building design or remodeling. Construction can affect fire and smoke spread, life safety, and the extent of fire damage that will occur within the building. A well-designed building will consider flame and fire spread, life safety, and smoke spread during the preplanning stage so that construction materials and design features can be selected to reduce or eliminate potential exposures. There are far too many instances of fatalities that have occurred because of the failure to recognize life safety concerns and fire and smoke spread as they relate to overall building fire safety. Fire in buildings adds to the fire problem for the following reasons (Brannigan 1971, 3):

1. The building itself may burn.
2. The contents of the building may be ignited.
3. Occupants of the building may be trapped by the fire.
4. The building structure may make it difficult to attack the fire.
5. The building may collapse in whole or part during the fire.
6. The fire may extend beyond the original point of origin to other buildings.
7. Firefighters may be injured or killed.

Fire protection systems to consider during the design phase of construction can be classified as either active or passive. Active fire protection systems actively engage the fire to reduce its spread, such as an automatic sprinkler system. Passive fire protection systems are part of the structure that reduce fire ignition, limit fire development, and/or reduce or eliminate the spread of smoke, such as a fire wall (Schroll 2002, 98).

BASIC TERMINOLOGY

Some basic terminology needs to be understood when discussing building construction. The following terms provide a better understanding of the concepts in this chapter:

- Fire resistance refers to the ability of the material, structure, or assembly to resist the effects of the heat and flame from the fire. Therefore, fire-resistant construction would reduce the ease of ignition and flame spread of the building structure (NFPA 2008, 6–129).
- Flame spread refers to the rate at which a fire will spread from the point of origin to involve an ever-increasing area of combustible material. This rate can be determined using a variety of testing methods such as the Steiner Tunnel Test or the radiant panel or flooring tests (NFPA 2008, 2–42).
- Noncombustible material is a material that in the form in which it is used and under the conditions anticipated will not ignite, burn, support combustion, or add appreciable heat to a fire (Brannigan 2008, 125).
- Dead load is the weight of the building itself and any equipment permanently attached to or built on the building (Brannigan 2008, 13).
- Fire load is the total amount of potential heat in the fuel available to a fire within the building (Brannigan 2008, 19).
- Safety factor is the ratio of the strength of a material prior to failure to the safe working stress. For example, a safety factor of ten will occur if the design load is only a tenth of the tested strength (Brannigan 2008, 21).

STRUCTURAL ELEMENTS

In a building fire, three separate elements can be identified (Brannigan 1971, 7):

1. The structural elements of the building. These structural elements are critical during a fire because if they fail it will affect the structural stability of the building.

2. The contents of the building which can have a significant impact on the fire load.
3. The nonstructural building elements. These elements include surface finishes, windows, interior vertical openings, decorative surfaces, air-conditioning systems, and the like.

The major structural elements involved with industrial building construction include beams, columns, trusses, and connectors. A beam is defined as a structural element that transmits force in a perpendicular direction to the points of support and is usually thought of as a horizontal member (although this is not always the case) (Brannigan 2008, 23). For example, a rafter is structurally a beam that is positioned vertically or diagonally. When a beam is loaded, it deflects or bends downward, with the initial load of its own deadweight plus any additional live load applied. Because of this fact, beams are built with a slight camber so that when the design load is superimposed, the beam will be level. The capacity of a beam increases by the square of its depth and in direct proportion to its width. However, the load capacity of the beam decreases in direct proportion to increases in span length, distance between support points. There are various types of beams. A simple beam is supported at two points. With simple beam construction, the load is delivered to the two end points, and the rest of the structure renders no assistance in cases of overload. When these two end points are rigidly held in place, the beam is referred to as a fixed beam. A continuous beam is a beam that is supported at three or more points (Brannigan 2008, 23–24).

A column is a structural member that conveys a compressive force along a straight path in the direction of the structural member. Columns are typically thought of as vertical structural members, but any structural member that is compressively loaded is in fact a column. For example, a strut is a diagonal column. The capacity of a column decreases by the square of the change in column length. For example, a 10-ft. column with a capacity of 10,000 lb. would have a capacity of only 2,500 lb. if its length were increased to 20 ft (Brannigan 2008, 26).

Walls transmit to the ground the compressive forces applied or received at any point on the wall, with the primary function of exterior walls being to protect the interior of the building from the elements. Walls are typically classified as either load bearing or non-load bearing. A load-bearing wall, as the name implies, carries not only the weight of the wall itself but also the load of some part of the structure. Non-load-bearing walls support their own weight only (Brannigan 2008, 28). Walls can be further classified based on their function within the structure and what follows are some examples (NFPA 2008, 19–10–11):

- A curtain wall is an exterior wall that is supported by the structural frame of the building. If the wall is only one story in height it will be referred to as a panel wall.
- An exterior wall is a boundary wall that protects the interior of the building from outside weather.
- A fire wall is a wall of sufficient fire resistance and stability to withstand the effects of a fire and remain standing despite the possible collapse of the structural framework to either side of the wall.
- A fire barrier wall is sometimes referred to as a fire partition and it is an interior wall that resists the spread of fire but not to the same level as a fire wall.
- A partition wall is a one-story, interior wall that separates two areas in a building but is not a fire barrier.

A truss is a framework of members that restrains the building through triangular formations which by their geometric shape are inherently stable. All parts of a truss, including the connectors, are critical to its stability and the failure of any part may cause the entire truss to fail (Brannigan 2008, 137138). The use of trusses is common with roof supports. However, trusses can also be used in large areas that must be column free.

Connectors transfer the load from one structural element to another, and the type of connectors is an important part of evaluating a building's overall stability during a fire. A pinned building uses simple connectors such as bolts, rivets, or welded joints to connect the components. This is different than a rigid frame building that has no hinged joints and the connectors are strong enough to reroute forces if a member is removed (Brannigan 2008, 34–40). In either case, connectors are often the weakest link in the structural assembly and are a critical component of structural stability during a fire.

CHARACTERISTICS OF BUILDING MATERIALS

Buildings are constructed of a variety of building materials, each of which influences how that building will be affected during a fire situation. Each material, even different forms of the same material, has certain physical and chemical characteristics that make it desirable for its intended function (Brannigan 2008, 52). All materials can be damaged by fire even if they do not burn because all structural materials used in building construction are adversely affected by the elevated temperatures caused by a fire. The degree and significance of this adverse behavior depends primarily on the function of the elements and the degree of protection afforded. In general, the mechanical properties of strength and stiffness decrease as the temperature rises. Other adverse behavior, such as excessive expansion and accelerated creep, also develop with increases in temperatures. What follows are some characteristics of common building materials that can influence their structural behavior in a fire.

Steel

In commercial construction, steel is the most common material used, especially regarding the structure of the building. Steel is noncombustible and does not contribute fuel to a fire. However, structural steel does have three characteristics that affect its performance when exposed to a fire. The first characteristic is that steel conducts heat, thereby aiding heat transfer. The second is its coefficient of expansion at elevated temperatures. This high coefficient of expansion affects the steel structure because the ends of the structural member axially restrained and the attempted expansion due to the heat causes thermal stresses to be transferred to the member. This stress, combined with those of normal loading, can increase the likelihood of a collapse. The last characteristic is that steel will lose its strength when subjected to high temperatures. The critical temperature of steel is 1,100°F, at which time its yield stress, or the point at which the steel will fail, is about 60 percent of its value at room temperature (Brannigan 2008, 55).

The aforesaid three characteristics of steel can create serious problems during a fire depending on four factors: (1) the function of the steel element, (2) its level of stress, (3) how the steel member is supported to other structural members, and (4) its surface area and thickness. The yield strength of steel decreases with increasing temperature; therefore, it is the most significant parameter in establishing steel's load-carrying capacity (NFPA 2008, 19–49). This decrease in yield strength at higher temperatures makes it critical that unprotected structural steel be protected. Often this protection is accomplished by fireproofing, which acts to insulate the steel from the heat produced by the fire. One of the most common methods of insulating structural steel is to apply a surface treatment that encases the steel. An example of a surface treatment that is sprayed on steel is a mixture of cementitous and mineral fiber coatings. The key to the effectiveness of these sprayed on surface treatments is their ability to adhere to the steel when exposed to the high temperatures of a fire and avoid spalling (NFPA 2008, 19–51).

Wood

Wood is combustible, and as it burns, it loses its structural integrity. The important factors that influence the fire endurance of wood are its physical size and its moisture content. As the moisture content in wood increases, so does its ignition temperature, and overall fire spread is reduced. In most cases, the burning of wood produces a char on its surface, which provides a protective coating that insulates the unburned wood from the flame and therefore reduces flame spread (NFPA 2008, 2–33). This charring is one of the reasons why large-dimension lumber (heavy timber) provides greater structural integrity over the same period of fire exposure than smaller-dimension lumber such as 2 × 4s. It has become increasingly difficult to harvest heavy timbers of such large dimensions, making the use of glued, laminated frames and beams increasingly popular in construction today, such as wooden I-joists. One of the advantages to these manufactured heavy-timber members are they provide dimensional stability over solid wood. However, a concern with their use is the stability of the glues when exposed to the higher temperatures of a fire. Wood frame construction, which is common in residential construction, utilizes structural members that are considerably smaller (2 × 4s) than heavy-timber construction; therefore, the fire resistance is considerably reduced. When this type of wood is exposed to fire, it offers little structural integrity and the fire spreads very quickly, hence its common name "quick burning." Figure 5.1 depicts wood frame construction.

Fire-retardant treatments may delay ignition and retard combustion when applied to wood. A common fire-retardant treatment of wood is to impregnate the wood with mineral salts. It is important to note that this treatment

BUILDING CONSTRUCTION

109

FIGURE 5.1
Wood Frame Construction

will reduce the wood's flame spread, but the wood is still combustible. In addition, some of the fire-retardant treatments have been shown to reduce the strength of plywood and structural lumber when applied. These members become brittle and have failed under loads well within their original capacity (Brannigan 2008, 52). Pressure treated lumber used to resist rot can also influence the fire risk of wood. Specifically, these treatments can make the wood easier to ignite, increase flame spread, and release toxic chemicals when burned.

Masonry and Brick

Masonry and brick products are quite fire resistant, but they can spall when subjected to elevated temperatures from a fire. With spalling, there is a loss of the surface of the brick and other masonry products. Hollow concrete blocks also generally retain their structural integrity when exposed to a fire but can crack at elevated temperatures. An important consideration regarding the stability of a non-load-bearing masonry or brick wall is the supporting structure of the wall. A steel or wood structure for such a wall could be damaged from a fire, putting the wall at risk of collapse.

Reinforced Concrete

Reinforced concrete is a composite material where steel rods are placed in the concrete. This composite material allows concrete, which naturally has compressive strength, to be combined with steel to provide excellent tensile strength. Concrete also has a low thermal conductivity, which helps to protect the steel from the heat produced by a fire. However, that does not mean that reinforced concrete is not affected during a fire because at elevated temperatures it does lose strength. The degree to which it is affected by these elevated temperatures is influenced by the type of aggregate used, its moisture content, and the fire loading. For example, lightweight concrete, which uses aggregates of vermiculite of perlite, performs better at elevated temperatures than traditional aggregates. In addition, when exposed to a fire, the concrete and steel bond can fail, which can result in failures of the reinforced member and spalling. However, as a rule, reinforced concrete is fire resistant, and it is rare to encounter the collapse of a reinforced concrete structure that has been exposed to a fire (NFPA 2008, 19–52).

Gypsum

Gypsum products are quite common in construction and include such products as wallboard and plaster, both of which have overall excellent fire-resistive properties. The wall board is formed using gypsum (calcium sulfate dehydrate), which is sandwiched between two sheets of cardboard. By its nature, gypsum has a high portion of chemically combined water, and when exposed to a fire, the evaporation of this water requires a great deal of heat energy. This is the primary reason why gypsum wallboard is such an excellent fire-retardant building material (NFPA 2008, 19–54).

FIRE RESISTANCE RATINGS

Fire-resistive barriers are evaluated in testing furnaces by exposure to a fire whose severity follows a time-varying temperature curve known as the standard time–temperature curve. The curve was adopted by the American Society for Testing and Materials (ASTM) in 1918 and has been the basis for almost all fire resistance testing ever since. Building materials are provided a rating on their ability to resist the effects of fire without failure, which is typically expressed in hours. There are a variety of tests used to evaluate the fire resistance of a material, some examples include: Underwriters' Laboratories

263 "Standard for Fire Tests of Building Construction and Materials," ASTM E119 "Standard Test Methods for Fire Tests of Building Construction and Materials," and NFPA 251 "Standard Methods of Test of Fire Resistance of Building Construction and Materials (NFPA 2012a, 4–5). These fire resistance tests contain detailed test procedures, a guide on the restraint required, if any, and a suggested format for reporting results. The standards also specify the preparation and conditioning of the test specimen and acceptance criteria, which are specific to the element tested. In general, the test continues until failure, which may include any of the following (NFPA 2008, 19–30):

- Failure of test specimen to support a load.
- An increase in the temperature on the unexposed surface of 250°F above ambient.
- Passage of heat/flame sufficient to ignite cotton waste.
- Excess temperature on steel members.
- Failure of walls and partitions when exposed to a hose stream.

Information on a building material's fire-resistance rating is available from a variety of sources such as UL's "Fire Resistance Directory," Factory Mutual's "Approval Guide," and Intertek Testing's "Directory of Listed Products." UL's Fire Resistance Directory provides hourly ratings for a variety of building components, such as beams, columns, and floors. The directory also contains a listing and classification for various building materials based on flame-spread ratings (NFPA 2008, 19–31).

MAJOR TYPES OF BUILDING CONSTRUCTION

The construction of a building has a significant influence on its fire and life safety capabilities. Building construction for life safety includes the layout of the facility, the traffic-flow patterns of the occupants, the types of construction materials used, and their fire-resistance ratings. The design of the facility can aid in preventing the fires from occurring and, once they do, limit the spread of the fire through containment and or active suppression systems.

The NFPA has developed a classification system for building types that is discussed in detail in NFPA 220 "Standard on Types of Building Construction." This standard classifies buildings based on their type of construction, with five basic types of construction designated by Roman numerals as Type I to V, with approved fire-resistance ratings. Table 5.1 illustrates the fire resistance for each of the five types of building construction discussed earlier (NFPA 2012a, 6).

Type I buildings, commonly called fire resistive, have structural members such as the frame, walls, floors, and roof that are all noncombustible with a minimum specified fire-resistive rating. There are two subclassifications within Type I: Type 442 and Type 332. The basic difference between these two subclassifications is the level of fire resistance specified for the structural frame. For example, Type 442 has exterior bearing walls with a 4-h fire resistance, a 4-h fire resistance of columns and a 2-h fire resistance for floor-ceiling assemblies (NFPA 2012a, 6). In general, these Type I buildings will withstand fire for several hours without structural failure and are the best type of construction from a fire safety perspective. Common building materials used in their construction include concrete, steel, and masonry.

Type II is a construction type in which the structural elements are made entirely of noncombustible or limited combustible materials, hence the common name of noncombustible used for this construction type. Although the building materials are noncombustible, they do not have a sufficient fire resistance rating to be classified as fire resistant. When exposed to a fire, the structure will not burn or contribute fuel to a fire involving contents, but it can still collapse due to structural steel failure. Common building materials include metal frame and metal clad, as well as steel, masonry, aluminum, and mineral fiber. There are three sub classifications of Type II construction. The first subclassification is Type 222. With this subclassification, the exterior bearing walls, interior bearing walls, columns, and beams and floors have a 2-h fire resistance. The next subclassification is Type 111. With this subclassification, the exterior bearing walls, interior bearing walls, columns, beams, and floors have a 1-h fire resistance. The third subclassification within Type II is Type 000, and as the name would imply this subclassification has exterior bearing walls, interior bearing walls, columns, and floors with no fire resistance (NFPA 2012a, 6). Type III, which is commonly called ordinary construction, is a construction type where the exterior walls are noncombustible with a 2-h fire resistance, but the interior may be constructed of noncombustible or approved combustible materials. Type III construction is further divided into two subclassifications. Type 211 has 2-h exterior walls and 1-h fire resistance for the interior bearing walls, columns, beams and floors. Type 200 has 2-h

Table 5.1. Fire Resistance Ratings for Type I through Type V Construction (h)

	Type 1		Type II			Type III		Type IV	Type V	
	442	332	222	111	000	211	200	2HH	111	000
Exterior Bearing Walls										
Supporting more than one floor, columns, or other bearing walls	4	3	2	1	0b	2	2	2	1	0b
Supporting one floor only	4	3	2	1	0b	2	2	2	1	0b
Supporting a roof only	4	3	1	1	0b	2	2	2	1	0b
Interior Bearing Walls										
Supporting more than one floor, columns, or other bearing walls	4	3	2	1	0	1	0	2	1	0
Supporting one floor only	3	2	2	1	0	1	0	1	1	0
Supporting roofs only	3	2	1	1	0	1	0	1	1	0
Columns										
Supporting more than one floor, columns, or other bearing	4	3	2	1	0	1	0	H	1	0
Supporting one floor only	3	2	2	1	0	1	0	H	1	0
Supporting roofs only	3	2	1	1	0	1	0	H	1	0
Beams, Girders, Trusses, and Arches										
Supporting more than one floor, columns, or other bearing Walls	4	3	2	1	0	1	0	H	1	0
Supporting one floor only	2	2	2	1	0	1	0	H	1	0
Supporting roofs only	2	2	1	1	0	1	0	H	1	0
Floor construction	2	2	2	1	0	1	0	H	1	0
Roof construction	2	1½	1	1	0	1	0	H	1	0
Interior nonbearing walls	0	0	0	0	0	0	0	0	0	0
Exterior nonbearing walls	0b	0b	0b	0b	0b	0b	0b	0b	0b	0b

Note: H = heavy timber members (see text for requirements).

exterior walls but no fire resistance for interior walls, columns, beams and floors. The interior construction is typically made of wood joist and studs; therefore, the entire interior is easily destroyed by fire (NFPA 2012a, 6).

Type IV is a construction type in which structural members are basically of unprotected wood with large cross-sectional areas, hence the common name of heavy timber. With this construction type the exterior bearing walls and interior bearing walls supporting more than one floor must have a 2-h fire resistance rating. All columns, beams, floor, and roof assemblies must meet specific minimum dimensions, for example, columns must be 8" × 8" when supporting floors, beams must be a minimum of 6" in width and 10" in depth when supporting floors, wooden floors must be a minimum of 3" tongue and groove, and wooden roofs must be a minimum of 2" tongue and groove (NFPA 2012a, 8–9). Characteristics of Type IV construction include slow burning with poor heat conduction; for that reason, they are sometimes superior to steel construction of the same load capacity. Although a few large trees of the necessary dimensions exist today, it is becoming more common to see wood laminates being used to provide such structural dimensions today.

Type V construction is a construction type where exterior walls and structural members are primarily made of wood or other combustible materials. As one might expect, Type V construction provides the lowest degree of fire protection. Type V construction is divided into two subclassifications: Type 111, which has a 1-h fire resistance for exterior and interior load-bearing walls, columns, beams, and floors. Type 000 has no fire resistance requirements. A wood frame residential home with 2" × 4" wood studs, wood siding, wood floors, and a wood roof would be a good example of Type 000 (NFPA 2008, 19–8).

As one might expect, mixed types of construction are common, and where two or more types of construction are used, it is generally recognized that the requirements for occupancy, size restrictions, or fire protection for the least fire resistant type of construction will apply. It should be noted, however, that in those cases where each building type is separated by adequate fire walls, each area may be considered as a separate building (NFPA 2008, 19–10).

FIRE PROTECTION FEATURES

Fire protection of building elements is typically done for two reasons. The first reason is to ensure the building frame or structure will not fail when exposed to a fire. A major focus here is the fire resistance of building components, which were discussed previously in the building construction classifications of Type I–V, but it is also influenced in a major way by local building codes. The second reason is to reduce or prevent the spread of fire within a building or from an outside exposure (NFPA 2008, 19–10). This fire spread is not only influenced by the building structure but also by building contents such as furnishings, interior finish, and both raw materials and finished goods.

Fire Spread

Fire can spread horizontally and vertically from the point of origin through spaces and compartments; however, this rarely occurs because of heat transfer through the wall, floor, or ceiling, or the structural failure of these assemblies. The most common reason for fire spread through a structure is through open doors, unenclosed stairways, unprotected penetrations of fire barriers, or along the outside surface of the building (NFPA 2008, 18–11). This is how fire most commonly travels down corridors and up open stairway shafts. Once a room becomes involved, a fire can travel quickly to adjacent rooms in the absence of fire-separation barriers. Flames can spread through various spaces and openings such as concealed spaces behind walls, above suspended ceilings, and in utility chases and attics. This is especially true of vertical concealed spaces that act as flues to spread smoke and gases upward, then mushroom out horizontally. In addition, fires in concealed spaces burn out of sight of the occupants and detection is usually delayed. Manual fighting of these fires can be difficult as well. Factors that influence the spread of fires along the outside of a building include the height of the building, the size and dimension of window openings, and air flow patterns along the outside of the building (NFPA 2008, 18–11).

Based on the aforesaid factors that influence fire spread, there are several building features that can be incorporated into the building design to reduce fire spread. These building features include fire walls/compartmentation, fire doors, fire stops, baffles, fire dampers, and parapets. When properly constructed, these fire barriers prevent the spread of fire, smoke, and toxic gases. An important part of the protection provided by these building features is their fire resistance. When designing the building, the fire-resistive qualities of the compartments can be determined. The fire rating of the construction can be achieved using specified building materials and the elimination of openings in the barriers where possible. What follows is an explanation of these building features to reduce fire spread.

Firestops are an important consideration in building design because they act as physical barriers that provide a specific fire resistance and prevent fire and smoke from spreading through concealed horizontal and vertical spaces. Examples of such spaces may include heating ventilation and air-conditioning, holes and openings in floors, and elevator shafts. Three factors are considered to determine the most appropriate firestop system: the fire barrier, the penetrating element, and the firestop product. Examples of fire stop products include sealants, gypsum board, and fire-retardant mineral wool. According to NFPA 5000 Building Construction and Safety Code, the firestop product must have "F" Rating (time the system will withstand the passage of flame) equal to the hourly rating of the fire barrier (NFPA 2008, 18–74).

Baffles are physical barriers that are fire resistant and extend between girders and the roof to prevent the spread of fire under the roof. Baffles also play an important role in containing smoke and aid in venting smoke out of the building.

Fire dampers are hinged fire-resistant panels placed inside heating, ventilation, and air-conditioning ducts. When activated, these fire dampers automatically close, reducing the spread of fire, smoke, and gases through the system (NFPA 2008, 8–14).

Parapets are extensions of an exterior load-bearing wall above the roof. Parapets have the same fire resistance rating as the exterior wall and extend at least 36 in. above the roofline. The primary purpose of a parapet is to reduce the spread of fire from adjacent buildings, especially in urban areas where buildings are close together.

Fire walls prevent the spread of fire to other adjacent areas. Typically, a fire wall will have a minimum three-hour fire resistance, extend from the foundation through the roof, and have as few openings as possible. Fire walls should not be confused with a fire barrier wall which is used to subdivide an area and the wall does not extend from the basement through the roof or have equivalent fire resistance. Using fire walls to create compartmentation is a great technique to prevent unlimited spread of a fire within a building. For example, compartmentation can be very important when isolating a high-risk fire area from the surrounding area and also minimize the risk of loss during evacuation as a result of fire and smoke in another space (Brannigan 2008,

80). Any openings in a fire wall must be protected. What follows are some strategies for protecting these openings.

Protection of Openings in Fire Walls

All openings in fire walls present a path for smoke, heat, and flames to travel to other areas of an occupancy and therefore must be protected. Ideally, we would prefer the protection for these openings to be equivalent to the wall itself. However, this is rarely the case and the rationale for this is based on the facts that easily ignitable materials are not normally stored against openings in a wall and if a fire were to occur it would be a small localized fire extinguishable by fire suppression forces (NFPA 2008, 18–15).

Fire doors, windows, and shutters are the most acceptable means of protecting openings in fire walls (NFPA 2008, 18–15). The NFPA and most building codes classify fire doors based on their fire-resistance rating. This rating is based on the testing procedures provided in NFPA 252 "Standard Methods of Fire Tests of Door Assemblies." The best doors, in terms of fire-resistance rating, are four- and three-hour fire doors. Four- and three-hour fire doors are required in openings in walls separating buildings or dividing a building into different fire areas. Most current codes require only a three-hour fire door when separation walls are required to have a three-hour resistance or more. However, some local municipalities require two three-hour doors, one on each side of the opening, or one four-hour door for walls required to have a four-hour fire resistance (NFPA 2008, 18–16).

Fire doors with 1.5-hour ratings are required in openings in two-hour enclosures for vertical openings, such as stairs. Some building codes permit the use of 1.5-hour fire doors to protect openings in walls separating buildings or dividing buildings into different fire areas when two-hour fire walls are required. Openings in exterior walls that can be subjected to severe fire exposures from outside are also commonly protected by 1.5-hour fire doors (NFPA 2008, 18–16).

One-hour fire doors are used for openings in one-hour enclosures for vertical openings in buildings such as stairs. Forty-five-minute fire doors are commonly used in openings in exterior walls subject to a moderate or light fire exposure from outside the building. These forty-five-minute doors are also used in openings in room partitions and walls around some hazardous areas and are also permitted by some building codes in partitions that subdivide the floors of a building. Thirty-minute and twenty-minute fire doors are primarily used for smoke control (NFPA 2008, 18–17).

Impact of Ventilation on Building Fires and Smoke Movement

The burning rates of materials depend on the properties of the combustible material, the available fuel surface area, as well as the air available for combustion. If sufficient air is available the fire will burn at a rate associated with fuel surface-controlled combustion, commonly referred to as stoichiometric combustion. Stoichiometric combustion for a fully developed fire requires considerable air (ventilating area) and if this is not available the fire will burn at its ventilation-controlled rate (NFPA 2008, 18–8). Most building fires burn at their ventilation-controlled rate at least during the time when containment is possible. The fire burning rate can change from the ventilation-controlled rate to stoichiometric combustion based on changes to ventilation (air), such as windows breaking out or a door failing during a fire (NFPA 2008, 18–8–9).

Ventilation can also play an important role in the movement and transfer of smoke throughout a building. As mentioned, smoke is the leading cause of death in fires. Therefore, managing smoke in a fire is important for minimizing deaths due to inhalation of toxic fumes and to improve visibility during a fire. In terms of ventilation, smoke management aims to do the following (Cote and Bugbee 2001, 160):

1. Maintain means of egress in a useable condition,
2. Contain smoke in the environment,
3. Maintain a condition outside of the fire area that will assist fire suppression, and
4. Assist with the protection of life and reduce property damage.

Interior Finish

An important consideration in building construction is the interior finish. NFPA 101 defines interior finish as those materials or assemblies of materials that form the exposed interior surface of walls, ceilings, and floors in a building (NFPA 2012b, 101–30). Examples of interior finish materials include wood, wood paneling, carpeting, drywall, plastics, fibrous ceiling tiles, and wall coverings. The interior finish materials have a tremendous influence on smoke and toxic gas generation, as well as the speed at which the interior of the building can become involved once ignition has taken place, flame spread. The most widely used test for flame spread is the Steiner Tunnel Test, which is included within NFPA 255: Standard Method of Test of Surface Burning Characteristics of Building Materials (ASTM E84). This test attempts to stimulate the spread of fire across a plane surface, where radiant heating of the sample from an external heat flux is applied (NFPA 2008, 2–42). Results from this test are expressed as a Flame Spread Index (FSI) which is a relative number based on the area under the flame spread distance versus the time curve of the test. Materials are tested based on two reference points for the FSI of asbestos cement board being 0 and red oak being 100. Interior finishes are categorized as Class A, B, and C. Class A materials have a flame spread rating from 0 to 25 and are the best in terms of flame spread; in other words, the flame does not propagate as far with these materials as it would with the other classes of materials. Class B materials have a flame spread index from 26 to 75, and Class C has a rating from 76 to 200 (NFPA 2008, 6–43–44).

Interior finish is also important due to its potential for smoke contribution. It is important not only to understand how much smoke is being generated but also to determine if the smoke contains any toxic gases. The interior finish can also increase the combustible load within a building and, thereby, increase the intensity of the fire. One last thing to consider about the fire hazard of interior finish is the combustibility of any adhesive or vapor seals used to apply the interior finish. In some cases, the adhesive used can greatly increase the flame spread of the basic interior finish material (NFPA 2008, 18–38).

Building Contents

In 1973, the U.S. Fire Administration (USFA) found that in terms of design and materials, the environment in which Americans live and work presents unnecessary hazards. The hazards of flames have been studied and regulated
to some extent, but recognition of the hazards of smoke and toxic gases came belatedly. The USAF's report *America Burning: The Report of the National Commission on Fire Prevention and Control* recommended that the impact of new materials, systems, and buildings on users and the community should be assessed during the design stages, well before use (USFA 1973, ix). However, more than forty years later, the toxic gases evolved from burning

contents and the increased fire loads they create still play a significant role in the extent of fire deaths in the United States.

Fire Loading

A relationship has been established between the severity of a fire in a structure and its fire load. Therefore, it is common to use a structure's fire load to predict the fire severity for various occupancies and to determine the fire resistance required of fire barriers as well as other structural components.

The fire load in a building is the total amount of potential heat energy that might evolve during a fire in the building. The fire load, expressed as BTU/sq. ft., is calculated by multiplying the weight of the fuel by the fuel's caloric value (BTU) divided by the floor area (Brannigan 2008, 19–20). The fire load in a building is influenced by several factors: the characteristics of the combustible material, such as heats of combustion, distribution of the fire load in the area, characteristic of the structure such as walls, floors and ceilings, and ventilation (air distribution) in the area. It is important to note that the fire load includes more than the combustible contents, it also includes combustible structural elements as well as interior finishes (NFPA 2008, 18–5).

Although fire load is a representation of the latent energy available in a fire, it is not the complete indication of the severity of a potential fire. The rate of energy release when the fuel burns is referred to as the heat release rate (HRR). The HRR is usually expressed in terms of kilowatts (kW). This rate is a critical factor in determining if flashover will occur in a compartment and is also used by fire protection engineers to describe the fire's size. Examples of HRR for common building contents follow (Brannigan 2008, 20–21):

Mattress (cotton): 40–970 kW
Wastebasket: 4–18 kW
Chair (cotton)
Chair (polyurethane)

OCCUPANCY AND COMMODITY CLASSIFICATIONS

In addition to building construction, it is also important to understand occupancy and commodity classifications because of their potential impact on fire spread. Occupancy and commodity classifications provide a convenient means for categorizing the fuel loads and fire severity associated with certain building operations. The proper occupancy hazard classification for a given building operation should be determined by carefully reviewing the descriptions of each occupancy hazard and by evaluating the quantity, combustibility, and heat-release rate of the associated contents. When developing building codes, it is quite common for the codes to classify buildings based on their use and occupancy. An example is the International Building Code (IBC), which classifies buildings into ten classifications: Assembly, Business, Educational, Factory and Industrial, High Hazard, Institutional, Mercantile, Residential, Storage, and Utility/Miscellaneous. Some of these classifications are further divided into groups. It is critical to select the correct classification and possibly group as the codes will vary based on these classification and groups. Refer to chapter 3 of the IBC for specific definitions for these codes and classifications (International Code Council 2012, 41).

This classification system is also critical for the design of fire suppression systems as it establishes a relationship between the burning fuels and the ability of a sprinkler system to control the associated types of fires. Specifically, the determination of the type of occupancy hazard influences system design and installation considerations, such as sprinkler-discharge criteria, sprinkler spacing, and water supply requirements. It cannot be overemphasized that proper classification of the occupancy is critical to the overall success of fire suppression systems. NFPA 13: Installation of Sprinkler Systems, Chapter 5, identifies three occupancy classifications (NFPA 2013, 13–25):

- Light hazard occupancies are occupancies or portions of other occupancies where the quantity or combustibility of contents is low and fires with relatively low rates of heat release would be anticipated. Light hazard occupancies represent the least severe fire hazard since the fuel loads associated with these occupancies are typically low and relatively low rates of heat release would be expected. Some examples of light hazard occupancies include schools, office buildings, and churches.
- Ordinary hazard occupancies are subdivided into two groups. Ordinary hazard Group 1 includes occupancies where combustibility is low, quantity of combustibles is moderate (stockpiles of combustibles do not exceed 8 ft.), and fires with a moderate rate of heat release would be expected. Examples of ordinary hazard Group 1 occupancies include bakeries and restaurant service areas. Ordinary hazard Group 2 includes occupancies

where the quantity and combustibility of contents is moderate to high (stockpiles do not exceed 12 ft. for contents with moderate rates of heat release and 8 ft. for high rates of heat release), and fires with moderate to high rates of heat release would be anticipated. Examples of ordinary hazard Group 2 occupancies include libraries, dry cleaners, and woodworking facilities.
- Extra hazard occupancies represent the potential for the most severe fire conditions and, therefore, present the most severe challenge to fire protection systems. The extra hazard occupancy classification has two subclassifications. Extra hazard Group 1 includes occupancies where the quantity and combustibility of contents is very high, and dust, lint, or other materials are present, introducing the probability of rapidly developing fires with high rates of heat release but little or no combustible or flammable liquids. Examples of extra hazard Group 1 occupancies include die casting, printing, and sawmills. Extra hazard Group 2 includes occupancies with moderate to substantial amounts of flammable or combustible liquids. Examples of extra hazard Group 2 occupancies include establishments involved with flammable liquid spraying, open oil quenching, and plastic processing.

In addition to understanding the hazards of the occupancies, it is also important to understand commodity classifications. Commodity classifications provide identification of the type, amount, and arrangement of combustibles and are essential to defining potential fire severity based on a commodity's burning characteristics. Commodity classifications are governed by the types and amounts of materials that are part of a product and its primary packaging. NFPA 13: Installation of Sprinkler Systems, Chapter 5, identifies four commodity classifications (NFPA 2013, 13–25–26):

- Class I commodities are noncombustible products that meet one of the following criteria: the noncombustible products are placed directly on wooden pallets, placed in a single layer, corrugated carton, or are shrink-wrapped or paper-wrapped as a unit load with or without pallets.
- Class II commodities are noncombustible products that are placed in slatted wood crates, solid wood boxes, multiple-layer corrugated cartons, or a packaging material of equivalent combustibility, with or without pallets.

- Class III commodities are products made of wood, paper, natural fibers, or Group C plastics with or without cartons, boxes, or crates and with or without pallets.
- Class IV commodities are defined as products with or without pallets that meet any of the following criteria: constructed partially or totally of Group B plastics, consist of free-flowing Group A plastic materials, or contain within themselves or their packaging an appreciable amount (5–15 percent by weight) of Group A plastics.

As one would expect, it is very possible within any facility to have mixed commodities of the above four classes. When mixed commodities are encountered, NFPA 13 §§ 5.6.1.2.2 requires that the storage area be protected by the requirements for the highest-classified commodity and storage arrangement within the facility (NFPA 2013, 13–25).

UNIQUE RISKS FOR FIRE IN HIGH-RISE BUILDINGS

The NFPA defines a high-rise building as a building more than 75 ft. in height as measured from the lowest level of fire department vehicle access to the floor of the highest occupiable story (NFPA 2012b, 101–28). The fire risk in high-rise buildings has been a special concern to the fire community for as long as there have been high-rise buildings. This interest increased after the collapse of the World Trade Center in 2001. There have been numerous studies and recommendations following this event, with one of the most comprehensive studies done by the National Institute of Standards and Technology (NIST). In 2005, NIST developed a report on this incident that included thirty recommendations addressing the following broad areas of high-rise construction (NFPA 2008, 20–79): Increase Structural Integrity, Enhanced Fire Resistance of Structures, New Methods for Fire Resistance Design of Structures, Improved Active Fire Protection, Improved Building Evacuation, Improved Emergency Response, Improved Procedures and Practices and Education and Training.

The risk of a significant hazardous event is higher in a high-rise structure for several reasons. First, a high-rise structure will have larger square footage and therefore may have more people exposed. Second, evacuation will take longer and the possibility of the event affecting more than one floor is greater. Third, a localized failure could impact other parts of the structure or adjacent structures. There are also some features of high-rise structures that affect firefighting (NFPA 2008, 20–80):

- High-rise buildings affect the fire department's access to the fire. Fire apparatuses have limitations in reaching the upper floors of the exterior of the building.
- The height of the fire affects the number of fire service personnel required to deliver adequate types and amounts of equipment to the fire. More time and energy are required to deploy forces and equipment to the fire; as a result, these resources could be exhausted before firefighting forces can mount an attack.
- Delays in deploying equipment and firefighters can indirectly affect fire growth, resulting in a fire of greater magnitude.
- The height and location of the building can restrict the fire department's ability to approach the fire at its origin from more advantageous locations.
- Due to building height, egress and people movement systems within a high-rise building are limited.
- Natural forces affecting fire and smoke movement are more significant in high-rise buildings than they are in lower buildings. Due to the height of high-rise buildings, significant stack effects can move large volumes of smoke and heat through a building.
- Due to their design, high-rise buildings significantly increase the occupant, equipment, and material load on a given building. Stacking floors increases the number of occupants and fuel load that could be exposed to a fire compared to lower-height buildings.

Based on the increased risk associated with high-rise structures and the unique challenges they present to firefighters, it is critical to address these increased risk in the design and construction of these structures. Local building codes as well as NFPA 5000 (Chapter 33) provide specific criteria for high-rise structures. Specifically, this chapter provides criteria on construction types, means of egress, voice communications for alarms, fire department communications, sprinkler systems, smoke proof exit enclosures, emergency lighting, stand-by power, and emergency command center (NFPA 2012c, 5000–323).

BUILDING CODES

When discussing building construction, one must also consider the local building codes. A building code can be defined as a law that sets forth minimum requirements for the design and construction of buildings and structures. It is important to note that a building code is not the same as a fire code, which regulates the use of a building after it is built (Brannigan 2008, 59 & 67). Building codes have been in existence since 1750 BC when King Hammurabi established a form of building codes in Babylonia. These building codes were very punitive in nature, with a focus on the prevention of building collapse. Specifically, if a building collapsed and killed the owner, the builder would be put to death. Some of the earliest fire codes in Europe started in England. In 1189, the Mayor of London created "Assize of Buildings" which regulated building construction and addressed fire spread using stone party walls (NFPA 2008, 1–52). In the United States during the early 1600s, similar building codes were developed by cities based on their own fire experiences. For example, New York City banned the use of wood and plaster for chimney construction, and Boston established building ordinances that required all homes be built of stone with roofs of tile or slate to reduce the spread of fire. The first national building code in the United States was published in 1906 by the National Board of Fire Underwriters and it was titled the National Building Code (Brannigan 2008, 66). This code represented the first model building code, which provided a complete building code that could be adopted by cities and states as written.

By 1950, there were three regional model code organizations: Building Officials and Code Administration (BOCA), Southern Building Code Congress International, and the International Conference of Building Officials (ICBO). BOCA published the National Building Code, ICBO published the Uniform Building Code, and the Southern Building Code Congress International published the Standard Building Code (Brannigan 2008, 67). In general, these three model building codes were a compromise between ideal safety and economic feasibility. Based on occupancy and construction type, fire suppression systems and other types of fire protection systems may be required. These model building codes may also address height and floor building limitations.

The code enforcement personnel for the municipality would typically review building plans for compliance with the model building codes when the building permit was issued. These same enforcement personnel would make periodic site visits to verify that the plans were being followed during construction. Construction features that were typically covered included: structural integrity, life safety, fire protection/suppression systems, mechanical and utility systems, enclosure of vertical openings, and interior finish. These model building codes were typically written as either specification or performance codes. A specification code spelled out in detail what materials can be used, the building size, and how components should be assembled. Performance codes indicated the objectives to be met and established criteria to determine whether the objectives had been met (NFPA 2008, 1–58).

The use of three model building codes in the United States was problematic for building architects, engineers, contractors, and for code enforcement. By the early 1990s, there was a major push from these groups, among others, to create a true national building code.

INTERNATIONAL BUILDING CODE

The International Code Council (ICC) was established in 1995 with representatives from the three regional model building code organizations discussed earlier. The purpose of the ICC was to develop one comprehensive set of regulations for building systems that was consistent with the three current model building codes (International Code Council 2003, iii). The ICC believed that combining the three codes into one would benefit code-enforcement officials, architects, engineers, designers, contractors, and manufacturers and provide uniform education and certification for building-code-enforcement officials.

The first building code published by the ICC was the IBC in 2000. This code is reviewed annually with input provided by a variety of sources such as code enforcing officials, industry representation, and design professionals. This code is published every three years, with the most recent edition published in 2012. The basic principles of the IBC are to develop codes that (International Code Council 2003, iii)

- adequately protect public safety, health, and welfare,
- do not unnecessarily increase building costs,
- do not restrict the use of new building materials, or methods of construction, and
- do not bias the use of particular types or classes of building materials or methods of construction.

In addition to the IBC, the ICC also has other codes related to building construction such as the International Energy Conservation Code, International Existing Building code, International Fire Code, International Fuel Gas Code, International Green Construction Code, International Mechanical Code, and International Plumbing Code. Although a relatively new building code, the IBC is increasingly being adopted by jurisdictions and it is only a matter of time that this will truly become a national building code adopted by the majority of jurisdictions.

For more information on building design, threats, and mitigations, see appendix.

CHAPTER QUESTIONS
1. What is fire resistance?
2. What are the three separate elements involved in a building fire?
3. What does a Type 222 subclassification for a building signify?
4. What are three characteristics of steel that can reduce its performance in withstanding a fire?
5. Why does gypsum have excellent fire-resistance characteristics?
6. Why is it important to consider the interior finish when evaluating the fire risk associated with a building?
7. Why do high-rise buildings pose a unique fire risk?
8. What is fire load and why is it important when assessing the fire risk of a building?
9. What are some examples of ordinary hazard occupancies?
10. Products that are primarily made up to Group B plastics on wooden pallets would be classified as what class of commodity?

REFERENCES

Brannigan, Francis L. (1971). *Building Construction for the Fire Service.* Quincy, MA: NFPA.

Brannigan, Francis L. (2008). *Building Construction for the Fire Service.* Quincy, MA: NFPA.

Cote, Arthur, and Percy Bugbee. (2001). *Principles of Fire Protection.* Quincy, MA: NFPA.

International Code Council. (2012). *International Building Code—2012*. Country Club Hills, IL: International Code Council.

National Fire Protection Association (NFPA). (2008). *Fire Protection Handbook*, 20th ed. Quincy, MA: NFPA.

NFPA. (2012a). *NFPA 220: Standard on Types of Building Construction*. Quincy, MA: NFPA.

NFPA. (2012b). *NFPA 101: Life-Safety Code*. Quincy, MA: NFPA.

NFPA. (2012c). *NFPA 5000: Building Construction and Safety Code*. Quincy, MA: NFPA.

NFPA. (2013). *NFPA 13: Standard for the Installation of Sprinkler Systems*. Quincy, MA: NFPA.

Schroll, Craig R. (2002). *Industrial Fire Protection Handbook*. Lancaster, PA: Technomic Publishing Co.

U.S. Fire Administration (USFA). (1973). *America Burning: The Report of the National Commission on Fire Prevention and Control*. Washington, DC: National Commission on Fire Prevention and Control.

USFA. (1987). *America Burning Revisited: National Workshop*. Tyson's Corner Virginia, November 30-December 2. Washington, DC: Federal Emergency Management Agency, USFA.

6

Life Safety in Buildings

LOSS OF LIFE IN BUILDINGS

From a fire loss control standpoint, no single element is more important than life safety efforts that prevent injury or death to occupants. The building structure and equipment can be replaced but a human life cannot. Life safety measures consider many things, such as building construction, provisions for exits and fire protection, occupant notification, and emergency response. The increased focus on fire prevention and life safety has resulted in a reduction in fires and fire injuries/deaths over the past ten years but we still have room for improvement. Take, for example, the year 2012 where public fire departments responded to 1,375,000 fires that caused 2,855 civilian deaths and 12,875 civilian injuries, an average of one fire injury every 32 minutes (Karter 2013). Although most of these fires were outside or residential, we still experience large-loss fires in commercial and industrial occupancies. Throughout its history, the United States has experienced some of the deadliest fires in these occupancies due to inadequate building and life safety codes, failure of organizations to comply with codes and standards, and as the direct result of terrorist activities. An example of one of these historical fires that had major impacts on the development of fire and life safety codes is the Triangle Shirtwaist Factory.

TRIANGLE SHIRTWAIST FIRE

One of the worst industrial fires in U.S. history occurred on March 25, 1911, when fire broke out in the Triangle Shirtwaist Company facility in New York City (Greer 2001). To keep the workers at their sewing machines, the company had locked the doors leading to the exits. As the fire spread rapidly, fed by thousands of pounds of fabric, workers rushed to the stairs, freight elevator, and fire escape. Many died when the rear fire escape collapsed, and many others jumped to their deaths to escape the burning building (Greer 2001). For all practical purposes, the ninth-floor fire escape in the Asch Building led nowhere, certainly not to safety, and it bent under the weight of the factory workers trying to escape the inferno. Other workers waited at the windows for the rescue workers only to discover that the firefighters' ladders were several stories too short and that the water from the hoses could not reach the top floors (Guthrie 2004). In total, 146 women died. The owners were acquitted of manslaughter charges and were ordered to pay $75 to the families of twenty-three victims. However, the public outcry from this tragedy did promote some of the nation's strongest changes in worker safety in the manufacturing industry, one of these being the forerunner to the NFPA 101 Life Safety Code.

Unfortunately, these large loss fires are not a thing of the past. For example, in the past fifteen years, some of the largest losses in terms of fatalities occurred in commercial properties such as the World Trade Center in 2001 and the Station Night Club in 2003.

The Station Night Club is a good example of inadequate fire and life safety codes as well as a lack of enforcement/implementation of these codes.

STATION NIGHT CLUB FIRE

The Station Night Club was in Warwick, Rhode Island. At 11:07 p.m. on the night of February 20, 2003, a band that was performing used pyrotechnics as part of its opening act. Unfortunately, the pyrotechnics ignited the polyurethane foam insulation lining the walls and ceiling of the platform. The nightclub was not equipped with sprinklers and the fire spread quickly along the walls and ceiling and smoke was visible in the exit doorways within 30 seconds. Shortly after this, the alarm sounded and the bulk of the crowd began to evacuate. Unfortunately, egress from the nightclub was hampered by overcrowding at the main entrance to the building. Club patrons and staff broke windows on the front of the nightclub so they could exit. Many of those who escaped attempted to extricate people who had been wedged in the front doorway. Within 5 minutes the flames were observed through the roof and extending out the windows and front doorway. Shortly after this, the first fire department arrived on site to initiate extinguishment. As a result of this fire, 100 people lost their lives (Bryner, Grosshandler, Kuntz, and Madrzykowski 2005, xvii).

This tragedy was investigated by a variety of agencies, one of them being the National Institute of Standards and Technology (NIST), an agency of the U.S. Commerce Department. A major role of the NIST is to develop and promote measurement, standards, and technology to enhance productivity, facilitate trade, and improve the quality of life. Based on their investigation, NIST urged all state and local governments to adopt and aggressively enforce national model building and fire safety codes for nightclubs. The NIST believed that strict adherence to the 2003 model codes available at the time of the fire would have gone a long way to preventing similar tragedies in the future. The NIST also believed that additional recommendations to the existing codes could strengthen occupant safety even further. A summary of these recommendations follows (Bryner, Grosshandler, Kuntz, and Madrzykowski 2005, xvii):

- The first recommendation urges all state and local jurisdictions to adopt a building and fire code covering nightclubs based on one of the national model codes and update local codes as the national standards are revised; implement aggressive and effective fire inspection and enforcement programs that address all aspects of these codes; and ensure that enough fire inspectors and building plan examiners—professionally qualified to a national standard—are on staff to carry out this work.
- The second and third recommendations addressed the use of automatic fire sprinkler systems for extinguishing fires in nightclubs and limiting the flammability of materials used as finish products. Specifically, the NIST recommends that NFPA 13 "Standard for the Installation of Sprinkler Systems" be adopted, implemented, and enforced for all new nightclubs regardless of size, and for existing nightclubs with an occupancy limit greater than 100 people.
- The fourth recommendation was to revise NFPA 1126 "Standard on the use of Pyrotechnics" by addressing the need for automatic sprinkler systems;

minimum occupancy/building size levels; the posting of pyrotechnic use plans and emergency procedures; and setting new minimum clearances between pyrotechnics and the items they potentially could ignite.
- The fifth recommendation was to revise all national model codes that increase the factor of safety for determining occupancy limits in all new and existing nightclubs. These include setting a maximum permitted evacuation time, calculating the number of required exits and permitted occupancies, increasing staff training and evacuation planning, and improving means for occupants to locate emergency routes when standard exit signs are obscured by smoke.
- The sixth recommendation addressed portable fire extinguishers, calling for a better understanding of the numbers, placement locations, and staff training required to ensure their effective use.
- The seventh recommendation was for state and local jurisdictions to adopt existing model standards on communications, mutual aid, command structure, and staffing.
- Recommendations 8 through 10 addressed the need for research to serve as the basis for further improvements in codes, standards, and practices. One of the specific areas recommended for additional research involved a better understanding of human behavior in emergency situations and to predict the impact of building design on safe egress in emergencies.

HUMAN BEHAVIOR DURING EMERGENCIES
It has long been known that the actions of people during a fire emergency play a critical role in their survival. Unfortunately, responding appropriately during an emergency is not a natural skill and some human behaviors, such as panic, can increase the likelihood of being a victim rather than surviving. There are a variety of factors that can influence how we react during a fire emergency and they include: actions of others, building characteristics such as marking of exits, perceived threat, and individual characteristics such as personality and previous experience/education (NFPA 2008, 4–3).

Time is of the essence during a fire emergency, and the faster an individual recognizes the perceived risk and acts the greater the probability of survival. Obviously, alarm systems play a major role in the recognition/awareness of a fire. Studies have shown the most effective type of alarm is one that combines an alarm bell with human voice directives. These human voice directives help to convey not only meaning but also a sense of urgency (NFPA 2008, 4-4-8).

Once the fire is recognized the individual will then attempt to determine the seriousness of the threat. Some of the factors that influence the perceived seriousness of the fire include the presence and intensity of smoke, flames, and heat. Based on the aforemenioned factors the individual will take an overt behavioral response, which could be a decision to extinguish the fire or evacuate. It should be noted that an individual's behavioral responses at this time are impacted in a major way by the actions of others as well as the familiarity with the building and emergency response plan. Examples of how human behaviors can be influenced by the actions of others include convergence clusters, panic behaviors, and reentry behavior. Convergence clusters occur when the occupants in a building that is on fire converge in specific rooms that they perceive as areas of refuge (NFPA 2008, 4-23). Panic behavior occurs when the occupants experience a sudden and excessive feeling of alarm or fear, leading them to undertake extravagant efforts to secure safety. This behavior is not common during a fire emergency and it must be avoided if at all possible as it reduces the escape possibilities of the group (NFPA 2008, 4-30). In reentry behavior, occupants who have successfully exited the building reenter for various reasons. Typically, they do so looking for loved ones, to assist others in exiting, and to assist with firefighting. Occupant firefighting behavior occurs more often with males, especially those who have economic or emotional ties to the building (NFPA 2008, 4-33).

In addition to human behavior there are some physical characteristics of people that can influence their survival during an emergency. These physical characteristics include age and disability. People with limited physical and cognitive abilities, especially children and older adults, are at a higher risk of death and injury from fire than other groups. These two age groups accounted for 46 percent of the fire deaths and 22 percent of estimated fire injuries in the United States in 2004 (USFA 2008)

ORIGIN AND DEVELOPMENT OF LIFE SAFETY CODES AND REGULATIONS

The National Fire Protection Association (NFPA) 101: Life-Safety Code had its origin in the work of the Committee on Safety to Life of the NFPA, which was appointed in 1913. This committee was a nongovernmental body formed

FIGURE 6.1
Fire Escape

in the days following the Triangle Shirtwaist fire in New York City to push for system-wide reforms for worker safety. The work of this committee led to the Building Exits Code in 1927, the forerunner to today's NFPA 101 Life Safety Code (NFPA 2013a). These early codes included standards for the construction of stairways, fire escapes (figure 6.1), and other egress routes for fire drills in various occupancies and for the construction and arrangement of exit facilities for factories, schools, and other occupancies. Major fire losses over the years have resulted in revisions of and additions to the life safety codes. Examples of major fire losses that resulted in significant life safety improvements include the Cocoanut Grove Night Club fire in Boston in 1942, Our Lady of the Angels school fire in Chicago in 1958 as well as more recent fires such as the Station Night Club fire in 2003.

In addition to the NFPA, there is also life safety standards included within the Occupational Safety and Health Administration (OSHA) as well as building codes. OSHA standards are included in 29 C.F.R. 1910 Subpart E: Means of Egress and they address life safety issues such as emergency lighting, exit signs, and means of egress. Some of these OSHA standards were adopted from existing NFPA life safety codes, while other standards were promulgated by OSHA. To provide some consistency between NFPA 101 and the standards included in Subpart E, OSHA now deems employers who are in compliance with the exit route provisions of NFPA 101: Life Safety Code to be in compliance with 29 C.F.R. § 1910.34, 36 and 37 (USDOL 2013a).

Building codes will also include standards on life safety. As discussed in chapter 5, the International Code Council (ICC) was created to develop a national building code. The ICC currently has two specific building codes directed at life safety, the International Building Code (IBC), and the International Fire Code. NFPA also has a building code that is included in NFPA 5000: Building Construction and Safety Code. NFPA 5000 covers a range of subjects, including allowable building heights and areas based on occupancy and construction; protection schemes for vertical openings; means of egress; and the rehabilitation of existing buildings (NFPA 2013b, 5000–1).

BUILDING OCCUPANCIES AND LIFE SAFETY

One of the most significant factors that will influence the life safety of the occupants as well as the overall building design and fire suppression systems is the occupancy classification. Occupancy classification is the principal use of the structure, which considers the relative hazards of processes/activities, types and quantities of regulated materials, and the capacities of occupants for self-preservation (Diamantes 2007, 41). The IBC, NFPA 5000 Building Construction and Safety Code, and the NFPA 101 Life Safety Code all classify buildings into general occupancy classifications which may be subdivided into groups. Chapter 3 of the IBC defines ten occupancy classifications and groups that are similar but not identical to NFPA 101 and 5000. The focus of this chapter is life safety so the occupancy classifications discussed in Chapter 6 of NFPA 101 will be used. A summary of these occupancy classifications follows (NFPA 2012, 101–6.1):

1. Assembly occupancy is an occupancy that is used for a gathering of fifty or more persons for deliberation, worship, entertainment, eating, drinking, amusement, awaiting transportation, or similar uses, or used as a special amusement building, regardless of occupant load. Examples of assembly occupancies include auditoriums, courtrooms, exhibition halls, and theatres. Specific life safety code requirements for these occupancies are included in Chapters 12 and 13 of NFPA 101 Life Safety Code.
2. Educational occupancy is an occupancy used for educational purposes through grade twelve by six or more persons for four or more hours per day or more than twelve hours per week. Specific life safety code requirements for these occupancies are included in Chapters 14 and 15 of NFPA 101 Life Safety Code.
3. Day-care occupancy is an occupancy where four or more clients receive care, maintenance, and supervision by individuals that are not their relatives or legal guardians for less than twenty-four hours per day. Examples of day-care occupancies include adult day care, child day care and nursery schools. Specific life safety code requirements for these occupancies are included in Chapters 16 and 17 of NFPA 101 Life Safety Code.
4. Health-care occupancy is an occupancy used for purposes of medical or other treatment or care of four or more persons, where such occupants are mostly incapable of self-preservation due to age or physical or mental disability, or because of security measures not under the occupants' control. Examples of such occupancies would include hospitals, limited care facilities, and nursing homes. Specific life safety code requirements for

these occupancies are included in Chapters 18 and 19 of NFPA 101 Life Safety Code.

5. Ambulatory health-care occupancy is an occupancy used to provide services or treatments to four or more patients on an outpatient basis any of the following: treatment or anesthesia for patients that would render them incapable of taking action for self-preservation under emergency conditions without the assistance of others, or emergency or urgent care where the injury renders the patient incapable of taking action for self-preservation under emergency conditions without the assistance of others. Specific life safety code requirements for these occupancies are included in Chapters 20 and 21 of NFPA 101 Life Safety Code.

6. Detention and correctional occupancy is an occupancy used to house one or more persons under various degrees of restraint or security and the occupants are incapable of self-preservation due to the security measures not under the occupants' control. Specific life safety code requirements for these occupancies are included in Chapters 22 and 23 of NFPA 101 Life Safety Code.

7. Residential occupancy is an occupancy that provides sleeping accommodations for purposes other than health care or detention and correction. Residential occupancy has several groups that include one- and two-family dwelling units, lodging or rooming houses, hotels, dormitories, and apartments. One- and two-family dwellings are buildings that contains not more than two dwelling units with independent cooking and bathroom facilities. Lodging or rooming houses are buildings that do not qualify as a one- or two-family dwelling units and provide sleeping accommodations for a total of sixteen or fewer people without separate cooking facilities for individual occupants. Hotels are buildings that provide sleeping accommodations for more than sixteen people and primarily used by transients for lodging with or without meals. Dormitories are building in which group sleeping accommodations are provided for more than sixteen people who are not members of the same family in one room or a series of associated rooms under joint occupancy without individual cooking facilities. Apartment buildings are buildings that contain three or more dwelling units that have independent cooking and bathroom facilities. Specific life safety code requirements for these occupancies are included in Chapters 24–31 of NFPA 101 Life Safety Code.

8. Residential board and care occupancy is an occupancy used for lodging and boarding four or more residents, not related to each other, for the purpose of providing personal-care services. Specific life safety code requirements for these occupancies are included in Chapters 32 and 33 of NFPA 101 Life Safety Code.
9. Mercantile occupancy is an occupancy used for the display and sale of merchandise. Examples of this occupancy classification would include department stores, restaurants with occupancy less than fifty people, and supermarkets. Specific life safety code requirements for these occupancies are included in Chapters 36 and 37 of NFPA 101 Life Safety Code.
10. Business occupancy is an occupancy used for the transaction of business other than mercantile. Examples of this occupancy include general offices, dentist offices, and college classrooms under fifty persons. Specific life safety code requirements for these occupancies are included in Chapters 38 and 39 of NFPA 101 Life Safety Code.
11. Industrial occupancy is an occupancy where products are manufactured or where processing, assembling, mixing, packaging, finishing, decorating, or repair operations are conducted. Examples of this occupancy include factories of all kinds, power plants, and sawmills. Specific life safety code requirements for these occupancies are included in Chapter 40 of NFPA 101 Life Safety Code. This is an extremely broad occupancy classification because of the variety of processes involved. It is for that reason; this occupancy classification was further divided into the following subclassifications (NFPA 2012, 101–40.1.2):
    a. General industrial occupancy is an occupancy which has low or ordinary hazard processes in buildings of conventional design suitable for various types of industrial processes. This subclassification also includes multistory buildings that are occupied by different tenants that are subject to a high density of employee population.
    b. Special purpose industrial occupancy is an occupancy that conducts low or ordinary hazard processes in buildings designed for particular types of operations. This occupancy is also characterized by a relatively low density of employee population, with much of the area occupied by machinery.
    c. High hazard industrial occupancy is an occupancy that conducts industrial operations that use high hazard materials/processes or store high hazard contents.

12. Storage occupancy is an occupancy used primarily for the storage or sheltering of goods, merchandise, products, or vehicles. Examples of such occupancies include cold storage facilities, warehouses, and grain elevators. Specific life safety code requirements for these occupancies are included in Chapter 42 of NFPA 101 Life Safety Code.
13. Multiple occupancy is an occupancy where two or more classes of occupancy exist within the structure. Within this classification there are two groups: mixed occupancy and separated occupancy. With mixed occupancy the two or more classes are intermingled, and the building must comply with the most restrictive requirements of the occupancies involved. An exception to this is where one of the occupancy classifications is considered incidental, such as a small storage area in any occupancy. With separated occupancy the different occupancies are separated by fire resistive assemblies.

GENERAL REQUIREMENTS FOR MEANS OF EGRESS

A means of egress is an unobstructed, continuous path from any point in a building to a safe location outside. The means of egress consists of three parts: exit access, the exit, and exit discharge (NFPA 2008, 4–78). The general requirements for the means of egress for all occupancies are included in NFPA 101: Life Safety Code Chapter 7 "Means of Egress," as well as 29 C.F.R. 1910 Subpart E Means of Egress. In addition to the general requirements included in chapter 7, specific requirements based on the type of occupancy are included in Chapters 12 to 42 of NFPA 101; see previous section for specific chapters. Most of the occupancy chapters are further divided into two chapters with one devoted to new occupancies and the other devoted to existing occupancies. It cannot be overemphasized the importance of consulting applicable building codes, 29 C.F.R. Subpart E, and both chapter 7 and the occupancy, specific chapter of NFPA 101 for the specific requirements of the means of egress for a particular occupancy.

Exit Access

The first part of the means of egress is the exit access which is the route one must take to go from anywhere in the building to reach the exit. Examples of exit access include aisle ways, corridors, or rooms. Important criteria for the exit access are an adequate width to accommodate the maximum number of people in each area, maintained clear and unobstructed, and meet the maximum travel distances established by codes. In terms of adequate width, the IBC and NFPA 101 both require a minimum of 36 in. for corridors serving fifty or less and a minimum of 44 in. for more than fifty people (Diamantes 2007, 118). This is a minimum width and can increase depending on the type of occupancy. The maximum travel distance is based on the type of occupancy and whether the building is sprinklered or unsprinklered. This travel distance ranges from a low of 75 ft. for an unsprinklered high hazard storage occupancy to a high of 400 ft. for a sprinklered ordinary hazard storage occupancy (NFPA 2012, 7.6).

Exit

An exit is that portion of the means of egress that is separated from the building and provides a protected pathway to the exit discharge. In a small building, it may be possible to go directly from the exit access to the exit discharge which would be a door leading to the outside. However, in larger buildings and in multistory buildings it may not be possible to meet the maximum travel distance requirements for exit access and an exit (protected pathways) will be necessary, such as a stairway. The term protected implies a specific fire resistance rating for the pathway. For buildings of three or fewer stories, the minimum is a 1-h fire resistance and for building with four or more stories a minimum fire resistance of 2 h (NFPA 2012, 7.1.3.2.1). The interior finish of the exits is also an important consideration in terms of its combustibility and flame spread. Specifically, NFPA 101 Table A10.2.2 limits the flame spreads of interior finishes and materials that can be used in the exits, exit access corridors, and other areas of the building based on the type of occupancy. For example, all interior wall and ceiling finishes used in exits in assembly occupancy must have a Class A Flame Spread Rating (NFPA 2012, 10.2.2).

The width of the exit will vary depending on the type of component, such as ramp or stairway as well as the occupant load, see next section on Capacity of Means of Egress. For example, the minimum stair width permitted by NFPA 101 for new stairs is 36 in. if the stair is serving fifty or fewer and this width will increase with an increase in the occupant load (NFPA 2012, 7.2.2.2.1.2).

The number of exits is influenced by the type of occupancy, number of occupants and the maximum travel distance for exit access. With few exceptions, most buildings are required to have at least 2 exits and this will increase as the occupant load increases, minimum of 3 exits for 501–1,000 occupants and a minimum of 4 exits for more than 1,000 occupants. In addition, exits must be remote enough from each other so that a fire in one area would not block access to several exits. To determine this distance, the maximum diagonal of the area is divided by 2 for unsprinklered and by 3 for sprinklered to determine the minimum separation distance between exits (Diamantes 2007, 114).

It is critical that all occupants traveling along any point in the exit access be able to identify the exit. This is accomplished by a combination of emergency lighting and marking all exits with clearly identifiable signs. The exit signs must be illuminated to a surface value of at least 5-ft.-candles and be distinctive in color. Self-luminous or electroluminescent signs that have a minimum luminance surface value of at least 0.06-ft. lamberts are permitted. Each exit sign must have the word "EXIT" in plainly legible letters not less than 6 in. high, with the principal strokes of the letters in the word "EXIT" not less than three-fourths of an inch wide (USDOL 2013b) (see figure 6.2).

FIGURE 6.2
Exit Sign

If the direction of travel to the exit or exit discharge is not immediately apparent, directional signs must be posted along the exit access indicating the direction of travel to the nearest exit. Each doorway or passage along an exit access that could be mistaken for an exit must be marked with the phrase "Not an Exit," or with a similar sign indicating its actual use (USDOL 2013b).

In addition to the sign, lighting must be appropriate so that an employee with normal vision can see along the exit route. During normal operations this lighting will be provided by the natural lighting in the facility. However, during an emergency there is a possibility of electrical failure thereby necessitating the need for emergency lighting throughout the means of egress. Emergency lighting should come on automatically and should last for at least one and a half hours in the event of the failure of normal lighting. Emergency lighting illumination levels should be not less than an average of 1-ft. candle and, at any point, not less than 0.1-ft. candle, measured along the path of egress at floor level (NFPA 2012, 101–7.9.2.1).

Doors are a critical component to any exit as they not only allow entry into the protected pathway, but they also help maintain the pathway's fire resistance. Examples of criteria related to doors include the width of a door, its swing direction, and its construction, such as panic hardware and construction material (fire rating). In new structures, the minimum width of the door is a clear width of 32 in., but this is also influenced by the type of occupancy and the occupant load (NFPA 2012, 7.2.1.2.1). In general, fire doors are categorized as either fire-rated doors, non-fire-rated doors, or smoke-resistant doors. The specific fire resistance of the door will be based in part on the fire resistance of the exit itself. Requirements in 29CFR 1910.36 indicate that each fire door, including its frame and hardware, must be listed, or approved by a nationally recognized testing laboratory for its fire resistance. Some other general requirements of fire doors are (USDOL 2013c):

- Exit doors must swing in the direction of travel if the room holds more than fifty people or is used in a high-hazard area.
- Exit doors must be self-closing and remain closed or automatically close in an emergency upon the sounding of a fire alarm or employee alarm system.
- Exit doors must be unlocked, allowing employees to open an exit door from the inside always without keys, tools, or special knowledge. The one exception to this is mental, penal, or correctional facilities.

LIFE SAFETY IN BUILDINGS 143

- Exit doors are permitted to use panic bars that lock only from the outside.

Exit Discharge

The exit discharge is the portion of the exit that separates the exit from the public area. A common exit discharge would be a doorway leading from the building to the outside public walkway. According to requirements in 29CFR 1910.36, each exit discharge must lead directly outside or to a street, walkway, refuge area, public way, or open space with access to the outside. If an exit discharges into a courtyard or other open space, the area must be large enough to accommodate all the building's occupants (USDOL 2013c).

Capacity of Means of Egress

It is critical that the capacity of the means of egress exceed the occupant load, which is the maximum number of persons the occupancy is intended to serve. Criteria used to determine this maximum capacity are based on the type of occupancy and the width of the means of egress components. For structures in the design stage, an occupant load factor is divided into the floor area to predict the occupant load. Table 7.3.1.2 of NFPA 101 provides the occupant load factor (ft.2/person) which is based on occupancy type. The square footage is calculated as either gross or net. Gross is the total square footage and net is the total square footage minus unoccupied areas such as closets and restrooms (NFPA 2012, 7.3). For existing structures, the maximum number of occupants in the structure is used as the occupant load.

The next step in the process is to measure the clear width for each component in the means of egress. This clear width is the narrowest point in the egress component. For example, a 36" exit door will typically have a clear width of 34" once the opening is measured between the door and jam assembly. NFPA 101 does allow some projections in the means of egress components if they do not extend more than 4.5" on either side to a height of 38", that is, handrails on a stairway. When these projections are encountered, they are ignored when measuring clear widths (NFPA 2012, 7.3.2).

Once we have obtained the clear widths for each component of the means of egress, we divide the width by capacity factors (in./person) provided in Table 7.3.3.1 in NFPA 101. These capacity factors range from a low of 0.2 in./person to a high of 0.7 in./person and are based on the occupancy classification and whether the egress surface is level or a stairway. For example, if a high hazard content occupancy had an exit door with a clear width of 34″ this would be divided by the capacity factor of 0.4 in./person for a capacity of eighty-five people. These calculations must be done for all components to each means of egress such as aisle ways, corridors, exit doors, stairways, etc. For each means of egress determine the "bottle neck" that is the most restrictive (least capacity) for the given means of egress. These "bottle neck" capacities for each means of egress are added together to provide the total means of egress capacity for the building which is then compared to the occupant load. It should be noted that in multistory buildings the egress capacity is determined for each story, and this capacity cannot decrease in the direction of egress travel (NFPA 2008, 4–82).

Detection, Alarm, and Communications Systems

Detection, alarm, and communications systems assist in identifying fires in their early stages, notifying the building occupants of the potential fire, and possibly notifying the fire department. In new construction, most building codes require the alarms signal to be both visible and audible alarms. Included in 29 C.F.R. Subpart E within Emergency Action Plans is a requirement for employers to install and maintain an employee alarm system for all employers with eleven or more employees. This alarm system must be capable of being perceived above ambient noise or light levels by all employees in the affected portions of the workplace. The system must also be distinctive and recognizable as a signal to evacuate or to perform actions designated under the emergency action plan. Maintenance and testing of the alarm systems are critical to ensuring their proper operations. For alarm systems that are not supervised, they must be maintained and tested every two months while all supervised systems must be tested annually (USDOL 2013d). Additional information on alarm and detection systems is included in chapter 8 as well as NFPA 101–9.6.1.

Building Services

The NFPA 101 Life Safety Code and the National Building Code establish requirements for building utilities and services such as heating, ventilating, air-conditioning equipment, smoke control, and utilities. With regard to ventilation, these codes address the ventilation aspects of the building including mechanical ventilation and pressurized stair enclosure systems, smoke-proof enclosure mechanical ventilation equipment, and boiler, incinerator, or heater rooms. The general requirements for these building services are identified in Chapter 9 of NFPA 101, "Building Service and Fire Protection Equipment." These general requirements typically refer to specific NFPA standards for design, installation, inspection, and testing, such as NFPA 92 "Standard for Smoke Control Systems." As mentioned in chapter 5, building construction such as the ventilation system and smoke control systems can have a major influence on the spread of fire, smoke, and toxic gases. Therefore, their proper design and installation can play a critical role in life safety (NFPA 2012, 9.2).

Emergency Action Plans

At the beginning of this chapter, it was indicated that one of the factors to consider as part of an overall life safety program is emergency response.

Human behavior plays a major role in the response of individuals during an emergency and this is not a natural skill. Two of the factors that can influence how an individual reacts during an emergency are the actions of others and previous experience/education. Both factors are addressed in the development and implementation of an effective emergency response plan.

The development of emergency response plans is included in NFPA 101, Chapter 4 and by OSHA in 29 CFR 1910 Subpart E Means of Egress. According to OSHA, an employer must have an emergency action plan whenever a specific OSHA standard requires one. An example of this is OSHA's "Process Safety Management of Highly Hazardous Chemicals" which requires employers who operate highly hazardous processes to establish and implement an emergency action plan that meets the requirements of 29 C.F.R. 1910.38. In general, these emergency response plans must be in writing, but they can be communicated verbally if the employers have ten or fewer employees. These written plans must be posted in the workplace for employee review. What follows is a summary of the OSHA requirements for a written emergency action plan (USDOL 2013e):

1. Procedures for reporting a fire or other emergency,
2. Procedures for emergency evacuation, including the type of evacuation and exit route assignments,
3. Procedures for employees to follow to determine who will remain to operate critical plant operations before evacuating,

4. Procedures to account for all employees after evacuation,
5. Procedures to be followed by employees performing rescue or medical duties, and
6. The name or job title of every employee who may be contacted by other employees who need more information about the plan or an explanation of their duties under the plan.

An important aspect of the emergency action plan is employee training. For an emergency action plan to be effective, the employer must designate and train employees to assist in a safe and orderly evacuation of other employees. OSHA requires the employer to review the emergency action plan when the plan is initially developed or a new employee is hired, an employee's responsibilities under the plan change or the plan itself changes (USDOL 2013e). A part of this training should be fire drills, which not only help in rehearsing the plan but also in identifying potential problems that need to be changed in the plan. Written documentation should be completed after each drill, such as date completed, time, and participants. Specific requirements for these drills are included in NFPA 101 Chapter 4 as well as specific occupancy Chapters 11–43.

Maintenance of Means of Egress

A critical component to life safety is the development and implementation of processes that ensure all components to the means of egress are maintained clear and unobstructed. Examples include blocked corridors or aisle ways, storage placed in exits, exit doors that are locked, damaged stair railings, damaged stair treads, or exit doors that are blocked open. A means of egress that is not maintained clear and unobstructed is ineffective and can increase the likelihood of loss of life during an emergency evacuation.

CHAPTER QUESTIONS
1. Identify four factors that can influence how an individual will react during a fire emergency.
2. What is occupancy under the life safety codes?
3. What three things comprise a means of egress?
4. What are the illumination requirements for emergency lighting?
5. When does OSHA require an emergency action plan?

6. What does the term "protected pathway" imply for an exit?
7. What criteria are used to determine the maximum capacity for a means of egress?
8. What additional research was recommended following the Station Nightclub fire?
9. Exhibition halls would fall under what type of occupancy according to NFPA 101?
10. In general, the number of exits in a building is influenced by what?

REFERENCES

Bryner, N., Grosshandler, W., Kuntz, K., and Madrzykowski, D. (2005). Report of the Technical Investigation of the Station Nightclub Fire, US Department of Commerce, Washington, DC, June 2005.

Diamantes, David. (2007). *Fire Prevention: Inspection and Code Enforcement*, 3rd ed. Clifton Park, NY: Delmar Cengage Learning.

Greer, M. E. (2001, October). 90 Years of Progress in Safety. *Professional Safety*, 46(10).

Karter, Michael J. (2013). *Fire Loss in the U.S. During 2012*. Quincy, MA: NFPA Journal, September 2013.

National Fire Protection Association (NFPA). (2008). *Fire Protection Handbook*, 20th ed. Quincy, MA: NFPA.

NFPA. (2012). *NFPA 101: Life-Safety Code*. Quincy, MA: NFPA.

NFPA. (2013a). 100 years ago, Frances Perkins urged NFPA to champion workplace safety. NFPA Today - Your On-line Journal, Quincy, MA, 6/13/2013.

NFPA. (2013b). *NFPA 5000: Building Construction and Safety Code*. Quincy, MA: NFPA.

USDOL. (2013a). *Occupational Safety and Health Standards for General Industry, Subpart E: Means of Egress: General Requirements, 29 C.F.R. § 1910.35 Compliance with Alternate Exit-Route Codes*. Washington, DC: USDOL.

USDOL. (2013b). *Occupational Safety and Health Standards for General Industry, Subpart E, 29 C.F.R. § 1910.37 Maintenance, Safeguards, and Operational Features for Exit Routes*. Washington, DC: USDOL.

USDOL. (2013c). *Occupational Safety and Health Standards for General Industry, Subpart E, 29 C.F.R. § 1910.36 Design and Construction Requirements for Exit Routes.* Washington, DC: USDOL.

USDOL. (2013d). *Occupational Safety and Health Standards for General Industry, Subpart L: Fire Protection: 29 CFR 1910.165 Employee Alarm Systems.* Washington, DC: USDOL.

USDOL. (2013e). *Occupational Safety and Health Standards for General Industry, Subpart E: Means of Egress: 29 CFR 1910.38 Emergency Action Plans.* Washington, DC: USDOL.

U.S. Fire Administration. (2008). *National Fire Data Center A Profile of Fire in the United States 1995–2004.* Washington, DC: Fourteenth Edition, February 2008.

7

Hazardous Processes

Hazardous processes can be found in industries as part of a variety of operations. The hazards of these processes can include flammable gases, flammable and combustible liquids, reactive gases, toxic gases, and liquefied compressed gases to name a few. The hazards can exist in the chemicals that are being used in the process, in the by-products of a reaction, or in the final process product. Hazards can also be observed when problems occur with the process such as in the case of a runaway chain reaction or improper maintenance and repair of process equipment. The result of any of these situations can include fires, explosions, and the release of hazardous materials.

Major fires and explosions associated with hazardous processes have unfortunately been part of the industrial work environment. During operations at the Phillips Petroleum Houston Chemical Complex in Pasadena, Texas, on October 23, 1989, an explosion and the ensuing fire resulted in 23 deaths and 123 injuries of varying severity. Metal and concrete debris were found as far as 6 miles away following the explosion (U.S. Fire Administration [USFA] 1990). The explosion occurred in the polyethylene plant when a flammable vapor cloud formed and was subsequently ignited, resulting in a massive vapor cloud explosion. Following this initial explosion, there was a series of further explosions and fires (Health and Safety Executive 2013). The day before the incident, scheduled maintenance work had begun to clear three of the six settling legs on a reactor. A specialist maintenance contractor was employed to carry out the work. A procedure was in place to isolate the leg to be worked on. During the clearing of a settling leg, part of the plug remained lodged in the pipework. A member of the team went to the control room to seek assistance. Shortly afterward, the release of flammable vapors occurred and approximately two minutes later, the vapor cloud ignited (Health and Safety Executive 2013). The company was fined $6.3 million by the Occupational Safety and Health Administration (OSHA) in 1990 but settled for $4 million.

PROCESSES INVOLVING FLAMMABLE AND COMBUSTIBLE LIQUIDS

Many industrial processes involve the use of flammable and combustible liquids. These processes pose potential fire hazards to those working nearby. Some of the more typical processes include spray finishing operations such as electrostatic spray painting, powder coating processes, dip-tank processes, and oil quenching. Because of the unique hazards of each of these processes, engineering controls and safe work practices must be followed. The National Fire Protection Association (NFPA) has established recommended practices, and OSHA has adopted safety standards designed to minimize the fire hazards associated with these processes.

Electrostatic Spray Operations

Spray finishing operations are found in several industries. In general, these operations apply materials and coatings to a surface. In many instances, these finishing materials consist of flammable liquids. A common example of a spray finish operation can be found in the automotive repair industry, when automobile bodies are spray painted. The process by which the paint is applied to the product can include the use of paint guns and electrostatic spray painting, when the spray material such as paint is negatively charged (while atomized or after having been atomized by the air or airless methods) through the connection of the spraying gun to a generator. The material being painted is positively charged; as a result, the difference in charge leads the particles toward the material to obtain the coating. This type of spray operation is used to reduce overspray, therefore material waste, and makes it possible to obtain very uniform and regular coating layers.

Electrostatic spray operations can be fixed or portable, as with a spray gun. In fixed operations, the parts are typically hung on a conveyor and transported through the spray area. Electrostatic spray equipment used in coating operations must meet design and use standards. The NFPA, Underwriter's Laboratories, and OSHA have established standards for the use, design, and installation of this type of equipment. Due to the hazards associated with the electrical components used to operate the electrostatic spray equipment, applicable standards should be followed to ensure that potential sources of electrical energy sufficient to ignite flammable vapors are located outside of the spraying area.

Electrostatic spray painting requires the use of high voltage and leads to electrodes. To prevent accidental damage that could result in electrocution, this type of equipment should be properly insulated and protected. Another hazard of using electrostatic atomizing heads is the accidental grounding of the equipment. To prevent this from happening, the electrostatic atomizing heads should be insulated from the ground and adequately supported. If the electrode system is deenergized, an automatic means shall be provided for grounding the electrode system (USDOL 2013b, 29 C.F.R. §1910.107).

In the application of the coating material, a safe distance of at least twice the sparking distance shall be maintained between goods being painted and electrodes or electrostatic atomizing heads or conductors. The safe distance should be posted near the assembly.

Workers can also be exposed to hazards created by the equipment. To prevent the workers from coming into contact with the equipment, means should be taken to guard the hazards using railings or guards suitable to maintain a safe distance between the workers and the equipment.

Electrostatic hand-spraying equipment involves equipment that uses electrostatically charged elements for the atomization or precipitation of materials for coatings on articles or for other similar purposes in which the atomizing device is handheld and manipulated during the spraying operation. Equipment used in electrostatic hand spraying should meet applicable design standards. The high-voltage circuits must be designed so as not to produce a spark of sufficient intensity to ignite any vapor–air mixtures or result in an appreciable shock hazard upon coming into contact with a grounded object under all normal operating conditions (USDOL 2013b, 29 C.F.R. § 1910.107). Transformers, power packs, control apparatus, and all other electrical portions of the equipment, with the exception of the handgun itself and its connections to the power supply, should be located outside of the spraying area or meet safety requirements for this type of environment (NFPA 2011a, 33–22). To prevent accidental electrocution of users and ignition of vapors, the guns and all equipment should be properly grounded. In addition, the objects being painted or coated must maintain metallic contact with the conveyor or other grounded support. The hooks should be regularly cleaned to ensure this contact and areas of contact should be sharp points or knife edges where possible (USDOL 2013b, 29 C.F.R. §1910.107).

Ventilation, also a critical element of any spraying operation, is used to keep the concentrations of flammable vapors well below their lower explosive limits. The electrical equipment should be interlocked in a manner that prevents the spraying equipment from operating unless the ventilation system is working (USDOL 2013b, 29 C.F.R. § 1910.107).

Spray Booths

Spray booths can be found in several industrial facilities. The spray booth is a power-ventilated structure provided to enclose or accommodate a spraying operation. Due to the potential hazards of the flammable vapor concentrations that can be generated in spraying operations, the booth is designed to confine and limit the escape of spray, vapor, and residue and to conduct or direct safely them to an exhaust system. An area containing dangerous quantities of flammable vapors or mists or of combustible residues, dusts, or deposits due to the operation of spraying processes is referred to as the spraying area.

To minimize the fire hazards that a spray booth poses for other activities in the industrial occupancy, the spray booth should be separated from other operations by 3 ft. or more, or it should be cordoned off with a partition or wall surrounded by a clear space of not less than 3 ft. on all sides that is kept free from storage or combustible construction (USDOL 2013b, 29 C.F.R. §1910.107).

To prevent the lighting used in a spray booth from becoming a source of ignition, only approved methods of lighting should be used. When spraying areas are illuminated through glass panels or other transparent materials, only fixed lighting units shall be used as a source of illumination (USDOL 2013b, 29 C.F.R. § 1910.107).

SOURCES OF IGNITION

Due to the fire hazards associated with the flammable and combustible vapors generated in spraying operations, no open flame or spark-producing equipment should be allowed in any spraying area or within 20 ft. of the spraying area unless it is separated by a partition (USDOL 2013b, 29 C.F.R. § 1910.107). Sources of ignition may include electrical sources, mechanical sources, and hot equipment surfaces. Due to the presence of flammable vapors in a sufficient concentration to ignite, electrical wiring on any equipment in or near the spraying areas must meet applicable electrical codes for the hazardous location. Electrical wiring and equipment not subject to deposits of combustible residues but located in a spraying area should be of the explosion-proof type approved for Class I, Group D locations (USDOL 2013b, 29 C.F.R. § 1910.107). Electrical wiring, motors, and other equipment outside of but within 20 ft. of any spraying area and not separated from it by partitions should not produce sparks under normal operating conditions and should meet construction requirements for Class I, Division 2 hazardous locations (USDOL 2013b, 29 C.F.R. § 1910.107). Portable lamps approved for Class I, Division 1 locations can be used only in conjunction with cleaning and drying operations. Spray booths, rooms, or other enclosures used for spraying operations should not be used alternately for the purpose of drying by any arrangement that will cause a material increase in the surface temperature of the spray booth, room, or enclosure.

VENTILATION

Spray booths protect the industrial occupancy by preventing harmful vapor concentrations from escaping the booth to areas where they can be accidentally ignited; it plays a big role in removing them from the inside of the booth to a safe location outside the industrial occupancy. The ventilation sweeps the air currents toward the exhaust outlet, where the vapors are ventilated to safe location outside of the booth. Mechanical ventilation is always required while spraying operations are being conducted and for a sufficient time afterward to ensure that vapors are reduced to a nonhazardous level. The exhaust system is intended solely for the spray booth and discharges the vapors to the exterior of the building. Thus, spray-booth ventilation is essential in spray-booth fire prevention. The NFPA's Standard for Blower and Exhaust Systems for Vapor Removal, NFPA No. 91-1961 establishes the criteria for ventilation systems used in spray booth operations. This standard has also been adopted by OSHA (USDOL 2013b, 29 C.F.R. § 1910.107).

To further prevent the accumulation of paint residues and aid in the movement of air inside the booth, the interior surfaces of spray booths should be smooth and continuous. Baffle plates are also used to ensure an even flow of air through the booth or to collect the overspray before it enters the exhaust duct. OSHA requires that air filters be designed, installed, and maintained so that the average air velocity over the open face of the booth (or booth cross-section during spraying operations) will not be less than one hundred linear feet per minute (USDOL 2013b, 29 C.F.R. § 1910.107). The accumulation of overspray materials can greatly decrease the air velocities in the spray booth; therefore, the air filters should be checked on a regular basis, cleaned, and disposed of properly.

Because the fan and its mechanical parts are a potential ignition source for the vapors, the rotating elements shall be non-sparking. Heat sources such as friction should be eliminated in the fan motor, belts, and other moving parts. The bearings should be of an approved type, typically self-lubricating or lubricated from the outside duct (USDOL 2013b, 29 C.F.R. § 1910.107).

Unless the spray booth exhaust-duct terminal is from a water-wash spray booth, the terminal discharge point shall be not less than 6 ft. from any combustible exterior wall or roof and shall not discharge in the direction of any combustible construction or unprotected opening in any noncombustible exterior wall within 25 ft. (USDOL 2013b, 29 C.F.R. § 1910.107). Air exhaust from spray operations shall not be directed so that it will contaminate makeup air being introduced into the spraying area or other ventilating intakes; nor shall it be directed to create a nuisance. Air exhausted from spray operations shall not be recirculated. Air intake openings to rooms containing spray finishing operations shall be adequate for the efficient operation of exhaust fans and shall be so located as to minimize the creation of dead air pockets.

FLAMMABLE AND COMBUSTIBLE LIQUIDS: STORAGE AND HANDLING

OSHA limits the quantity of flammable or combustible liquids kept in the vicinity of spraying operations to the minimum required for operations and should ordinarily not exceed a supply for one day or one shift. The liquids should be stored in approved containers or pumped into the spray finishing room using approved piping and pumping systems. As discussed in chapter 3, transferring flammable and combustible liquids from one container to another can create fire and explosion hazards. Therefore, transferal of the liquids shall be done using approved equipment and procedures, including the use of bonding wires, adequate grounding, and adequate ventilation. Precautions should also be taken to prevent spills and the presence of unwanted sources of ignition.

In many spraying operations, it is common practice to draw the spray liquid from containers. Provisions for the containers supplying spray nozzles should meet the following requirements (USDOL 2013b, 29 C.F.R. § 1910.107):

1. They should be of closed type or provided with metal covers that are kept closed.
2. Containers not resting on floors should be on metal supports or suspended by wire cables.
3. Containers supplying spray nozzles by gravity flow should not exceed a 10-gallon capacity.
4. Original shipping containers should not be subject to air pressure for supplying spray nozzles.
5. Containers under air pressure supplying spray nozzles should be of limited capacity, not exceeding that necessary for one day's operation; they should be designed and approved for such use and provided with a visible pressure gauge and a relief valve.
6. Containers under air pressure supplying spray nozzles, air storage tanks, and coolers should meet applicable pressure vessel standards.

FIRE PROTECTION

OSHA requires spray-booth areas, including the interior of the booth, to be protected with either automatic sprinklers or other approved automatic extinguishing equipment (USDOL 2013b, 29 C.F.R. § 1910.107). The sprinkler systems should be properly maintained. To ensure they work properly, sprinklers protecting spraying areas should be kept free from deposits and buildup of paint and spray material. The sprinkler heads should be cleaned on a regular basis to ensure their proper operation, which may mean cleaning as often as daily. In addition to a fixed extinguishing system, an adequate supply of suitable portable fire extinguishers shall be installed near all spraying areas, and "NO SMOKING" signs in large letters on a contrasting color background shall be conspicuously posted in all spraying areas and paint storage rooms.

OPERATIONS AND MAINTENANCE
To reduce the potential for accidental fires involving spraying operations, safety measures should be incorporated into the spraying operations and maintenance procedures. Fires can start from residue material being ignited. To prevent this, housekeeping practices should be followed to clean up overspray residue and the accumulation of deposits of combustible residues. Spraying should only take place in designated areas and never outside of the spray booth. Tools and equipment used in the cleaning of spray booths should be non-sparking. Approved metal waste cans should be provided wherever rags or waste are impregnated with finishing material, and all such rags or waste should be deposited therein immediately after use (USDOL 2013b, 29 C.F.R. § 1910.107). The waste cans should be emptied at least once daily or at the end of each shift.

Because solvents used in the cleaning of the spray booth and equipment can also create a fire hazard, they should have flash points not less than 100°F. Solvents used for cleaning spray nozzles and auxiliary equipment may have flash points not less than the material normally used in spray operations. To remove potentially flammable vapors from the work area during cleaning, ventilating equipment should be operated during cleaning.

DRYING, CURING, AND FUSION APPARATUSES
The drying process and associated equipment can pose potential fire hazards in spray-booth areas. This drying and curing equipment can present potential heat sources and sources of electricity sufficient to ignite vapors that may still be present in the spray booth following the application of coating materials. To prevent potential fires from these sources of ignition, drying, curing, and fusion equipment used in connection with the spray application of flammable and combustible finishes must conform to NFPA 86A: Standard for Ovens and Furnaces (1969) and "drying, curing, or fusion equipment" (USDOL 2013b, 29 C.F.R. § 1910.107).

OSHA requires that drying, curing, and fusion units utilizing a heating system that has open flames or that may produce sparks shall not be installed in a spraying area. This type of equipment may be installed adjacent to spraying areas equipped with an interlocked ventilating system.

The heating components on drying equipment shall be prevented from starting until the drying space is thoroughly ventilated. The ventilating system should also maintain a safe atmosphere at any source of ignition, and the heating system should automatically shut down if the ventilating system fails.

Spray booths and enclosures used for automobile refinishing may also be used for drying with a portable, electrical, infrared drying apparatus if the following conditions are met (USDOL 2013b, 29 C.F.R. § 1910.107):

1. The interior (especially floors) of spray enclosures is kept free of overspray deposits.
2. During spray operations, the drying apparatus and electrical connections and wiring thereto should not be located within the spray enclosure or in any other location where spray residues may be deposited.
3. The spraying apparatus, the drying apparatus, and the ventilating system of the spray enclosure should be equipped with suitable interlocks so arranged that:

- The spraying apparatus cannot be operated while the drying apparatus is inside the spray enclosure.
- The spray enclosure will be purged of spray vapors for a period of not less than three minutes before the drying apparatus can be energized.
- The ventilating system will maintain a safe atmosphere within the enclosure during the drying process, and the drying apparatus will automatically shut off if the ventilating system fails.
- All electrical wiring and equipment of the drying apparatus conforms with electrical standards, and only equipment of a type approved for Class I, Division 2 hazardous locations is located within 18 in. of floor level.
- All metallic parts of the drying apparatus are properly electrically bonded and grounded.
- The drying apparatus contains a prominently located, permanently attached warning sign, indicating that ventilation should be maintained during the drying period and that spraying should not be conducted in a vicinity where spray will deposit on the apparatus.

Aerated Powder Coating Operations

An aerated powder is any powdered material used as a coating material that is fluidized within a container by passing air uniformly from below. It is common practice to fluidize such materials to form a fluidized powder bed and then to dip the part to be coated into the bed in a manner like liquid dipping. Such beds are also used as sources for powder spray operations. Because the process suspends powder materials in air, a major fire hazard associated with aerated powder coating operations involves explosions from igniting the suspended particles.

To prevent these explosions, exhaust ventilation shall be sufficient to maintain the atmosphere below the lowest explosive limits for the materials being applied. All non-deposited, air-suspended powders shall be safely removed via exhaust ducts to the powder recovery cyclone or receptacle. Housekeeping measures should be established and followed to prevent the accumulation of powder coating dusts, and measures should be taken to eliminate sources of ignition from electrical equipment, equipment that produces heat, and from smoking.

ELECTROSTATIC FLUIDIZED BEDS

An electrostatic fluidized bed is a container that holds powder coating material, which is aerated from below so as to form an air-supported, expanded cloud of such material, which is electrically charged with a charge opposite to the charge of the object to be coated; this object is transported through the container immediately above the charged and aerated materials in order to be coated. Fire hazards with electrostatic fluidized beds involve the ignition of the powder coating, which could result in an explosion. In addition, the equipment charges the powder material with high voltage, which serves as an additional source of ignition.

To minimize the potential ignition of the powder coating, electrostatic fluidized beds and associated equipment should be of approved types. The maximum surface temperature of this equipment in the coating area should not exceed 150°F (USDOL 2013b, 29 C.F.R. § 1910.107). The high-voltage circuits should be designed so as not to produce a spark of sufficient intensity to ignite any powder–air mixtures or result in an appreciable shock hazard upon coming into contact with a grounded object under normal operating conditions.

Transformers, power packs, control apparatuses, and all other electrical portions of the equipment, with the exception of the charging electrodes and their connections to the power supply, should be located outside of the powder coating area and meet applicable standards for use in hazardous areas

(USDOL 2013b, 29 C.F.R. § 1910.107). All electrically conductive objects within the charging influence of the electrodes shall be adequately grounded. The powder coating equipment is required to have a permanently installed warning regarding the necessity for grounding these objects. Objects being coated shall be maintained in contact with the conveyor or other support to ensure proper grounding. Hangers shall be regularly cleaned to ensure effective contact, and areas of contact shall be sharp points or knife edges where possible. Electrical equipment should be interlocked with the ventilation system so that the equipment cannot be operated unless the ventilation fans are in operation.

Dip Tanks and Coating Operations

Dip-tank and coating operations can pose a fire hazard due to the flammability characteristics posed by the coating material. The fire hazards exist around the tank as well as in the areas where the coating material is draining off the dipped objects and in the drying areas (see figures 7.1 and 7.2). Examples of operations that can pose fire hazards include dip-tank operations in which objects are painted, dipped, electroplated, pickled, quenched, tanned, degreased, stripped, roll-coated, flow-coated, and curtain-coated. OSHA standards for dip tanks and coating operations apply when a dip tank containing a liquid other than water is used and the liquid in the tank or its vapor is used to do any of the following (USDOL 2013a, 29 C.F.R. § 1910.123):

1. Clean an object,
2. Coat an object,
3. Alter the surface of an object, and
4. Change the character of an object.

The NFPA also identifies electrical equipment used in areas around dip-tank operations that contain flammable or combustible vapors as a common ignition source for fires involving dip-tank operations. Typically, the electrical equipment found in dip-tank areas are not suited for use in hazardous environments. The vapors from dip-tank operations can also be ignited from other sources, such as open flames and heating equipment located in the dipping coating process area (NFPA 2008, 9–45–46). Welding, cutting, and other spark-producing operations can also serve as an ignition source for fires

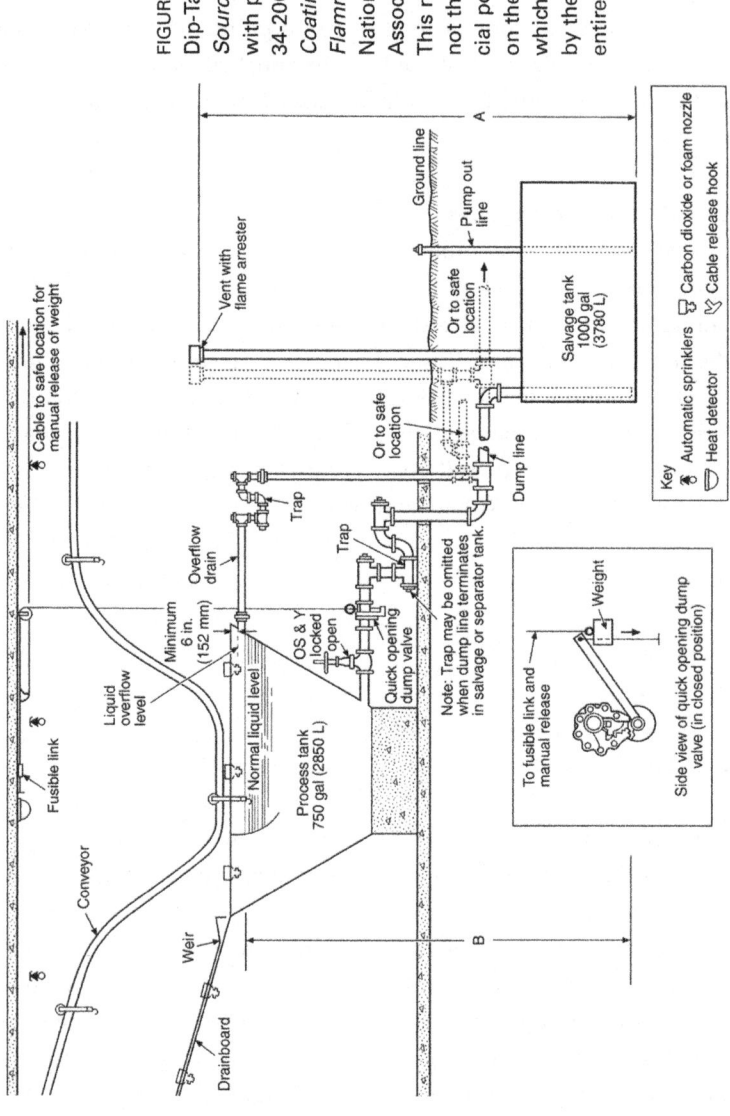

FIGURE 7.1
Dip-Tank Operation
Source: Reprinted with permission from 34-2003, *Dipping and Coating Processes Using Flammable Liquids*, 2003, National Fire Protection Association, Quincy, MA. This reprinted material is not the complete and official position of the NFPA on the referenced subject, which is represented only by the standard in its entirety

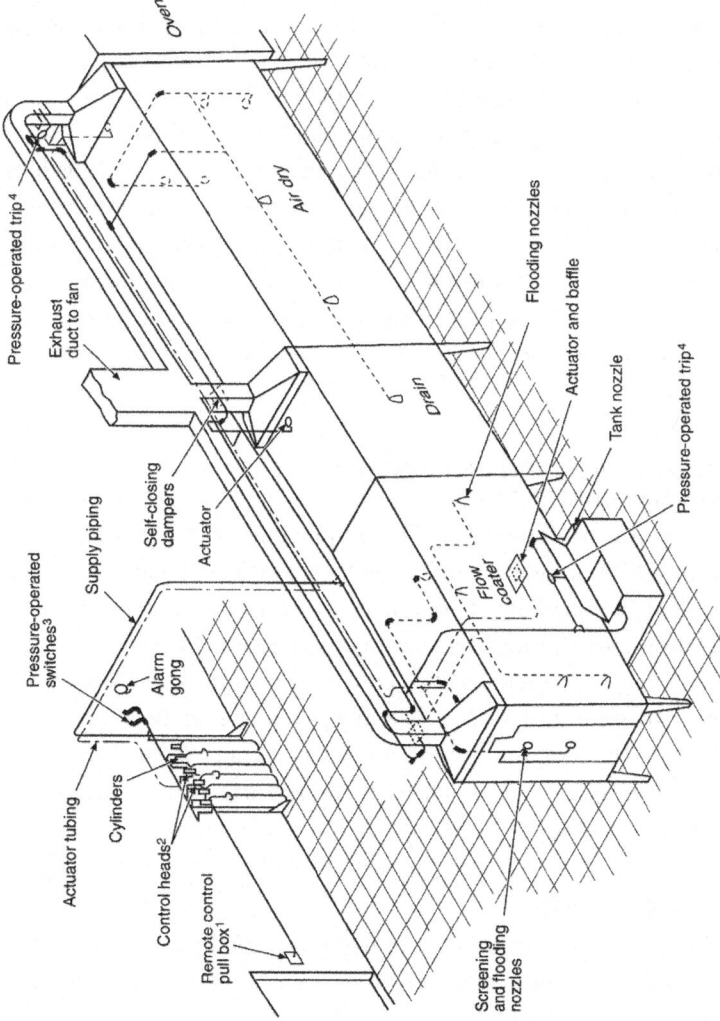

FIGURE 7.2
Flow-Coating Operation. Source: Reprinted with permission from 34-2003, Dipping and Coating Processes Using Flammable Liquids, 2003, National Fire Protection Association, Quincy, MA. This reprinted material is not the complete and official position of the NFPA on the referenced subject, which is represented only by the standard in its entirety

Notes:
1. Normal manual pull-box control and cable.
2. Pneumatic control heads with emergency manual controls.
3. Pressure-operated switches to shut down exhaust fan, pumps, conveyor, and so on, and sound alarm.
4. Pressure-operated trips to release self-closing dampers in exhaust duct and self-closing cover on paint tank.

involving dip-tank operations (NFPA 2008, 9–47). As a precaution, hot work should not be permitted in or adjacent to dipping or coating operations areas unless hot work safety precautions are used.

As is the case when handling flammable and combustible liquids, static electricity can be generated when the liquid from a container is transferred to the dip tank. The static electricity can contain enough energy to ignite the liquids. Proper grounding and bonding should be followed when transferring the liquids. Finally, carelessly discarded smoking materials or matches have been another common ignition source involving dip-tank operations. To prevent fires, smoking should be prohibited around dip tank, draining, and drying areas, and "NO SMOKING OR OPEN FLAMES" signs should be posted at all dipping areas, coating areas, and paint storage rooms.

Proper layout of dip-tank processes is the first step in preventing fires. Dipping and coating processes shall be separated from other operations, materials, or occupancies by location, fire walls, fire partitions, or other acceptable means. Separating dip-tank processes from other parts of the facility will aid in reducing the possibility that sources of ignition will come into contact with the tank and, if a fire does occur, will reduce its potential to spread to other areas.

DESIGN AND CONSTRUCTION OF DIPPING AND COATING EQUIPMENT AND SYSTEMS

Dip-tank fire hazards are controlled by fire-prevention design features on the tanks. Both NFPA standards and OSHA standards stipulate the design and construction features of dip- tank processes. Many of the adopted OSHA standards pertaining to dipping and coating operations are NFPA standards. Dipping and coating equipment shall be constructed of steel, reinforced concrete, masonry, or other noncombustible material and shall be securely and rigidly supported. The supports for dipping and coating tanks that exceed either a 500-gallon capacity or 10 sq. ft. of liquid surface must have a fire-resistance rating of at least one hour.

OVERFLOW PREVENTION

To prevent the overflow of burning liquid from the dipping or coating tank if a fire in the tank actuates automatic sprinklers, one or more of the following should be done (NFPA 2008, 9–48):

HAZARDOUS PROCESSES

1. Drain boards should be arranged so that sprinkler discharge will not flow into the tank.
2. Tanks should be equipped with automatically closing covers.
3. Tanks should be equipped with overflow pipes.

LIQUID LEVEL CONTROL

Control of the liquid level of a dip tank is crucial to preventing fires. Accidental overfilling of the tank or release of the liquid from the tank could result in rapidly spreading fire. The liquid in the dipping or coating tank shall be maintained at a level that is at least 6 in. below the top of the tank to allow effective application of extinguishing agents in the event of fire. To prevent accidental overfilling and subsequent spilling of the liquid, dip tanks exceeding a 150-gallon capacity or 10 sq. ft. of liquid surface should be equipped with a trapped overflow pipe leading to a safe location. Depending on the area of the liquid surface and the length and pitch of pipe, overflow pipes for dipping or coating tanks that exceed a 150-gallon capacity or 10 sq. ft. of liquid surface shall be capable of handling either the maximum rate of delivery of process liquid or the maximum rate of automatic sprinkler discharge, whichever is greater. The overflow pipe shall have a diameter of at least 3 in. (NFPA 2008, 9–48).

Bottom drains are used on dip tanks to drain the tank in the event of a fire, thus removing the fuel from the fire. The drains automatically open and release the flammable or combustible liquid from the tank and discharge it to a closed, vented salvage tank or a safe location. Dipping or coating tanks that exceed a 500-gallon capacity shall be equipped with bottom drains arranged to drain the tank in the event of fire. Exceptions to this requirement include tanks that are equipped with automatic closing covers and tanks using liquids in which the viscosity of the liquid at normal atmospheric temperatures makes this impractical. The NFPA has established that acceptable bottom drain pipe diameters capable of emptying the dipping or coating tank within five minutes are required on these dip-tank operations (NFPA 2011b, 34–9). When the liquid is evacuated from the dip tank to a salvage tank, the salvage tank must meet design requirements. Some of these requirements include means to pump the released contents from the salvage tank and minimum capacity requirements.

CONTROL OF LIQUID TEMPERATURE

Where dipping or coating liquids are heated, either directly or by the workpieces being processed, procedures shall be followed to prevent excess temperature, vapor accumulation, and possible autoignition (NFPA 2011b, 3410). Excess temperature means any temperature above which the ventilation cannot safely confine the vapors generated; in no case shall the temperature exceed the boiling point of the liquid or a temperature that is 100°F less than the autoignition temperature of the liquid. To prevent accidental overheating of the liquid, the dipping or coating tank shall be equipped with approved equipment designed to limit the temperature of the liquid. If excessive temperatures are reached, interlocks designed to shut down the equipment, including conveyor systems, shall be provided.

ELECTRICAL AND OTHER SOURCES OF IGNITION

Dipping and coating process areas where Class I liquids are used, or where Class II or Class III liquids are used at temperatures at or above their flash points, are considered hazardous locations; as a result, electrical equipment used in these areas must meet applicable codes. The hazardous environment classifications can include the dip-tank area, drain area, and drying areas. Sources of ignition such as open flames and spark-producing equipment should be prohibited from areas classified as hazardous. Light fixtures used around dip-tank operations shall be approved for use in that area. In order to prevent discharges or sparks from the accumulation of static electricity, all persons and all electrically conductive objects, including any metal parts of the process equipment or apparatus, containers of material, exhaust ducts, and piping systems that convey flammable or combustible liquids, shall be electrically grounded (NFPA 2011b, 34–13).

VENTILATION

As with spraying operations involving flammable and combustible liquids, ventilation plays an important role in reducing fire hazards in dipping and coating operations as well. Ventilation is intended to maintain the flammable and combustible vapors at safe levels without the ventilation equipment itself serving as a potential ignition source. Ventilating and exhaust systems used with dipping and coating operations should be installed in accordance with NFPA 91: Standard for Exhaust Systems for Air Conveying of Vapors, Gases, Mists, and Noncombustible Particulate Solids (NFPA 2010).

The goal of ventilation is to confine the vapor of the dipping and coating process to an area not more than 5 ft. from the vapor source and to remove the vapors to a safe location. If the ventilation system fails, it should be interlocked with the process so that the dipping or coating process automatically shuts down and an alarm sounds. Vapors from the process should be exhausted to a safe location following applicable codes. The exhaust equipment, including ducts, motors, bearings, and belts, should be approved for use with dipping and coating operations as to not serve as a potential ignition source.

To prevent accidental ignition of vapors from freshly dipped or coated pieces, materials should be dried only in spaces that are ventilated to prevent the concentration of ignitable vapors from exceeding 25 percent of the lower flammable limit (NFPA 2011b, 34–14). Areas serving as drying areas should be well ventilated in a manner that keeps the concentration of vapors below 25 percent of the lower flammable limit.

STORAGE, HANDLING, AND DISTRIBUTION OF FLAMMABLE AND COMBUSTIBLE LIQUIDS

The storage, handling, and transference of flammable and combustible liquids shall follow accepted practices. Chapter 3 covers flammable- and combustible-liquid handling, transference, and storage. Procedures limit the maximum quantities to be stored in various containers and cabinets and describe approved containers. The NFPA recommends the maximum quantity of liquid to be located in the vicinity of the dipping or coating process area—but outside of a storage cabinet, inside storage room, cut-off room or attached building, or other specific process area cut off by at least a 2-h fire-rated separation from the dipping or coating process area—should not exceed the greater of the quantities given below (NFPA 2011b, 34–15):

1. The amount for one continuous 24-h period
2. The aggregate sum of the following:
    a. 25 gallons of class 1A liquids in containers
    b. 120 gallons of Class 1B, 1C, or Class III liquids in containers
    c. 1,585 gallons of any combination of the following:
       Class 1B,1C, II, or IIIA liquids in portable metal tanks not exceeding 793 gallons
       Class II or Class IIIA liquids in nonmetallic intermediate bulk containers not exceeding 793 gallons
    d. 20 portable tanks or intermediate bulk containers not exceeding 793 gallons of Class IIIB liquids.

LIQUID PIPING SYSTEMS

In some instances, flammable and combustible liquids used in dip-tank and coating operations are transferred to the tank through a pump and piping system. Components of the system, including the piping, pumps, and meters, used for transferring liquids through this system must be approved for such use. When the piping system fills the tank from the top, the free end of the fill pipe shall be within 6 in. of the bottom of the tank (NFPA 2011b, 34–16). If the dip tank uses Class I liquids, the tank and fill pipe shall be electrically connected through a bond wire. Also, for Class I liquids, the tank, piping system, and storage tank shall be bonded and grounded to reduce the potential for fires due to static electricity generated during the filling process.

The pumps used to fill the tank must be approved for their use and designed in a manner that prevents the pump from exceeding the design pressures of the system. To prevent the tanks from accidental overfilling, automatic shut-offs must be provided in the tank system. In the event of the activation of the fire-detection system or automatic fire-extinguishing system, the pumping system should be designed to shut down the pump automatically.

FIRE PROTECTION

Fire protection for dip-tank and coating operations includes the use of automatic extinguishing systems and fire-detection systems. Where required in the fire codes, areas in which dipping, or coating operations are conducted shall be protected with an approved automatic sprinkler system. Dipping and coating processes can be protected with an approved automatic fire-extinguishing system. The following systems may be permitted (NFPA 2011b, 34–16):

1. An approved water-spray extinguishing system,
2. An approved foam extinguishing system,
3. An approved carbon dioxide system,
4. An approved dry-chemical extinguishing system,
5. An approved gaseous-agent extinguishing system,
6. An approved sprinkler system for tanks containing liquids having flash points above 200°F, and

7. For tanks equipped with a tank cover arranged to close automatically in the event of fire, a sprinkler system that meets NFPA 13 requirements.

The requirements for fire protection involving dip-tank processes are based on the size of the process. Small dip-tank processes are those tanks that do not exceed a 150-gallon capacity or do not exceed 10 sq. ft. of liquid surface area. Large dip-tank processes exceed 150 gallon or 10 sq. ft. of liquid surface area. On open, small dip tanks, automatically closing process tank covers or extinguishing systems shall be provided. The tank covers shall close automatically in the event of a fire. The covers and their components shall be constructed of metal and noncombustible materials. When the tanks are not in use, the covers should be kept closed.

On large dip-tank processes, an automatic extinguishing system shall be provided for enclosed processes, for open processes with peripheral vapor containment and ventilation, and for process tanks of a 150-gallon capacity or more or of 10 sq. ft. in liquid surface area or greater. The systems shall be designed to protect the following areas (NFPA 2011b, 34–17):

1. For dip tanks, the system shall protect the tank, its drain board, freshly coated work pieces or material, and any hoods and ducts.
2. For flow coaters, the system shall protect open tanks, vapor drying tunnels, and ducts. Pumps circulating the coating material shall be interlocked to shut off automatically in the event of fire.
3. For curtain and roll coaters or similar processes, the system shall protect the coated work pieces or material and open troughs or tanks containing coating materials. Pumps circulating the coating material shall be interlocked to shut off automatically in the event of fire.
4. Approved, automatic closing process tank covers or fire-protection systems shall be permitted for enclosed systems that do not exceed a 150-gallon capacity or 10 sq. ft. in liquid surface area and for open processes with peripheral vapor containment and ventilation.

OPERATIONS AND MAINTENANCE

In addition to the safety requirements pertaining to the design of dip tanks and coating processes, there are procedural requirements intended to minimize the potential for fires and explosions. Examples of these procedures include the following (NFPA 2011b, 34–17):

1. Areas in the vicinity of dipping and coating operations, especially drain boards and drip pans, should be cleaned on a regular basis to minimize the accumulation of combustible residues and unnecessary combustible materials.
2. Use of combustible coverings (e.g., thin paper, plastic) and strippable coatings should be permitted to facilitate cleaning operations in dipping and coating areas.
3. If excess residue accumulates in work areas, ducts, duct discharge points, or other adjacent areas, then all dipping and coating operations should be discontinued until conditions are corrected.
4. Approved waste containers should be provided for rags or waste impregnated with flammable or combustible material, and all such rags or waste should be deposited therein immediately after use.
5. The contents of waste cans should be disposed of at least once daily or at the end of each shift.
6. Portable fire extinguishers should be provided and located in accordance with NFPA 10: Standard for Portable Fire Extinguishers.

INSPECTION AND TESTING

Inspections and tests of all process tanks, including covers, overflow pipe inlets, overflow outlets and discharges, bottom drains, pumps and valves, electrical wiring and utilization equipment, bonding and grounding connections, ventilation systems, and all extinguishing equipment shall be made monthly. Any defects found shall be corrected.

TRAINING

Personnel involved in dipping or coating processes should receive documented training on the safety and health hazards associated with dip tanks, the operational, maintenance, and emergency procedures required, and the importance of constant operator awareness (NFPA 2011b, 34–19). Training should also include the use, maintenance, and storage of all emergency, safety, or personal protective equipment that employees might be required to use in their normal work performance. Employees responsible for handling flammable and combustible liquids should be trained.

CHAPTER QUESTIONS
1. How does electrostatic spray painting work?
2. What are acceptable methods of overflow protection on dip tanks?
3. Why is liquid-level control important in dip-tank operations?
4. How does an aerated-powder coating operation work?
5. What does the NFPA recommend for the maximum quantity of flammable or combustible liquids stored in a coating operation area?
6. When are dipping and coating operation areas considered hazardous environments?
7. Describe the type of training workers in dipping and coating operations should receive.

REFERENCES

Health and Safety Executive. (2013). *Hazardous Installations Directorate.* Sudbury, Suffolk: Health and Safety Executive. www.hse.gov.uk/comah/ sragtech/casepasadena89.htm

National Fire Protection Association (NFPA). (2008). *Fire Protection Handbook*, 20th ed. Quincy, MA: NFPA.

NFPA. (2010). *NFPA 91, Standard for Exhaust Systems for Air Conveying of Vapors, Gases, Mists, and Noncombustible Particulate Solids.* Quincy, MA: NFPA.

NFPA. (2011a). *NFPA 33: Standard for Spray Application Using Flammable or Combustible Materials.* Quincy, MA: NFPA.

NFPA. (2011b). *NFPA 34: Standard for Dipping and Coating Processes Using Flammable or Combustible Liquids.* Quincy, MA: NFPA.

U.S. Department of Labor (USDOL). (2013a). *Occupational Safety and Health Standards for General Industry, Subpart H: Hazardous Materials: Dipping and Coating Operations, 29 C.F.R. § 1910.123.* Washington, DC: USDOL.

USDOL. (2013b). *Occupational Safety and Health Standards for General Industry, Subpart H: Spray Finishing Using Flammable and Combustible Materials, 29 C.F.R. § 1910.107.* Washington, DC: USDOL.

U.S. Fire Administration (USFA). (1990). *Technical Report Series: Phillips Petroleum Chemical Plant Explosion and Fire, Pasadena, Texas.* USFA Technical Report Series TR-035. Washington, DC: FEMA, USFA, National Fire Data Center.

8

Alarm and Detection Systems

A fire alarm system monitors and announces the status of the alarm or supervisory signal-initiating devices, then initiates the appropriate response to the signal. In general, fire alarm systems have three primary functions: (1) to provide an indication and warning of abnormal fire conditions, (2) to alert building occupants and summon appropriate assistance in adequate time to allow egress to a safe place and for rescue operations to begin, and (3) to control fire safety functions (Richardson and Roux 2010, 4).

Based on the aforesaid three functions, a fire alarm system may provide three types of signals: alarm, supervisory, and trouble. An alarm signal will indicate the presence of a fire. A supervisory signal will indicate the operational status of fire protection systems being monitored, such as an automatic sprinkler system. The trouble signal will indicate a problem or fault with a component or circuit of the alarm system, such as a smoke detector.

An excellent reference for alarm and detection systems is National Fire Protection Association (NFPA) 72: National Fire Alarm Code. This code provides the minimum installation, test, maintenance, and performance requirements for fire alarm systems. It also provides requirements for the application, location, reliability, and limitations of fire alarm components. Examples of equipment addressed in this code include fire alarm systems, power supplies, monitoring systems, and notification appliances (Richardson and Roux 2010, 75). Equipment that meets the requirements of NFPA 72 must be listed for its specific purpose. This listing will normally identify the permitted use, required ambient conditions, mounting orientation, voltage tolerances, and compatibility. The plans for a fire alarm system should be developed by a qualified person who has working knowledge of NFPA 72 and is experienced in the proper design, application, installation, and testing of fire alarm systems.

NFPA 72 CLASSIFICATIONS FOR FIRE ALARM SYSTEMS

The National Fire Alarm Code has developed a classification system for alarm systems. This classification includes the following four types of alarm systems (NFPA 2010, 72–24):

1. Combination systems are the systems in which components are used, in whole or part, in common with non-fire signaling systems.
2. Household fire warning systems are alarm systems installed in dwelling units to warn the occupants of a fire emergency so that they can immediately evacuate the building.
3. Municipal fire alarm systems are public emergency alarm reporting systems.
4. Protected premises fire alarm systems are designed to warn building occupants to evacuate the premise, actuate the building's fire protection features, and provide environmental protection.
5. Building fire alarm systems are protected premises fire alarm systems that include systems such as manual alarm systems, automatic alarm systems, and monitoring systems and serve the general fire alarm needs for a building.
6. Dedicated function fire alarm systems are systems that provide protection for buildings where fire alarms are not required.
7. Releasing fire alarm systems are protected premises systems that are part of the fire suppression system and provides inputs to the suppression systems sequence of operations and outputs.

POWER SUPPLIES FOR ALARM SYSTEMS

Fire alarm systems need to have at least two independent and reliable power supplies that have an adequate capacity for their specific application. The first

ALARM AND DETECTION SYSTEMS

power source is commonly referred to as the primary power source and is usually provided by electrical service from a commercial electrical distributor. The primary power service must be on a dedicated branch circuit(s) that is mechanically protected, and all electrical wiring and equipment must be in accordance with the NFPA 70: National Electrical Code.

The secondary power supply serves as a backup to the primary supply and must automatically supply electrical energy to the system within thirty seconds after failure of the primary supply. This secondary supply must also have a capacity that will allow the alarm system to operate for twenty-four hours (Richardson and Roux 2010, 84). Two examples of secondary supplies include storage batteries and an engine-driven generator. An excellent reference for generators as secondary power supplies is NFPA 110: Standard for Emergency and Standby Power Systems.

INITIATING DEVICES

An initiating device includes all types of sensors, ranging from manually operated fire alarm boxes to switches that detect the operation of a fire

FIGURE 8.1
Fire Alarm Pull Station

suppression system (NFPA 2010, 72–25) (see figure 8.1). These initiating devices do not respond to the fire itself but to some change in conditions because of the fire such as elevated temperatures or smoke. Important considerations for initiating devices include installation location, selection of the type of device, and temperature classifications for heat-sensing devices.

BASIC CONSIDERATIONS FOR INSTALLATION

In general, where possible, initiating devices should be installed throughout the entire facility, including uninhabited spaces such as closets and storage areas. Possible references for installation of devices include the device manufacturer, local building codes, and NFPA 72. In some situations, the term total complete coverage is used regarding coverage of initiating devices. Total complete coverage implies coverage in all rooms, halls, storage areas, basements, attics, lofts, spaces above suspended ceilings, other subdivisions and accessible spaces, the inside of all closets, elevator shafts, enclosed stairways, dumbwaiter shafts, and chutes (Richardson and Roux 2010, 250). It is important to note that inaccessible areas need not be protected by detectors when total complete coverage is applied. Another common term applied regarding coverage is partial coverage. Partial coverage, also referred to as selective coverage, implies that initiating devices are placed in all common areas, workspaces, and other unoccupied spaces where the environments are suitable for proper detector operation. In all cases, initiating devices must be installed so that they are accessible for periodic maintenance and testing (Richardson and Roux 2010, 252).

An important consideration regarding the placement and spacing of smoke and heat detectors is the transfer of combustion products, such as heat and smoke, from the fire to the detector. This transfer of smoke, aerosol, or heated combustion product gases and air can be described through a set of physical principles generally called a fire plume (NFPA 2008, 2–5355). Specifically, a fire plume is the buoyant column of flame and heated combustion products rising above the fuel. This buoyancy occurs because of density differences. Density is inversely proportional to temperature; therefore, the hotter gases associated with the plume cause an upward force on the hot gases relative to the cooler surrounding air. This basic principle is behind the requirement that most detectors be placed on the ceiling or within 12 in. of the ceiling.

ALARM AND DETECTION SYSTEMS

Selection of Initiating Devices

The two most important considerations in selecting an alarm system are the need for speed and accuracy of response to a fire with minimal chances for false alarms. Therefore, it is critical to select the proper type of initiating device for each application. This selection process requires a thorough understanding of how the detector operates and the conditions it will operate under. Factors to consider include the kinds of potential fires expected, the quantity and types of fuel sources, potential ignition sources, and the range of ambient air temperatures (NFPA 2008, 14–26). There are three general types of fire detectors to choose from smoke, heat, and flame. In general, heat detectors have the lowest rate of false alarms, but they also have the slowest response time. In contrast, flame detectors offer the fastest response time but the highest rate of false alarms.

Heat-Sensing Fire Detectors

A heat detector responds to an increase in the ambient temperature in its immediate vicinity. Specifically, the increase in temperature of the sensing element of the heat detector is due to the absorption of heat from fire. There are three basic principles for detecting heat: fixed temperature, rate compensation, and rate of rise. Fixed temperature detectors are set to activate when the operating element reaches a set temperature. A rate-of-rise detector activates in response to a rate of temperature change such as 15°F in one minute. Rate compensation detectors activate when the surrounding air reaches a set temperature (NFPA 2008, 14–17).

The performance of a heat detector depends on two parameters: its temperature classification and its time-dependent thermal response characteristics. It is critical to select a detector that will be stable in the environment in which it is installed. For this reason, all heat-sensing fire detectors must be marked with their operating temperature and thermal-response coefficient. As a rule, the temperature rating of the detector should be at least 20°F above the maximum expected temperature at the ceiling. A temperature difference less than this may cause false alarms, while selecting a temperature significantly higher will increase response time. Traditionally, the temperature classification has been the principal parameter used in selecting the proper detector for a given area. Heat-sensing fire detectors of the fixed temperature or rate compensated type are classified by the temperature of operation and

Table 8.1. Detector Temperature Classification and Color Code

Temperature Classification and Color Code	Temperature Rating Range (°F)
Low: uncolored	100–134
Ordinary: white	135–174
Intermediate: blue	175–249
High: red	250–324
Extra high: green	325–399
Very extra high: orange	400–499
Ultra-high	500–575

marked with a color code, as specified in table 8.1 (Richardson and Roux 2010, 258).

LOCATION OF HEAT-SENSING FIRE DETECTORS: SPOT VERSUS LINE DEVICES

Each heat-sensing fire detector can be installed as either a spot type or line type device. As the name implies, spot-type devices occupy a specific spot or point, while line devices extend over a distance, sensing temperature along their entire length.

NFPA 72 requires that spot-type heat detectors be located on the ceiling not closer than 4 in. from the sidewall or on the sidewalls between 4 and 12 in. from the ceiling. Line-type heat detectors are generally considered to be equivalent to a row of spot type detectors for the purposes of spacing and location. However, different manufacturers have different mounting requirements unique to their products (Richardson and Roux 2010, 266).

The distance between the detectors must not exceed their listed spacing. The listed spacing for a heat detector is determined using several variables, such as anticipated fire size, fire growth rate, ambient temperature, ceiling height, and thermal response coefficient.

Smoke-Sensing Fire Detectors

A smoke detector responds to the presence of smoke in the air in its immediate vicinity. The sensitivity of a smoke detector in responding to smoke is based on the percentage of obscuration required to produce a signal. This sensitivity is expressed as the percentage per foot obscuration and must be identified on the detector (Richardson and Roux 2010, 284). For most fires,

ALARM AND DETECTION SYSTEMS

smoke detectors respond much faster than either automatic sprinklers or heat detectors because smoke does not dissipate as quickly as heat in large, open areas. The difference in the speed of response becomes even more dramatic with low-energy smoldering fires.

LOCATION AND SPACING OF SMOKE DETECTORS

In general, the placement of smoke detectors is based on the premise that for a smoke detector to respond, the smoke must travel from the point of origin to the detector (see figure 8.2). This flow of smoke is based on the fire plume, discussed previously, as well as on building and site characteristics, such as ceiling shape and surface, ceiling height, configuration of contents in the area to be protected, burning characteristics of the combustible materials present, and ventilation (Richardson and Roux 2010, 260–261).

The location and spacing of smoke detectors are based on the effect of the aforementioned site-specific characteristics on the flow of smoke from these early-stage, low-energy-output fires. The design process begins by locating detectors so that they will provide general area protection with additional detectors added to take into account known building characteristics that may affect smoke movement. Some general guidelines provided by NFPA 72 for the spacing of smoke detectors are that spot type smoke detectors be located on the ceiling not closer than 4 in. from a sidewall or, if placed on a sidewall,

FIGURE 8.2 SMOKE DETECTOR

between 4 and 12 in. from the ceiling. On smooth ceilings, spacing of 30 ft. is permitted to be used as a guide and all points on the ceiling shall have a detector within a distance equal to 0.7 times the selected spacing (Richardson and Roux 2010, 287–288). It is important to note that these are general guidelines, and the manufacturer's recommendations or the local building codes, whichever are more stringent, should take precedent for smoke detector placement.

Radiant-Energy-Sensing Fire Detectors

A radiant-energy-sensing fire detector responds to the influx of radiant energy that has traveled from the fire to the detector. The radiant-energy-sensing fire detector can be classified as either a flame detector or a spark or ember detector. In each case, heat, smoke, or light travels from the fire to the detector before the device initiates the alarm signal. When using radiant-energy-sensing detectors, the designer must match the detector to the radiant energy emissions, or signature of the flame or spark/ember to be detected (Richardson and Roux 2010, 325). Some extraneous sources of radiant emissions that have been identified as interfering with the stability of flame detectors include sunlight, lightning, X-rays, and ultraviolet radiation from arc welding. Radiant-energy-sensing fire detectors are commonly used in environments that are unsuitable for heat or smoke detectors, such as the following:

- Open-spaced buildings with high ceilings, such as warehouses;
- Outdoor areas where winds can prevent smoke and heat from reaching a heat or smoke detector; and
- Areas where rapidly developing, flaming fires can occur, such as aircraft hangers.

There are three basic types of flame detectors. First, there are ultraviolet flame detectors that use a vacuum photodiode Geiger–Muller tube for detecting ultraviolet radiation from a flame (Richardson and Roux 2010, 326). The second type of flame detector is a single-wavelength, infrared flame detector, which uses different photocell types to detect the infrared emissions in a single-wavelength band that are produced by a flame. Because these flame detectors respond to specific, single wavelengths, they can be fuel specific. A similar type is the multiple-wavelength infrared flame detector, which, as the

ALARM AND DETECTION SYSTEMS

179

name implies, senses radiation at two or more narrow bands of wavelengths. The third type of flame detector is the ultraviolet/infrared flame detector, which is a combination of the first two types. Specifically, this detector senses ultraviolet radiation with a vacuum photodiode tube and a selected wavelength of infrared radiation with a photocell and uses this combined signal to indicate a fire (NFPA 2008, 14–25).

The second broad classification of radiant-energy-sensing fire detectors includes spark or ember detectors, which use a solid-state phototransistor to sense the radiant energy emitted by embers. The embers are typically between 0.5 and 2.0 in normally dark environments. As a rule, these detectors are extremely sensitive and can have response times in the microseconds (Richardson and Roux 2010, 328).

LOCATION AND SPACING OF RADIANT-ENERGY-SENSING FIRE DETECTORS

Radiant-energy-sensing fire detectors must be located and spaced consistently with their listing or manufacturer's recommendation. Simply stated, a flame detector cannot detect what it cannot "see." Therefore, the number of detectors should be based on the detectors being positioned so that no point requiring detection in the hazard area is obstructed or outside the field of view of at least one detector. Some general considerations for the location and spacing of radiant-energy-sensing fire detectors are size of the fire that is to be detected, the fuel involved, the sensitivity of the detector, the distance between the fire and the detector, and the field of view of the detector.

The field of view of the detector is based on the fact that the greater the angular displacement of the fire from the optical axis of the detector, the larger the fire must become before it is detected. This principle establishes the field of view of the detector (Richardson and Roux 2010, 330).

Sprinkler Water Flow Alarm-Initiating Devices

In many facilities, the sprinkler system is used as both a suppression system and a detection system. Specifically, a sprinkler water flow alarm initiates an alarm signal when the flow of water through the system is greater than that from a single sprinkler head of the smallest orifice size. NFPA 72 recommends that the water flow alarm be adjusted so that it is initiated within ninety seconds of a sustained flow equal to or greater than that of a single sprinkler (smallest orifice) that can be installed in a system (Richardson and Roux 2010, 345).

In most sprinkler systems, this water flow alarm will provide a signal locally, either in the immediate vicinity of the riser or throughout the entire facility. The signal can also be transmitted off premise to a supervising station.

In a wet pipe sprinkler system, the water flow alarm-initiating device is either a hydraulically operated alarm check valve commonly called a water motor gong or an electrically operated, vane type water flow alarm. Water flow alarms that can be used for dry pipe, preaction, and deluge sprinkler systems include hydraulically operated water motor gongs and drop-in-pressure water flow alarms. An electrically operated, vane type water flow alarm is not permitted on these types of sprinkler systems (NFPA 1999, 5–3.2)

In a large wet pipe system with large sprinkler risers, the flow from a single head can be hard to detect, depending on the amount of trapped air in the piping. The air in the piping acts as a gas cushion, allowing pulsating variations in water pressure within the riser when a single head discharges. These variations in water pressure can prevent either the vane of a vane-type water flow switch or the clapper of an alarm check valve from opening for long enough to initiate the alarm signal. Therefore, it is not uncommon to install an excess pressure pump and pressure drop alarm-initiating devices in large wet pipe systems to reduce the effects of pressure variations in the system and to minimize response time (Richardson and Roux 2010, 345).

Signal Annunciation

The primary purpose of fire alarm system annunciation is to enable responding personnel to identify the location of a fire quickly and accurately and to indicate the status of emergency equipment or fire safety functions that might affect the safety of the occupants in a fire situation (Richardson and Roux 2010, 102). NFPA 72 requires all protected premise systems to indicate fire alarm, supervisory, and trouble signals distinctively.

Supervisory signals can be used to monitor the functions of a sprinkler system, such as the water control valve and the water pressure and level in the system. Trouble signals with the alarm system indicate a fault in a monitored circuit, system, or component (NFPA 2010, 72–29).

The fire alarm annunciation system must be specific enough to identify the origin of a fire alarm signal as quickly as possible to reduce response time. For

ALARM AND DETECTION SYSTEMS

alarm annunciation, each floor of the building must be considered as a separate zone. However, if a floor area exceeds 22,500 sq. ft., additional zoning should be considered. In addition, if the alarm system serves more than one building, each building must be indicated separately on the signal annunciation panel (Richardson and Roux 2010, 105).

NOTIFICATION DEVICES

The primary purpose of a notification device is to convey information to building occupants during a fire emergency. This may be as simple as notifying building occupants to evacuate or alerting fire and emergency brigades of appropriate extinguishment strategies. In general, notification devices should be sufficient in quantity, audibility, intelligibility, and visibility to reliably convey the intended information. To guide the designers and installers of fire alarm systems so that the system will deliver audible and visible information with appropriate intensity, the device nameplate must state the capabilities of the appliance, as determined through tests conducted by the listing organization. This may include electrical requirements and rated audible or visible performance (Richardson and Roux 2010, 358). The proper location of notification devices is critical to ensure that they operate as intended. The notification devices should be mounted in accordance with manufacturers' published instructions (NFPA 2010, 72–101).

Audible and Visible Criteria of Notification Devices

A notification device is only effective if it is heard or seen. Therefore, specific audible and visual criteria have been established by NFPA 72. Audible notification devices intended for operation in the public mode must have a sound level of at least 75 dBA at 10 ft. or a sound level at least 15 dBA above the average ambient sound level, or 5 dBA above the maximum sound level, having a duration of at least sixty seconds (Richardson and Roux 2010, 368–370). These sound levels should be measured 5 ft. above the floor in the area in which the device is installed. Emergency voice/alarm communications systems used as part of the alarm system must be capable of reproducing prerecorded, synthesized, or live messages intelligibly. Visible signaling notification devices are often used to supplement audible devices; however, they are required when the average sound pressure level exceeds 105 dBA because it

FIGURE 8.3
Fire Alarm

could be harmful to occupants to try and overcome such a high sound level with audible fire alarm signals. Figure 8.3 depicts a typical fire alarm.

In general, there are two methods of visible signaling. The first involves the identification of an emergency condition by direct viewing of an illuminating appliance, such as a strobe light. At present, both NFPA 72 and NFPA 101, Life Safety Codes recognize only strobe lights as visible signaling devices. NFPA 72 requires that the flash rate not exceed two flashes per second and that the maximum pulse duration be 0.2 seconds. In addition, the light source color should be clear or nominal white and not exceed 1,000 candlepower. The second method of visible signaling involves illumination of the surrounding area. With this method, an illumination of 0.0375-ft. candles in all occupied spaces requiring visible notification meets the minimum light intensity requirements according to NFPA 72 (NFPA 2010, 72–103–104).

REPORTING SYSTEMS

The primary purpose of a fire alarm reporting system is to send a signal or communication to an outside agency regarding a fire emergency. In general,

ALARM AND DETECTION SYSTEMS

these reporting systems can be classified as public fire alarm systems, proprietary supervising systems, central station systems, and remote supervising systems.

Public Fire Alarm Systems

A public fire alarm reporting system allows the public to initiate a fire alarm signal from manual fire alarm boxes. Such an alarm system may also be used to transmit other signals or calls of a public emergency nature if the transmission does not interfere with the transmission and receipt of fire alarms. The manual fire alarm boxes must be located where they are conspicuous and accessible. One way of accomplishing this is to mount the boxes on support poles that are identified by distinctive colors or signs placed at least 8 ft. above the ground and visible from all directions, wherever possible. Indicating lights of a distinctive color can also be used to make the manual fire alarm boxes more conspicuous (NFPA 2010, 72-99-100).

Central Station Fire Alarm Systems

A central station fire alarm system receives fire alarm, supervisory, and trouble signals from a protected premise. These stations are controlled and operated by a person, firm, or corporation whose business is the furnishing of such systems. The company will have obtained a specific listing from a nationally recognized testing laboratory as a provider of central station service (NFPA 2008, 14-9).

Central station components typically include the central station physical plant, exterior communications channels, subsidiary stations, and signaling equipment located at the protected premises. At the protected premise, the central station will be involved with the installation, testing, maintenance, and runner servicing of the alarm system. At the actual central station, the monitoring of the system and signals will occur, as will the retransmission of signals to the appropriate emergency response service and record-keeping. Today, it is not uncommon for computer-based automation systems to manage the receipt and retransmission of signals. When such systems are used, the central station must automatically record the time and date of the retransmission of signals.

The central station must always have a minimum of two trained operators on duty to monitor signals (Richardson and Roux 2010, 589). According to

NFPA 72, all alarm signals must be treated as positives. This standard requires that upon receiving an alarm, a supervisory, or a trouble signal by the remote supervisor station, the operator on duty at the central station should notify the owner or the owner's designated representative immediately (Richardson and Roux 2010, 590).

Proprietary Supervising Station Systems

A proprietary supervising station is like a central station system, except this system is located at the protected property or at another property belonging to the same owner. Such a system is owned and operated by the property owner and receives signals from one or more properties under the same ownership. Where more than one building exists, the alarm signal must identify the specific building in which the signal originates. For large facilities, the floor, section, or other subdivision of the building must be designated at the proprietary supervising station or at the building that is protected (Richardson and Roux 2010, 577). These systems transmit fire alarm signals, as well as supervisory and trouble signals, to the station. The actual proprietary station must be in a fire-resistive, detached building or in a cutoff room not exposed to the hazardous areas of the properties that are to be protected. This station must have an automatic emergency lighting system that has a power source independent of the primary lighting source and capable of providing illumination for a minimum of twenty-six hours. There must be two different means for alerting the operator when each signal is received. The maximum elapsed time from sensing a fire alarm at an initiating device circuit, until it is recorded or displayed at the proprietary supervising station, must not exceed ninety seconds (Richardson and Roux 2010, 580).

All communications and transmission channels between the proprietary supervising station and the protected premise control panel must be tested either manually or automatically once every twenty-four hours to verify operation. In addition, all operator controls at the proprietary supervising station must be tested at each change of shift.

All personnel at the proprietary supervising station must be trained, competent, and in constant attendance so that appropriate action can be taken when necessary. At least two operators should always be on duty, with one of the operators acting as a runner to the alarm location if an alarm is activated.

Remote Supervising Station Fire Alarm Systems

A remote supervising station system provides a means for transmitting alarm, supervisory, and trouble signals from the protected premises to a remote supervising station. Such systems normally transmit alarm signals to a public fire service communications center, community emergency response center, or other constantly attended location acceptable to the authority having jurisdiction. These types of systems are typically selected when management does not believe it needs the level of protection offered by a central station or proprietary supervising system.

The remote supervising station must always have a minimum of two operators on duty to monitor signals. When a signal is received at the station, the operator on duty must notify the owner or the owner's designated representative immediately. All controls at the remote supervising station must be tested at the beginning of each shift or change in personnel. The signal-receiving equipment at the remote station must indicate each signal both audibly and visibly. In addition, a trouble signal must be received when the system or any portion of the system at the protected premises is placed in a bypass or test mode. As with the other systems, a record must be maintained of all signals received, including time, date, location, and any system restorations (Richardson and Roux 2010, 584).

INSPECTION, TESTING, AND MAINTENANCE OF FIRE ALARM SYSTEMS

The owner of an alarm system is responsible for the testing and maintenance of the system. In general, this inspection, testing, and maintenance should follow the requirements specified in NFPA 72, conform to the alarm manufacturer's recommendations, and be completed in such a way as to verify the proper operation of the alarm system.

All personnel working on the system must be qualified and experienced in the inspection, testing, and maintenance of fire alarm systems. An example of a qualified person would be one who is factory-trained and certified by the state or local municipality or by an agency such as the International Municipal Signal Association (Richardson and Roux 2010, 79).

A specification for the inspection, testing, and maintenance of an alarm system will typically identify the procedure to be performed; how it is to be performed; the type of data to be recorded; and the frequency of testing, inspection, and maintenance. An example of a procedure is notifying all persons and facilities receiving alarm, supervisory, or trouble signals and all building occupants of testing to prevent unnecessary response.

Examples of documents recommended by NFPA 72 for testing and maintaining a system include the fire alarm system record of completion, point-to-point wiring diagrams, individual device interconnection drawings, as-built (record) drawings, copies of original equipment submittals, operational manuals, and manufacturer's proper testing and maintenance requirements (Richardson and Roux 2010, 111–141).

NFPA 72 also provides a detailed listing of the recommended frequencies for testing alarm system components. In almost all cases, testing is required initially after installation and then at intervals ranging from weekly to yearly. In situations where automatic testing is performed at least weekly by a remotely monitored fire alarm control unit, the manual testing frequency is permitted to extend to yearly. The following are some examples of NFPA 72 recommended testing frequencies after installation (NFPA 2010, 71–75):

- Annual testing: emergency voice/alarm communications equipment and remote annunciators;
- Semiannual testing: batteries in fire alarm systems;
- Quarterly testing: control equipment used in building systems not connected to a supervising station;
- Monthly testing: engine-driven generators and batteries used in central station facilities.

All records on the maintenance, inspection, and testing of alarm systems must be retained until the next test and for one year following. This record must identify, at a minimum, the following information (Richardson and Roux 2010, 211–234):

- Date;
- Test frequency;
- Name and address of property;
- Name of person performing inspection, maintenance, tests, or combination thereof, as well as affiliation, business address, and telephone number;
- Name, address, and representative of approving agencies;

ALARM AND DETECTION SYSTEMS

- Designation of the detector(s) tested;
- Functional test of detectors; and
- Functional test of required sequence of operations.

CHAPTER QUESTIONS

1. List and describe the three functions of an alarm system.
2. Explain what factors need to be considered when selecting an alarm system.
3. What is the primary purpose of signal annunciation?
4. Maintenance, inspection, and testing records must be retained until the next test and for one year thereafter. What information must be provided in these records?
5. What is a zone, and what are its criteria?
6. What are the three basic principles used for detecting heat from a fire?
7. What characteristic of smoke influences the density of a smoke detector in responding to a fire?
8. Which NFPA standard establishes the requirements for fire alarm notification systems?
9. What is the difference between a proprietary system and a central station monitoring system?

REFERENCES

National Fire Protection Association (NFPA). (2008). *Fire Protection Handbook*, 20th ed. Quincy, MA: NFPA.

NFPA. (2010). *NFPA 72: National Fire Alarm and Signaling Code*. Quincy, MA: NFPA.

Richardson, Lee, and Richard Roux (2010). *National Fire Alarm and Signaling Code Handbook*, 6th ed. Quincy, MA: NFPA.

9

Fire Extinguishment

The four components of the fire tetrahedron—fuel, oxygen, a heat source, and a chain reaction—were discussed in detail in chapter 2, "Chemistry and Physics of Fire." Since fires can only occur if all four components of the fire tetrahedron are present in sufficient concentrations, removal of any one of the four components will result in fire extinguishment. All fire-extinguishing methods used in firefighting apply techniques designed to attack one or more of the four components of the fire tetrahedron. The equipment used for fire extinguishment can range from a portable fire extinguisher to a fixed sprinkler system. Examples of the various extinguishing agents/methods and the component of the fire tetrahedron they influence are presented in table 9.1.

CLASSIFICATIONS OF FIRES

The appropriate type of extinguishing media is based in part on the material that is burning, the fuel. Some extinguishing media work very well in extinguishing certain fuels while with other fuels they are ineffective. For example, water is an effective extinguishing media for fires involving wood but is ineffective for flammable liquids that are insoluble in water. It is for this reason fires can be classified by the fuel involved in the combustion process. National Fire Protection Association (NFPA) 10 "Standard for Portable Fire Extinguishers" classifies fires into five groups based on the fuel (NFPA 2013a, 5.2):

Table 9.1. Extinguishing Methods

Extinguishing Method	Tetrahedron Component
Water spray	Heat (cooling) and to a lesser extent the water spray remove the oxygen from the fuel surface
Carbon dioxide	CO_2 removes the oxygen from the fuel surface, and, to a lesser extent, the carbon dioxide cools the surface of the burning material
Dry chemical fire extinguisher	This interrupts the chemical reaction
Fire break cut in a forest fire	This removes fuel from the fire

- Class A: ordinary combustible materials (wood, paper)
- Class B: flammable and combustible liquids/gases and petroleum greases (oils, solvents, alcohols, and flammable gases)
- Class C: electrical fires
- Class D: combustible metals (sodium, titanium, magnesium)
- Class K: combustible cooking media (vegetable oils, animal oils, and fats)

In Class C, electrical fires, the fuel involved may be from any other class (A, B, D, or K). The special classification for electrical fires is necessary to provide an extinguishing agent that is not electrically conductive. In addition to classifying fires based on the type of fuel, NFPA 10 also classifies fire hazards based on the amount of fuel or fire load. These three fire hazard classifications are critical in determining the size and placement of fire extinguishers (NFPA 2013a, 5.4):

- Light hazard: quantity and combustibility of Class A and B materials are low and fires with low rates of heat release expected.
- Ordinary hazard: quantity and combustibility of Class A and B materials are moderate and fires with moderate rates of heat release expected.
- Extra hazard: quantity and combustibility of Class A and B materials are high and fires with high rates of heat release expected.

EXTINGUISHING AGENTS

The agent used to extinguish a fire can range from water to dry chemical compounds to carbon dioxide. Some extinguishing agents are preferred over others based on their characteristics and the characteristics of the burning material. What follows is a summary of common extinguishing agents.

Water

Of all the extinguishing materials available, water is the most widely used and available due to many reasons: it is inexpensive, abundant, and effective in fire suppression. Water is considered a good extinguishing agent because of its physical properties (NFPA 2008, 17–32):

1. Water is a heavy, stable liquid that is nontoxic, and it is not flammable or combustible.
2. Water has a high specific heat (1 BTU/lb°F) allowing it to absorb a great deal of heat from a fire and the burning fuel.
3. When water is converted to a vapor, its volume increases about 1,600 times, displacing an equal volume of oxygen.
4. Water has a high heat of evaporation, 970 BTU/lb, allowing it to be an effective cooling agent.

Water can extinguish a fire by a combination of methods. The most common method of extinguishment when using water is cooling. In chapter 2 it was indicated that to sustain combustion, the surface of the material must be heated to its ignition temperature. Water extinguishes a fire primarily by cooling the surface of the materials below its ignition temperature. Water can also extinguish a fire by removing oxygen from the surface of the burning material, thereby smothering the fire. Steam can be generated from the water hitting the burning fuel; thus, the steam displaces the air around the fuel.

Water can also extinguish a fire through dilution. Water-soluble materials, such as ethyl and methyl alcohol, can be extinguished by dilution with water. The last method of extinguishment by water is emulsification. With emulsification, water is applied to certain vicious liquids that form froth/bubbles at the surface reducing the release of flammable vapors (NFPA 2008, 17–32–34).

Water does have a few physical characteristics that reduce its effectiveness as an extinguishing agent. At temperatures below 32°F, water freezes and it does conduct electricity. Water also has a high surface tension which acts to hold the surface film of water together. This high surface tension reduces its ability to penetrate burning combustibles, as well as its spread through tightly packed or baled materials (NFPA 2008, 17–36). To improve the performance of water, man-made additives have been designed to prevent the water from freezing or corroding piping and improve its ability to penetrate burning material.

There are many instances in which water is used as an extinguishing agent where environmental conditions expose the agent to possible freezing. This is most common when using water in fixed extinguishing systems in areas such as building attics and uninsulated storage areas. Additives like glycerin or propylene glycol are used to lower the freezing point of water in the sprinkler system, thus preventing the water from freezing at 32°F and breaking the piping. For portable fire extinguishers using water, alkali-metal salt solutions are added to the water to protect it from freezing to temperatures as low as −40°F (NFPA 2008, 17–42).

Wetting agents are added to water to lower its surface tension and improve its ability to penetrate materials. Water combined with these wetting agents is commonly referred to as "wet water." Experience with these wetting agents has shown they increase the effectiveness of water extinguishing a fire by reducing both the time for extinguishment as well as the amount of water needed. Wet water should not be used on electrical equipment because of the increased conductivity of the solution. The use of wetting agents must be approved by the authority having jurisdiction and standards regarding their use and testing are included in NFPA 18 Standard of Wetting Agents (NFPA 2011a, 4.1).

Another characteristic of water that limits its use as an extinguishing agent is its ability to conduct electricity. When water is applied to electrical equipment, a continuous circuit is formed that can conduct the electrical charge back to the person applying the water. Some factors that influence the conductivity of water when it is used as an extinguishing agent include the voltage and current of the electrical equipment, the nozzle design, the purity of the water, the length and area of the water stream, and the resistance of the person's body to ground (NFPA 2008, 17–38). The electrical conductivity of water in firefighting is most important to firefighters applying water streams in a fire. Numerous studies have shown that a shock hazard exists in these situations if adequate distances are not maintained. Unfortunately, based on current data these adequate distances can only be estimated and therefore wherever possible it is always recommended to use water spray rather than a solid stream where electricity is present. When water spray is used, the safe distance is influenced by the type of nozzle used and the voltage/current of the equipment. Safe distances between water used in fixed systems, such as sprinkler systems, and electricity are available in NFPA 15 Standard for Water Spray Fixed Systems for Fire Protection (NFPA 2008, 17–40).

WATER USE ON SPECIAL HAZARDS

While water has been found to be an ideal extinguishing agent for many types of fires, in some situations water may make the fire problem worse. For example, certain chemicals, such as carbides and peroxides, can react with water, releasing flammable gases and heat. Combustible metals such as titanium, magnesium, and sodium also react with water, releasing energy. Radioactive metals pose a hazard with water, not from a fire standpoint but from a contamination standpoint in which the water can become irradiated and spread the hazardous material. While water can be effective in cooling the surface of flammable and combustible liquid fires, it can also cause the fire to spread due to the burning liquid floating on the surface of the water (NFPA 2008, 17–41).

Carbon Dioxide

Carbon dioxide is a noncombustible, nonreactive gas that has been used for many years as an extinguishing agent. It extinguishes fires primarily by displacing the oxygen surrounding the surface of the burning material. As a result, the oxygen levels are reduced below the point required to sustain combustion, usually 10 percent by volume in air. Other physical properties of carbon dioxide that increase its effectiveness as an extinguishing agent include its ability as a gas to penetrate the surface of burning materials, it is not electrically conductive, and it leaves no residue. While it is commonly identified as the extinguishing agent for electrical fires, carbon dioxide will also work as an extinguishing agent on most materials except oxidizers that provide their own oxygen as they burn and some reactive metals. Specifically, reactive metals such as magnesium and titanium will decompose the carbon dioxide, releasing oxygen. Carbon dioxide can be used for Class A fires but is limited due to its low cooling capacity and its ability to reach sufficient carbon dioxide concentrations in a given atmosphere. Another concern with using carbon dioxide with deep-seated Class A materials, such as a mattress, is the possibility of re-ignition after initial extinguishment (NFPA 2008, 17–6).

There are some potential health hazards that should be noted when using carbon dioxide. When discharged, carbon dioxide is both a gas and dry ice, which is often referred to as snow. The extremely low temperatures of the dry ice can cause burns to the skin if contact is made. Carbon dioxide concentrations above 6 percent cause harmful effects to humans, with concentrations above 9 percent causing unconsciousness. These levels are far below the minimum design concentrations of 34 percent for carbon dioxide extinguishment systems. Therefore, as a rule, total flooding carbon dioxide extinguishment systems should not be used in occupied areas. If used in other extinguishment systems, it is critical to design adequate safety precautions, such as audible and visual alarms, to allow predischarge evacuation of all humans (NFPA 2008, 17–8).

Halogenated Extinguishing Agents

Halogenated materials use a hydrocarbon material with an atom of the halogen series: fluorine, bromine, chlorine, or iodine (NFPA 2008, 17–93).

Halogenated agents (halons) extinguish a fire primarily by interrupting the chemical chain reaction. These agents have been around since the early 1900s, with Halon 104, or carbon tetrachloride, one of the first agents used in portable fire extinguishers. Unfortunately, carbon tetrachloride is toxic and therefore was replaced with new forms of halogenated extinguishing agents, specifically Halon 1301 (Bromotrifluoromethane) and Halon 1211 (Bromochloro-difluoromethane). The U.S. Army Corp of Engineers developed the numbering system for halogenated agents (Cote 2001, 206). The digits represent the chemical composition of the agent as follows:

- Digit 1: number of carbon atoms in the chemical
- Digit 2: number of fluorine atoms
- Digit 3: number of chlorine atoms
- Digit 4: number of bromine atoms
- Digit 5: number of iodine atoms if any (if omitted, none are used)

Halogenated agents are commonly used in hand-held fire extinguishers and fixed extinguishing systems protecting electrical equipment and electronics. These agents vaporize quickly in a fire, leave no corrosive or abrasive residue, and therefore do not damage electrical components. In addition, unlike carbon dioxide they are effective at lower concentrations in air that allow sufficient oxygen supply. For these reasons as well as the overall speed of extinguishment, halons became extremely popular as an extinguishing agent for sensitive electrical and computer equipment prior to 1990. In 1993, Halon 1301, 2402, and 1211 were identified in the Montreal Protocol on Substances that Deplete the Ozone Layer as ozone-depleting substances. As a result, the production of these halons was phased out in developed countries starting

January 1, 1994. However, halon systems still exist with recycled halon being used to replenish existing systems.

To replace halon, new extinguishing agents have been developed with the same firefighting characteristics. A standard that addresses these halon replacements is NFPA 2001 "Standard on Clean Agent Fire Extinguishing Systems." This standard defines a clean agent as one that is not electrically conductive and leaves no residue after evaporation. Clean agents are divided into two chemical groups: inert or halocarbon. Examples of two halocarbon clean agents identified in Table 1.4.1.2 of NFPA 2001 are HFC-125 Pentafluoroethane and HFC-23 Trifluoromethane (NFPA 2012, 1.4.1.2). An example of an inert clean agent is IG-541 which is a blend of three naturally occurring gases, nitrogen (52%), argon (42%), and carbon dioxide (8%), also known as Inergen (NFPA 2008, 17–129).

Dry Chemical Extinguishing Agents

Dry chemicals are the predominant extinguishing agent used in fire extinguishers, and they are especially effective in extinguishing fires caused by flammable liquids. There are several different types of dry chemicals used but the most popular is sodium bicarbonate because of its lower cost. Other dry chemicals used include potassium bicarbonate, potassium chloride, and monoammonium phosphate. Except for monoammonium phosphate, the other dry chemicals are listed for Class B and C fires. Monoammonium phosphate is a multipurpose extinguishing agent that is listed for Class A, B, and C fires. All these dry chemicals have a tendency to absorb moisture and pack so they will be mixed with additives such as metallic stearates and silicones to avoid packing and improve flow of the chemical (NFPA 2008, 17–8). The dry chemicals are stored in pressure containers and are expelled with an inert gas, most commonly nitrogen.

Dry chemical extinguishing agents extinguish a fire primarily by interruption of the chemical reaction and reducing the concentration of free radicals in the flame. However, these agents also extinguish to some level through cooling and smothering. For example, sodium bicarbonate, when heated by fire, releases CO_2 and monoammonium phosphate forms a sticky film both of which act to smother the fire.

Dry chemical agents are extremely popular because of their versatility, especially the multipurpose chemical agents. However, there are some

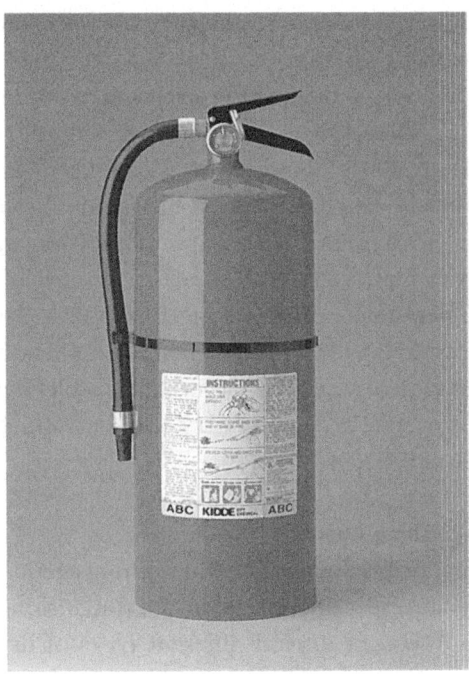

FIGURE 9.1
Multipurpose Dry Chemical Extinguisher. Source: Photo courtesy of Kidde Products

downsides to their use that must be considered. Generally, the materials are nontoxic; however, they can cause minor skin irritation and when inhaled in heavy concentrations can be respiratory irritants. Multipurpose dry chemical agents (see Figure 9.1) can be used on combustible materials, on electrical fires because they are nonconductive, and on flammable liquid fires. The dry chemical agents can be slightly corrosive, which can damage delicate electrical equipment. Therefore, these agents should not be used on delicate electrical equipment such as computers, and they may not be compatible with certain foams (NFPA 2008, 17–20).

Wet Chemical Extinguishing Agents

Wet chemical agents are used primarily for fixed fire suppression systems involving cooking equipment. These agents are typically a mixture of water with organic or inorganic salts such as potassium citrate, potassium acetate, or potassium carbonate. These wet chemicals are typically stored in plastic containers and are expelled using an inert gas under pressure. These agents extinguish cooking fires primarily through a combination of cooling and smothering that occurs when the agent foams when contacting the surface of the grease. Due to the presence of water, these agents cannot be used around electrical equipment or any other material that is water reactive (NFPA 2008, 17–24).

Foam Extinguishing Agents

Firefighting foam is an aggregate of gas-filled bubbles from a water-based solution. The gas used in the bubbles is typically an inert gas. Since the foam is lighter than liquids, it will float on top of flammable and combustible liquid. This foam layer extinguishes a fire by a combination of excluding oxygen, cooling, and suppressing fuel vapors. The foam agent is produced by mixing a concentrate with water, then aerating the mixture. Foams can be classified based on their expansion ratio, how they are generated, and by the type of concentrate. For example, low-expansion foam is a concentrate that produces foam-to-solution volume ratios of less than 20:1, while medium and high expansion foam produces foam-to-solution volume ratios of 20:1 to 1,000:1 (NFPA 2010, 3.3.12.6). When classified based on generation, the two most common methods of generation are blower type or aspirator type. When classified based on the foam concentrate, common types include: alcohol-resistant foam concentrates, aqueous film-forming foam (AFFF) concentrates, fluoroprotein foam concentrates, and protein foam concentrates. Each of these concentrates have their advantages and disadvantages and must be selected based on the type of application and the type of liquid they are protecting.

Foams are primarily used to extinguish Class B flammable or combustible liquid fires in tanks or fires associated with spills. However, AFFF, which has a low surface tension can be used on some Class A fires. As an extinguishing agent, foams do have some disadvantages. Foams are electrically conductive and therefore cannot be used on Class C fires. Their effectiveness is based on their ability to cover the surface of the liquid. Therefore, the volume and rate at which the foam is applied over the surface will also determine the successful extinguishment of the fire. Foams are also much more effective on horizontal surface fires (NFPA 2008, 17–46).

Combustible Metal Extinguishing Agents

Fires involving combustible metals, such as sodium, magnesium, and potassium, pose special fire hazards. These metal fires typically generate very high temperatures, thus increasing the cooling time. In addition, some of these metals are reactive with water and nitrogen and common dry chemical extinguishers are ineffective on fires involving these materials. To extinguish these combustible metal fires, specially formulated dry powder agents have been developed. These dry powder agents are not the same as the dry chemical agents discussed previously. Two of the most used dry powder extinguishing agents are sodium chloride and G1 powder, which is a combination of granular graphite and phosphorous. Sodium chloride will be combined with additives to prevent "caking" and it can be applied either through an extinguisher or by hand. G1 powder may be applied to the fire by hand via a scoop or shovel. In general, these dry powders extinguish a combustible metal fire by removing oxygen (smothering) and there is some cooling as well (NFPA 2008, 17–81).

PORTABLE FIRE EXTINGUISHERS

The first fire extinguisher was patented by Alanson Crane on February 10, 1863 (http://inventors.about.com/library/inventors/blfiresprinkler.htm) and has since become the most common method of extinguishing fires. The fire extinguisher has evolved in terms of design and function over its 150 years of existence. Early versions of portable fire extinguishers worked using a pump to generate pressure inside the cylinder. Other fire extinguishers, referred to as inverting type, required the user to tip the extinguisher upside down, causing baking soda to create pressure from the generated carbon dioxide inside the cylinder. Many of these early types of fire extinguishers are now considered obsolete and as of 1969, inverting extinguishers are no longer manufactured in the United States. Today's modern portable fire extinguishers use either a stored pressure cylinder or a cartridge to expel the extinguishing agent. One of the earliest standards on portable fire extinguishers was NFPA 10 Standard for Portable Fire Extinguishers that was published in 1921 (Cote 2001, 218).

Because different classes of materials can be involved in a fire, portable fire extinguishers are designed to extinguish materials based on their class. The fire extinguishers are classified according to the types of materials they may be used on. The classification of fires (Classes A, B, C, D, and K) used in NFPA 10 was discussed earlier in this chapter under "Classification of Fires." Fire extinguishers that are effective on more than one class of fire have multiple-letter classifications. Table 9.2 summarizes the various classes of fire extinguishers and the extinguishing agents more commonly used in ordinary building protection.

In addition to the class of fire, it is also important to consider the fire load, which is a measure of the amount of combustibles. NFPA 10 has three classifications based on fire load: light hazard, ordinary hazard, and extra hazard. These three classifications were discussed earlier in chapter 5 "Occupancy Classification." These classifications based on fire load are important in determining the size and placement of portable fire extinguishers.

Labeling

Fire extinguishers are labeled so that users can quickly identify the classes of fire on which the extinguisher will be effective. The marking system combines pictographs of both recommended and unacceptable extinguisher types on a single identification label.

FIGURE 9.2
Standard Labeling for Fire Extinguishers. Source: Kidde Products

Fire Extinguisher Rating Systems

In addition to the classification letter, fire extinguishers used for Class A and B fires will also have a rating number preceding the classification letter.

This rating number reflects the relative effectiveness of the extinguisher. For example, a fire extinguisher that is rated and classified 4-A:20-B:C, implies it should extinguish approximately twice as much Class A fire as a 2-A fire extinguisher and approximately twenty times as much Class B fire

Table 9.2. Summary of the Types of Portable Fire Extinguishers

Fire Extinguisher Class	Types of Materials	Extinguishing Agents
Class A	Ordinary combustibles: wood, paper	Water, aqueous film-forming foam, multipurpose dry chemical (ammonium phosphate), halogenated agents[a]
Class B	Flammable and combustible liquids, oils, grease	Carbon dioxide, aqueous film-forming foam, multipurpose dry chemical (ammonium phosphate), halogenated agents[a]
Class C	Electrical fires	Carbon dioxide, multipurpose dry chemical (ammonium phosphate), halogenated agents[a]
Class D	Metal fires	Special dry powder agents such as Met-L-X and NA-X
Class K		

[a]Halogenated agents may include Halon 1301 and Halon 1211. "Halogenated type agents" include Haltron® I (American Pacific Corporation) and FE-36TM (DuPont).

as a 1-B-rated fire extinguisher. The current classification numbering system used for Class A and B includes the following (NFPA 2013a, 10–56): Class A: 1-A, 2-A, 3-A, 4-A, 6-A, 10-A, 20-A, 30-A, and 40-A; Class B: 1-B, 2-B, 5-B, 10-B, 20-B, 30-B, 40-B, 60-B, 80-B, 120-B, 160-B, 240-B, 320-B, 480-B, and 640-B. Although the numbering systems provide a guide for the size of fire that can be extinguished, it should be recognized that the size of fire that can be extinguished by a particular extinguisher is also related to the degree of training and experience of the operator. This numbering system is also used as one criterion to determine the maximum coverage area per extinguisher or the maximum travel distance to an extinguisher. For example, the maximum travel distance for a 10-B fire extinguisher in an ordinary hazard occupancy is 30 ft. (NFPA 2013a, 10–13).

No classification numbering system is used for Class C or D fire extinguishers. For Class C fires, the rating of the extinguisher should be based on the size and extent of the Class A or Class B components, or both, of the electrical hazard being protected. For Class D fires, the relative effectiveness of the Class D extinguishers is identified on the fire extinguisher nameplate (NFPA 2013a, 10–56). For a complete listing of UL Fire Extinguisher Classifications/Ratings as well as extinguishing agents, their method of operation, capacity, and horizontal range see NFPA 10 Annex H, Table H-2 "Characteristics of Extinguishers."

FIRE EXTINGUISHMENT

Fire Extinguisher Use in the Workplace

Employers have a critical decision to make regarding the use of fire extinguishers in the workplace. Firefighting can be a dangerous activity, and some employers may decide that immediate evacuation is the best option for employees in the event of a fire. Other employers may find themselves in situations in which fire services in the area are limited or the response time to the facility is not acceptable. In those cases, some employers may decide to have employees use fire extinguishers and standpipe hose systems to fight the fire; while others may establish a fire brigade. Depending on the company's policy, Occupational Safety and Health Administration (OSHA) standards will be applied differently; however, OSHA does not require any employer to assign firefighting duties to an employee. As mentioned earlier, the employer may simply choose to adopt as its policy that all employees are required to evacuate the building immediately. In that case, the policy must be implemented as part of a comprehensive emergency action plan and a fire prevention plan, see chapter 10. Where extinguishers are provided, but are not intended for employee use, and the employer has an emergency action plan stating so, then fire extinguisher training is not required. In situations where the employer provides fire extinguishers for general employee use, OSHA standards specify requirements for their distribution, placement, design, testing, and maintenance, and for employee training in their use. A discussion on each of these areas follows.

Fire Extinguisher Distribution and Mounting

According to OSHA, portable fire extinguishers should be mounted, located, and identified in such a manner that they are readily accessible. Any signs used to identify the location of fire extinguishers should be as close to the extinguisher as possible and visible from normal paths of travel. In terms of mounting height, NFPA 10 recommends that extinguishers weighting 40 lbs. or less be no more than 5 ft. from the floor to the top of the extinguisher. For extinguishers weighting more than 40 lbs. the height should be no more than 3 ft. above the floor (NFPA 2013a, 6.1.3.8).

OSHA requires that portable fire extinguishers be distributed so that maximum travel distances to get to the extinguishers are met. These travel distances are based on the class of fire. For Class A fires, the travel distance for employees to an extinguisher should be 75 ft. or less. Instead of fire extinguishers, the employer may provide uniformly spaced standpipe systems or hose stations. For Class B fires, portable fire extinguishers should be distributed so that the travel distance to any extinguisher is 50 ft. or less. For Class C fires, the employer shall base the travel distances upon the existing Class A or Class B hazards. Class D fire extinguishers or powder containers shall be distributed so that the travel distance from the combustible metal working area to any extinguishing agent is 75 ft. or less (USDOL 2013). In terms of distribution, NFPA 10 also considers the size of the extinguisher, hazard occupancy type, and maximum square feet of floor area. Specific requirements for distribution of fire extinguishers based on these criteria are outlined in NFPA 10 Chapter 6 Installation of Portable Fire Extinguishers (NFPA 2013a, 6.2 and 6.3).

Maintenance, Inspection, and Testing

The employer is responsible for ensuring that all portable fire extinguishers are maintained in a fully charged and operable condition and kept in their designated places. To accomplish this, the employer must inspect, maintain, and test all portable fire extinguishers. According to OSHA, a visual inspection of fire extinguishers must be conducted monthly. This inspection is a quick visual check to ensure the extinguisher is in the designated location and will operate. In addition, the employer must assure that portable fire extinguishers are subjected to a maintenance check on at least an annual basis. The annual maintenance checks should be recorded and include a thorough examination of each extinguisher including examining the mechanical parts of the extinguisher, cleaning the extinguisher, replacing defective parts, and recharging/pressurizing the unit where necessary. For a complete listing of recommended maintenance checks for fire extinguishers, see NFPA 10 Annex I "Maintenance Procedures." These maintenance check records must be kept for one year after the last recorded inspection.

In addition to the monthly visual inspections and annual maintenance checks, fire extinguishers must also go through a hydrostatic test. Hydrostatic testing of portable fire extinguishers is done to evaluate the integrity of the shell and protect against unexpected in-service failure. Shell failure can be caused by internal corrosion, external corrosion, and/or damage to the shell itself. The employer shall ensure that fire extinguishers go through the appropriate hydrostatic testing at the prescribed intervals based on the type of extinguisher. This time interval ranges from every five years for carbon dioxide extinguishers to twelve years for dry chemical extinguishers with mild steel shells. Additional hydrostatic testing on cylinders is required in the following circumstances (USDOL 2013):

1. The extinguisher has been repaired by soldering, welding, brazing, or use of patching compounds.
2. The cylinder or shell threads are damaged.
3. There is corrosion that has caused pitting, including corrosion under removable nameplate assemblies.
4. The extinguisher has been burned in a fire; or
5. When a calcium chloride extinguishing agent has been used in a stainless-steel shell.

Any extinguisher that fails the hydrostatic test must be taken out of service immediately. Certification records pertaining to hydrostatic testing of fire extinguishers shall be maintained by the employer until the next hydrostatic test or until the extinguisher is taken out of service.

Training

If employees are required to use fire extinguishers in the event of a fire, then they must receive annual education and training. The educational program must at a minimum address the general principles of extinguisher use and the hazards associated with incipient stage firefighting. The training program should cover topics such as proper selection of an extinguisher, how to use the extinguisher, and the hazards of fires and firefighting. Ideally, training on the use of fire extinguishers should be done through actual operation of the extinguisher on an actual fire. NFPA 10 Annex D "Operation and Use" provides some additional topics that can be used for both the fire extinguisher education and training program. The training should be conducted when employees are initially assigned to a job where they may have to use an extinguisher and then at least annually thereafter (USDOL 2013).

WATER-BASED SPRINKLER SYSTEMS

Water-based sprinkler systems are devices that automatically distribute water over a fire to either extinguish the fire or prevent the fire from spreading. The key word here is automatic, as the system does not require human intervention for the system to discharge the water. Sprinkler systems are typically activated by the heat of the fire, and the water distributed from the sprinkler heads primarily extinguishes the fire via cooling. In the United States, the first automatic sprinkler system was patented by Philip W. Pratt of Abington, Massachusetts, in 1872. Henry S. Parmalee of New Haven, Connecticut, is considered the inventor of the first practical automatic sprinkler head.

Parmalee improved upon the Pratt patent and created a better sprinkler system that he installed in his piano factory in 1874. Prior to the 1940s, most sprinkler systems were installed in commercial buildings. Many of the owners of these buildings were able to offset the cost of the systems through savings in insurance premiums (Automatic Sprinkler Company 2014). Since their introduction, sprinkler system design and construction have changed with both their performance and reliability significantly better. With such improvements, their use has expanded both within commercial and residential properties. It is not uncommon for local building codes to require their use in hospitals, schools, hotels, and other high-rise properties.

Impact of Sprinkler Systems on Fires

Sprinkler systems have an excellent record of fire suppression. Recent statistical data has shown that 96 percent of the time they have controlled fires when activated (Klinoff 2012, 405). In addition to their effectiveness in fire suppression, they also offer benefits in terms of life safety. Not only does the system reduce the spread of fire, it also lowers the smoke level in an area, and it sounds an alarm providing occupants more time to evacuate. These systems also typically cause less water damage than a hose stream. Data from the NFPA shows that in 81 percent of fires in a wet-pipe system two or fewer heads were activated (NFPA 2008, 16–32). The insurance industry has long recognized the value of sprinkler systems and for that reason reduces premiums for those buildings that are fully sprinklered. These systems are extremely reliable but when they do fail it is typically related to a human error such as shutting off a water control valve.

Types of Sprinkler Systems

Sprinkler systems can be classified according to their design and operation. In terms of design, there are two basic classifications: pipe schedule and hydraulically designed. Pipe schedule design is based on a prescribed schedule for the number of sprinkler heads based on the pipe size. Pipe schedule design is restricted by NFPA 13 Automatic Sprinkler Systems as it is impractical to use in most cases (NFPA 2008, 16–47). Hydraulically designed systems select pipe sizes based on a pressure loss basis to provide a prescribed water density over a specified area (NFPA 2013c, 13–18). Most sprinkler systems designed today are hydraulically designed because they are more cost effective and are more flexible with regard to the selection of piping and type of sprinkler.

From an operation perspective, there are four general classifications of sprinkler systems: wet-pipe systems, dry-pipe systems, preaction systems, and deluge systems. The selection of the type of sprinkler system is based primarily on the type of environment in which it will be used but other possible considerations may be the type of building construction and the hazards of occupancy. For example, wet-pipe systems are not recommended when the system could be exposed to an environment where temperatures are below 40°F (NFPA 2008, 16–34).

In general, wet-pipe sprinkler systems are preferred over the other types because they are the least expensive systems to install and maintain, they are simpler to design, the most versatile system, and are the most reliable. In a wet-pipe system, water is under pressure throughout all the piping, from the riser to the sprinkler heads at all times. As the sprinkler heads open, water is discharged immediately. Water flowing through the pipe will also activate an alarm warning occupants of the presence of a fire and the need to evacuate.

Dry-pipe systems are used in areas of buildings that are susceptible to freezing. In a dry-pipe system, water is only present in the system up to the check valve at the riser. Between the check valve and the sprinkler, air or nitrogen under pressure is in the system. As the sprinkler heads are opened from the fire, the air or nitrogen escapes, and the check valve opens allowing water to fill the piping and ultimately discharging at the opened sprinkler head. These systems are more complex than a wet-pipe system and they require a reliable air supply source to maintain the pressure in the piping above the activation pressure for the dry-pipe valve. The delay in water delivery to the fire may limit the size of these systems as well as the need for accelerators to aid in the discharge of the air from the piping (NFPA 2008, 16–35).

In a preaction system, the piping between the check valve at the riser and the sprinkler head is filled with air that may or may not be under pressure.

With a preaction system, detection system sensors are used to detect the fire and to open the preaction valve, allowing water to flow through the piping. However, water will not flow through the sprinkler heads until sufficient heat is generated from the fire to open the sprinkler heads. These systems are popular in areas where the contents and/or equipment are sensitive to water damage, such as computer equipment, because they reduce the likelihood of inadvertent water discharge from the system (NFPA 2008, 16–37).

A deluge system is essentially a preaction system where all sprinkler heads in the system are always open. When the detection system sensors are activated by a fire, they open the deluge valve and water flows through the piping and is discharged through all sprinkler heads. These systems discharge large quantities of water very quickly over a specified area. Therefore, these systems are commonly used in areas where large quantities of high combustible materials are present and a rapidly growing fire is likely.

Sprinkler System Components

The typical components of a sprinkler system are illustrated in figure 9.3. For such a system to work properly and control or extinguish a fire, an adequate, unobstructed water supply must be provided in sufficient volume. This section describes the components of a typical water-based sprinkler system.

Water Supply and Distribution

The water supply system is the source where the water is found. The water distribution system is the portion of the water system that delivers the water to the sprinkler system. No single component of a sprinkler system is more important than the water supply. No matter how well designed a system is, it will fail if the water supply is not adequate in terms of water flow and pressure. All sprinkler systems are required to have at least one automatic water supply. According to Chapter 23 of NFPA 13, the acceptable sources of the water supply include public and private water supplies, fire pumps, pressure tanks, gravity tanks, and naturally occurring water supplies such as lakes and rivers. All water supplies must be reliable and flow tests must be completed on a routine basis to verify the condition of the water, its flow, and pressure.

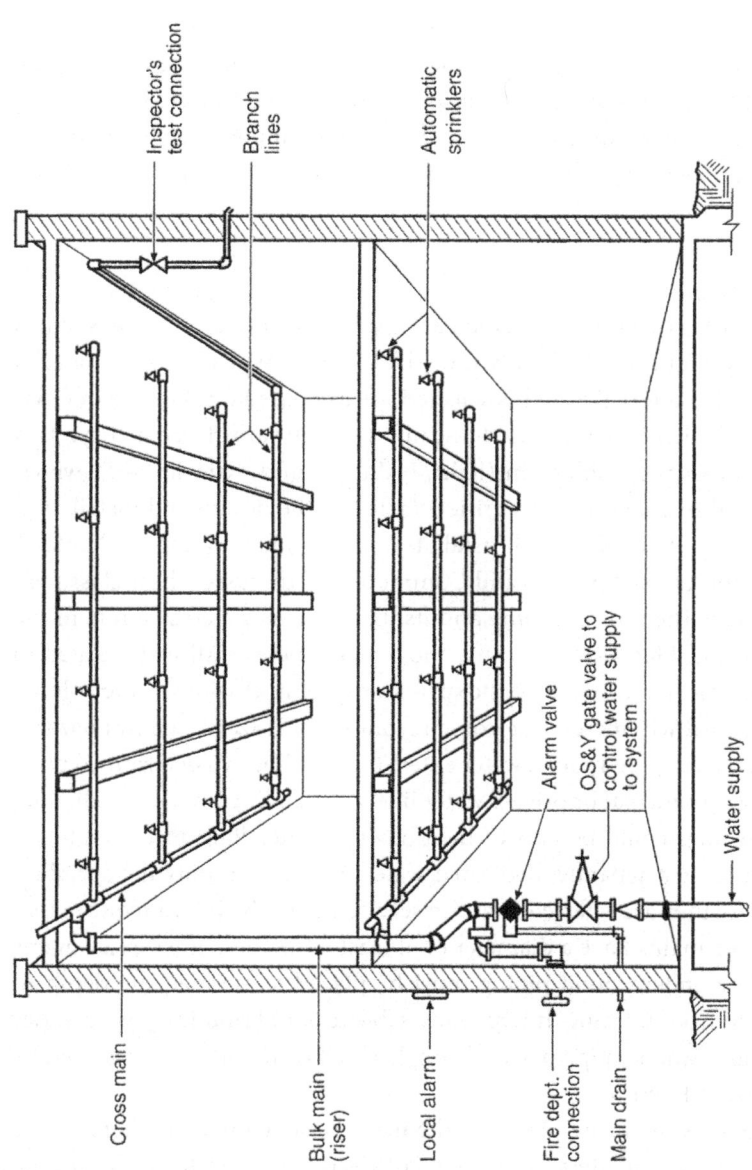

FIGURE 9.3
Basic Components of a Wet Pipe Sprinkler System. Source: Reprinted with permission from the National Fire Protection Handbook, 19th ed., 2003, National Fire Protection Association

Two major types of water distribution systems are gravity systems and direct pumping systems. A true gravity system provides water to the source without the use of pumping equipment. A gravity system is extremely reliable because it does not require the use of mechanical equipment to provide the flow of water. Pumping systems are located at the source of the water. They are used to overcome the friction loss in the supply system and to provide adequate working pressures in the distribution system (Klinoff 2012, 380).

Sprinkler Piping

Various types of piping materials are used in sprinkler systems such as steel, copper, and nonmetals (PB and CPVC). Each type of piping material has unique characteristics and is selected for use based on the type of environment in which it will be used and the type of hazard occupancy. This hazard occupancy classification (light, ordinary, and extra hazard), which was discussed in chapter 5, is a critical factor in the design and installation of a sprinkler system. NFPA 13 Installation of Sprinkler Systems, Chapter 5 "Classification of Occupancies and Commodities" discusses these classifications in detail. The type of piping and its diameter play a critical role in the design of a sprinkler system because the pipe diameter is directly related to the volume of water available to the sprinkler system. Obviously, the volume of water needed will increase as you go from Light Hazard to Extra hazard.

In a typical wet-pipe system, water enters the building at a determined pressure through municipal or private water lines. Depending on the jurisdiction, these water lines could be part of the domestic water lines to the building, or they could be a separate, dedicated water line serving only the sprinkler system where they are referred to as fire service mains. NFPA 13 requires that all fire service mains have a diameter of at least 6 in. These fire service mains and all other underground piping must be placed at an adequate depth to protect them from freezing. It is also advisable to avoid running piping under buildings, and when a pipe passes through a wall or foundation, it should be protected from fracture.

Backflow preventers are placed in the fire service main somewhere before the sprinkler riser. If there are significant decreases in water pressure, the backflow preventers prevent water from being sucked out of the sprinkler system into the municipal water system. If this occurred, water from sprinkler system piping could contaminate water supplies. The installation of backflow preventers is required by NFPA 13 for all systems using antifreeze, but they may also be required on other systems by local ordinances and the Environmental Protection Agency.

The vertical piping that is connected to the fire service main is referred to as the riser. The riser contains the main drain or test valve, the sprinkler control valve as well as any water-flow alarm devices for the system. The riser typically ends at or close to the roof/ceiling where it connects with the horizontal pipe called a feed main. The feed main connects with the cross mains that lead to the branch lines. Sprinkler heads are typically located on the branch lines. To maintain water pressure, the pipe diameters will decrease as they go from the riser to the branch lines. At the end of the branch line containing the sprinkler head that is the furthest from the riser is the inspector's test valve. This valve is used to simulate the opening of the most remote sprinkler head in the system. Opening this valve should activate the water-flow alarm in the same manner as an opened sprinkler head would. On some systems, a water gauge is present so that the inspector can verify the water pressure at this most remote sprinkler head. The inspector's test, also referred to as a trip test, will identify problems with the water-flow alarm and inadequate water pressure due to poor water supply and blocked or restricted pipes. If the water pressure is adequate at the inspector's test valve, it will be adequate throughout the entire system.

It is critical to size all sprinkler piping to meet hydraulic design calculations. Sprinkler piping must also be maintained to prevent corrosion and blockage. Foreign material can get into the sprinkler piping for a variety of reasons and this will affect water flow. System piping that is not inspected and flushed routinely can develop silt and sludge that can totally obstruct the smaller diameter branch lines and sprinkler heads.

Sprinkler System Valves

It may be necessary to shut off the flow of water at the fire service main for a variety of reasons such as maintenance on the system. The sprinkler system is typically provided with outside stem and yoke (OS&Y) valves or using a post indicator valve (PIV) (see figure 9.3). These valves allow the building owner to shut off the water to the sprinkler system. To close the OS&Y valve, a handwheel is turned on the valve, causing the stem, or screw, portion of the valve to lower and seal the valve closed. An OS&Y valve is in the open position when the stem is visible. To prevent unwanted closing of the OS&Y valves and thus shutting off of the water to the sprinkler system, the handwheels may be locked and chained in the open position or equipped with a valve supervisory switch that will sound an alarm and may alert the monitoring service when the valve is closed.

In a wet-pipe system, a clapper valve is located on the riser with water pressure gauges positioned above and below the valve. These pressure gauges will alert the inspector of potential problems in the water lines, such as a loss in pressure of water entering the system. Located below the sprinkler system water flow check valve is a 2-in. pipe and valve, which allows the inspector to conduct a 2-in. main drain test. During this test, the valve located at the base above the water flow alarm check valve is closed to ensure the water above the main clapper valve remains in the system. Opening the 2-in. main drain valve allows water to flow from the municipal supply through the 2-in. pipe and into a drain. The water pressure gauge located below the clapper valve indicates the water pressure that would be available to the sprinkler system should it be activated. The 2-in. main drain test will alert the inspector of potential water supply problems, such as inadequate water pressure and the presence of debris or obstructions in the water supply line (NFPA 2008, 16–43).

In a dry-pipe system, air or nitrogen is under pressure in the system above the dry-pipe valve. An air compressor keeps the system under pressure above the valve and a supervisory signaling device monitors this air pressure. The air pressure above the valve must be 20 psi above the trip pressure of the dry-pipe valve. When a sprinkler head opens, air rushes out reducing the pressure above the dry-pipe valve and it opens allowing water to flow through the piping. In a dry-pipe system, it is important that the clapper be maintained in its seated position so that water does not enter the piping. Once discharged, it is important to drain all the water out of the system and reseat the main clapper in the dry-pipe valve. In older systems, reseating the clapper valve may require removing the cover from the valve and manually closing the valve inside.

In preaction and deluge systems, the sprinkler control valve is opened due to the activation of a detection system sensor or automatic sprinklers.

FIRE EXTINGUISHMENT

Specifically, if it is a single interlock system the valve opens by the single operation of the detection sensor. With a non-interlock system, the valve will open with either the detection sensor being activated or the operation of a sprinkler head. The last type, double interlock requires both the detection system and the sprinkler system to operate before the valve will open allowing water to flow (NFPA 2013b, 7.3.2.1).

Water Flow Alarms and Supervisory Initiating Devices

As mentioned earlier, a water-flow check valve is attached to the riser piping. This water flow check valve serves two purposes. First, in a wet-pipe system, if there is a drop in water pressure below the check valve, the clapper will remain seated and keep the water in the sprinkler system piping above the clapper valve. The second purpose of the check valve is to activate a water flow alarm in some systems. The water-flow alarm alerts the building occupants that water is flowing in the sprinkler system. When water begins to flow, the clapper valve will be raised from its seated position. The actual activation of a water flow alarm may be accomplished by various means. A mechanical water gong alarm can be activated by the moving water rotating a wheel inside the riser pipe. This rotating wheel causes a striker to hit a gong. Water flow alarms may also be electrically operated. With this type of water flow alarm, the water flow may be detected by electrical switches incorporated into a pressure or water flow device. The water flow alarm must be local, but it may also be transmitted off site to a central station (NFPA 2008, 14–41).

Supervisory initiating devices are used to supervise the operation of critical operating features of a sprinkler system. Examples of features that may be supervised include closing of sprinkler water control valve, air pressure in a dry-pipe system, level of water in a storage tank, and integrity of a fire pump. A local alarm is sounded when these devices sense a problem with any of these critical features, and the signal may also be sent to a monitored central station as well.

Sprinkler Heads

The termination point of a sprinkler system is the sprinkler head. In a typical wet-pipe system, sprinkler heads are designed to open when the temperatures around the head exceed a predetermined level. There are two major mechanisms used to release the cap over the orifice of the sprinkler head: bulb sprinkler heads and fusible sprinkler heads. With bulb sprinklers, there is a frangible bulb, usually glass, with a liquid and air bubble inside. As the liquid heats up, it expands, the air bubble disappears and pressure rises causing the frangible bulb to shatter, releasing the cap that was holding back the water in the branch line. In a fusible link sprinkler head, a piece of metal (lever, strut, or link) with a predetermined melting point is used to hold the cap in place over the orifice. When the temperature at the head reaches the predetermined melting point the metal part melts, falls away, and the cap is released allowing the water to flow.

The temperature at which the sprinkler head will operate is stamped on the sprinkler head itself and it is also identified by a color code. The color used for the various temperature ranges is specified in Table 6.2.5.1 of NFPA 13 Installation of Sprinkler Systems (see table 9.3). For the fusible sprinkler head this color will be painted on the frame arm, deflector, or coating material. For frangible bulb sprinklers, the color of the liquid in the bulb is used to indicate the temperature rating of the head. For ornamental and decorative sprinkler heads, there is no color coding but as mentioned earlier its operating temperature will be stamped on the sprinkler head (NFPA 2013b 6.2.5.1).

The selection of the appropriate sprinkler head based on temperature rating is for the most part based on the maximum ceiling temperature in the area it will be located. In general, most locations will use ordinary or intermediate temperature sprinklers, which have a maximum ceiling temperature of 150°F. High-temperature sprinklers are common in ordinary and extra hazard occupancies and they are required in other areas as specified in NFPA 13 section 8.3.2.5.

Table 9.3. Ratings of Fire Sprinkler Heads

Temperature Rating (Degrees Fahrenheit)	Temperature Classification	Color Code	Glass Bulb Color
135–170	Ordinary	Uncolored or black	Orange or red
175–225	Intermediate	White	Yellow or green
250–300	High	Blue	Blue
325–375	Extra high	Red	Purple
400–475	Very extra high	Green	Black
500–575	Ultra-high	Orange	Black
650	Ultra-high	Orange	Black

In addition to the temperature at which a sprinkler head is designed to operate, sprinkler heads also differ in their design. Early sprinkler heads were nothing more than holes drilled in piping. Today, sprinkler heads are specially engineered components designed to ensure uniform water coverage throughout the area they protect. Variations in sprinkler head design include the size of the sprinkler orifice, mounting position, intermediate level sprinklers, and early suppression fast response sprinklers. In terms of mounting, upright sprinklers are mounted on top of the branch line while pendant sprinklers are mounted on the bottom of the branch line. Pendant sprinklers can be recessed in the ceiling or they can be flush-type sprinklers where the deflector drops below the ceiling when activated. Sidewall sprinklers as the name implies are installed along a wall and they have specially designed deflectors that discharges the water along the wall. Intermediate level sprinklers are another name for in-rack sprinklers, which provide protection for rack storage. These sprinklers have specially designed shields that protect the head from water that may be discharged from overhead sprinklers. In terms of the orifice size, the standard is ⅝" and anything smaller is considered a small orifice sprinkler and anything larger a large orifice sprinkler. Orifice size determines water flow, which is critical in the design of a sprinkler system. Large orifice sprinklers will get a required flow at a lower pressure than a standard sprinkler and offer the advantage of producing better size water droplets (NFPA 2008, 16–25). The early suppression fast response (ESFR) sprinkler head was designed primarily for high challenge storage occupancies. These heads are designed to operate earlier than standard heads and provide adequate water discharge to suppress the fire before a severe fire plume develops.

Paramount to ESFR sprinkler performance is the ability of the sprinkler to provide large amounts of water, in a specific discharge pattern, to the fire source in the incipient phase of fire development. Obstruction of the sprinkler discharge pattern could greatly affect the ability of the ESFR sprinkler to achieve fire suppression, according to a study published by the Fire Protection Research Foundation in 2020 (Report number: FPRF-2020-10).

As obstructions created by ceiling structural members, lighting, piping, or cable trays, could hinder ESFR sprinkler performance, the FPRF conducted a six-year study (2014-19) including full-scale testing, to assess the situation. Its findings were:

- The obstruction created by an open web steel truss 22–36 in. in depth, located a minimum of 6 in. horizontally from an ESFR sprinkler (K14 or K17), will not significantly decrease sprinkler performance.
- The obstruction created by a bridging member or other obstruction 1½ × 1½ inches in size or less, located a minimum of 12 in. directly below an ESFR sprinkler (K14 or K17), will not significantly decrease sprinkler performance. This also applies to a bridging member attached to open web steel truss.
- The obstruction created by a flat or round obstruction less than or equal to 12 in. in width, located a minimum of 6 in. horizontally from an ESFR sprinkler (K14 or K17), will not significantly decrease sprinkler performance.
- The obstruction created by a flat or round obstruction less than or equal to 24 in. in width, located a minimum of 12 in. horizontally from an ESFR sprinkler (K14 or K17), will not significantly decrease sprinkler performance.

A critical component in the design of a sprinkler system is the area of coverage (ft^2) of a sprinkler head. This area of coverage is influenced by the spacing of the individual sprinklers on the branch lines as well as the spacing between the branch lines. The maximum area of coverage (ft^2) as well as the maximum spacing between heads and branch lines is identified in NFPA 13 Chapter 8 and is based on the type of sprinkler head and the hazard occupancy. This standard also identifies the maximum and minimum distance from walls. It is critical to stay within these spacing requirements to help ensure uniform, overlapping water coverage on the floor. In general, sprinklers should be placed throughout the entire building and the heads should be uniformly spaced throughout the area.

To ensure their proper operation, sprinkler heads should not be painted. Doing so could seal the cap shut, preventing the flow of water, or increase the temperature at which they will activate. Maintaining clearance below the sprinkler head is also important to help ensure adequate water coverage. OSHA standards for general industry require a minimum clearance of 18 in. below sprinkler heads in areas where ordinary combustible materials are stored and 36 in. below sprinkler heads where flammable and combustible liquids are stored.

FIRE EXTINGUISHMENT

FIGURE 9.4
Sprinkler Head.

Fire Department Connections

In addition to the water supply coming into the building, a sprinkler system may also be equipped with an additional fire department connection. The purpose of this connection is to allow the fire department to add additional water sources to the sprinkler system, thus increasing the water flow or to provide water flow that may be missing to the sprinkler system. To ensure that fire department connections are in proper working order, the caps should be in place. Inspections should be conducted to ensure that foreign objects have not been placed in the piping, posing a potential clog in the sprinkler system water supply line. Figure 9.4 shows an example of a fire department connection.

Sprinkler-System Inspections

The inspection of a sprinkler system should consider three separate aspects of the system. The first aspect of the system is its mechanical components, the second is the adequacy of the system for the hazards being protected, and the third is the devices used by the fire department to support the system. A guide for the inspection of the mechanical components of the system is NFPA 25: Standard for the Inspection, Testing, and Maintenance of Water-Based Fire Protection Systems. This standard covers the inspection, testing, and maintenance of the sprinkler systems, but also the water supply, standpipes, fire pumps, and water spray systems (NFPA 2013c, 260). Sprinkler heads are one of the many items that must be inspected. For example, if the sprinklers heads were installed prior to 1920 they must be replaced and any heads older than fifty years must be tested every ten years. In addition to this standard, both OSHA and local building/fire codes exist for the inspection and testing of these systems. All three of these standards should be considered when developing a program for the inspection, testing, and maintenance program for a sprinkler system. The procedures included in this program must identify the specific sprinkler system components, criteria for evaluation, documentation, and the frequency.

The second aspect of the sprinkler system inspection is the adequacy of the system for the hazards protected. Any changes made to the building, processes, or material stored could influence the adequacy of the sprinkler system. For example, if a system was designed for a light hazard occupancy and it was changed so that now it was an ordinary hazard occupancy, the system would no longer be adequate. It is for this reason, NFPA 25 indicates the owner must evaluate the adequacy of the sprinkler system before changes are made to the building, processes, or materials stored (NFPA 2011b, 25–11).

The third aspect of the sprinkler system inspection is the inspection of the devices used by the fire department to support the sprinkler system. Devices that should be included in these inspections include fire department connections, standpipe systems yard hydrants, and PIVs. Standpipe systems and hydrants are discussed later in this chapter and the inspection and testing of the fire department connection should follow the requirements outlined in Chapter 13 of NFPA 25.

There are a variety of tests that must be completed on a sprinkler system to ensure that it is operating properly. For a complete listing of the testing requirements consult the current edition of NFPA 25 Standard for the Inspection, Testing, and Maintenance of Water-Based Fire Protection Systems. If the sprinkler system has remote supervision, it is important to notify them prior to any testing that may activate an alarm. What follows is a brief discussion on two of the most common tests completed on a sprinkler system, the inspectors test connection (trip test) and the 2-in. main drain test.

The trip test is performed to test the flow and pressure at the most remote location on the system and to test the water flow alarm. To run the trip test, the inspector opens the inspector's test valve and notes the water pressure reading. A slight decrease in pressure should be observed. The water flow alarm should be activated, and the fire department should receive a signal that the fire alarm has been activated if there is a remote connection. There should be a short delay between the time when the water begins to flow and the activation of the alarm, but the delay should not be any greater than ninety seconds. If the alarm activates immediately after the valve is opened, the valve may be susceptible to false alarms due to slight changes in water level in the riser. When the test has been completed, the inspector's valve is closed and the alarm system is reset.

The purpose of the main drain test, which is located on the riser, is to test the water supply to the sprinkler system. To run the 2-in. main drain test, the inspector closes the main supply valve below the water flow alarm check valve. The inspector opens the 2-in. drain fully and checks and records the water pressure. There should not be a significant drop in water pressure when the 2-in. drain is open. If there is, there may be a blockage in the supply line. The inspector then closes the 2-in. drain valve. The pressure should return to approximately normal. The inspector then restores the water supply to the sprinkler system (NFPA 2008, 16–210).

It is the responsibility of the building owner to ensure the sprinkler system is thoroughly inspected, tested, and maintained. Records of these activities must be in writing and available for review on site. These records should identify the date, name of person performing activities, procedures performed for inspection, maintenance and testing, and the findings. These records must be maintained for one year following the last inspection, test or maintenance activity (NFPA 2011b, 25–11).

FIRE HYDRANTS

There are two major types of fire hydrants used in the United States: the dry barrel and the wet barrel (see figure 9.5). The dry-barrel fire hydrant is used when temperatures get below freezing. The dry-barrel hydrant has its valve located below the frost line. A drain valve located at the base of the hydrant allows residual water to drain out. The wet, or "California," barrel is used in warmer climates. These hydrants have valves at each outlet.

FIGURE 9.5
Barrel Type Fire Hydrant.

Fire hydrants should be tested and maintained annually for proper operation and water flow. The hydrants should be opened, and water flow allowed for at least one minute. After completing the flow test the dry-barrel hydrants should be observed for proper drainage and the hydrant should be checked for water leaks. As part of the maintenance, the hydrant stems, caps, plugs, and threads should be lubricated and the hydrants themselves should have clear access. The hydrant outlets should be at least 18 in. above ground (NFPA 2013c, 239). There is no single standard for the placement of fire hydrants. However, a general rule for their placement is 250 ft. between hydrants in

FIRE EXTINGUISHMENT 219

compact mercantile and manufacturing areas, 400 ft. in residential areas, and at least 40 ft. from buildings (NFPA 2008, 15–32).

Standpipe and Hose Systems

Standpipe hose systems provide a fixed piping system that transports water from a reliable water supply to designated areas of a building where hoses can be used for manual firefighting. Standards for the design of standpipe systems are included in NFPA 14 "Standard for the Installation of Standpipe and Hose Systems" and their design is based in part on their expected use. Uses of s standpipe system can be broken down into full-scale firefighting, first aid firefighting, or both. These three uses correspond with the three design classes of standpipe hose systems as follows (NFPA 2008, 16–192):

1. Class I standpipe systems have 2.5-in. hose connections at designated locations in buildings for full-scale firefighting by fire department personnel. Class I systems are generally required in buildings more than three stories high. These systems reduce the need for firefighting personnel to lay extended lengths of hose.
2. Class II Standpipe Systems have 1.5-in. hose connections that are intended for first-aid firefighting by trained fire brigades before the fire department gets to the scene. These systems typically include a hose, nozzle, and a rack installed on each hose connection. Class II systems are common in large unsprinklered buildings.
3. Class III Standpipe Systems are provided for both first aid and full-scale firefighting. With these systems, both Class I and Class II connections are available, either using separate connections or the use of removable 2.5 in. to 1.5 in. adapters. Class III systems are common in buildings that require both Class I and II systems.

Because of the hazards posed by untrained personnel and occupants' handling hoses, the installation of Class II and III systems is declining. In addition to the above design classifications, standpipes are also subdivided into wet and dry systems. The wet systems always have water in the piping while the dry systems have piping filled with air.

There are two methods for determining the location of standpipe hose systems in a building. The actual-length method requires that standpipes be arranged in a manner so that 100-ft. lengths of hose with an average water discharge of 30 ft. will reach all sections of the building. The actual-length method is only allowed to be used for 1.5 in. hose connections. Today, the most common method for determining the location of standpipe hose systems for 2.5 in. connections on Class I and III systems is the exit location method. With this method, standpipe stations are located at exit stairs, horizontal exits, and in exit passageways according to the local building code. The assumption here is that since exits must be reasonably distributed, hoses will be adequately distributed. An advantage to this method is that it allows the fire department to attach their hoses to the stations before entering the fire area (NFPA 2008, 16–194).

The basic components of the standpipe system are the hose, the piping, and the hose case or station. The piping in a standpipe system is commonly made of steel. The water for the standpipes can come from a variety of sources with the most common being the municipal water supply when available. Some standpipe systems are "manual systems" meaning that the fire department must connect its water supply to the standpipe system, typically through connections located outside the building. These "manual systems" are not allowed in high rises or for use in Class II and III systems. The hoses used on standpipe systems must be lined and meet the standards for standpipe use as outlined in NFPA 14 "Standard for the Installation of Standpipe and Hose Systems." The preconnected hoses used on Class II and III systems typically have a maximum length of 100 ft. to minimize problems untrained people may have using them. For Class I systems, the 2.5 in. hose attached to the standpipe will be brought on site by the responding fire department (NFPA 2008, 16–199).

As with any water-based fire protection system, the maintenance, testing, and inspection of a standpipe system is critical. Chapter 6 of NFPA 25 "Standard for the Inspection, Testing, and Maintenance of Water-Based Fire Protection Systems" outlines the requirements for standpipe and hose systems.

CHAPTER QUESTIONS
1. Why is a two-inch main drain test completed on a wet-pipe sprinkler system?
2. What is the purpose of completing a trip test on a wet-pipe sprinkler system?

3. What is a Class B fire, and what type of extinguishing medium is appropriate for such fires?
4. Explain how a bulb-style sprinkler head activates.
5. In what type of situations are dry-pipe sprinkler systems the most suitable?
6. Describe the three different design classes of standpipe hose systems.
7. Supervisory initiating devices are used to supervise the operation of critical operating features of a sprinkler system. Give an example.
8. What are the two major types of fire hydrants used in the United States?
9. What is the purpose of the water-flow check valve in a sprinkler system?

REFERENCES

Automatic Sprinkler Company. (2014). A History of the Fire Sprinkler Systems. http://www.automaticfire.ie/history.htm, Accessed 1/18/2014.

Cote, Arthur, and Percy Bugbee. (2001). *Principles of Fire Protection*. Quincy, MA: NFPA.

Klinoff, Robert. (2012). *Introduction to Fire Protection*. Clifton Park, NY: Delmar Learning.

National Fire Protection Association (NFPA). (2008). *Fire Protection Handbook*, 20th ed. Quincy, MA: NFPA.

NFPA. (2010). *NFPA 11 Standard for Low, Medium and High Expansion Foam*. Quincy, MA: NFPA.

NFPA. (2011a). *NFPA 18: Standard of Wetting Agents*. Quincy, MA: NFPA.

NFPA. (2011b). *NFPA 25: Standard for the Inspection, Testing, and Maintenance of Water-Based Fire Protection Systems*, 2011 ed. Quincy, MA: NFPA.

NFPA. (2012). *NFPA 2001: Standard on Clean Agent Fire Extinguishing Systems*. Quincy, MA: NFPA.

NFPA. (2013a). *NFPA 10: Standards for Portable Fire Extinguishers*. Quincy, MA: NFPA.

NFPA. (2013b). *NFPA 13 Standard for the Installation of Sprinkler Systems*. Quincy, MA: NFPA.

NFPA. (2013c). *Fire and Life Safety Inspection Manual*, 9th ed. Quincy, MA: NFPA.

U.S. Department of Labor (USDOL). (2013). *Occupational Safety and Health Standards for General Industry, Subpart L: Fire Protection: Portable Fire Extinguishers, 29 C.F.R. § 1910.157*. Washington, DC: USDOL.

10

Fire Program Management

INTRODUCTION

In today's highly competitive marketplace, few companies can survive a major loss from a fire or other emergency incident. Over the years, it has become clear that organizations can substantially reduce these losses by developing and implementing effective fire risk management programs. Such programs focus on the identification, evaluation, and control of hazards to protect employees, the public, the environment, and company assets from loss due to fire or other emergency incidents. This process includes the following primary steps (Schneid and Collins 2002, 1–4):

1. Identification of the fire and emergency hazards or events that could lead to significant loss.
2. Quantification of the risk (probability of a fire or emergency event's occurrence and loss consequences).
3. Development and evaluation of alternative prevention and protection strategies to reduce fire and emergency risk.
4. Measurement to determine the effectiveness of the strategies in reducing the fire and emergency risk associated with the implemented alternatives.

A thorough risk assessment, using these four steps, will provide decision makers with a fundamental knowledge of the potential risks present at a facility and its survivability following a fire or other emergency. It is understood that every organizational structure and its culture are different; therefore, the actual responsibilities of a safety professional in this process will vary. These four areas will address some of the safety professional's more common responsibilities.

HAZARD IDENTIFICATION

Hazard identification is the process of recognizing hazards that can pose significant, undesirable losses. This identification of hazards should start during the preplanning stages with the evaluation of new materials, processes, and production modifications and should continue during the inspection of existing facilities. The safety professional will be responsible for providing the technical knowledge related to the fire codes and standards that may be used to identify actual or potential fire hazards. These fire codes could be the National Fire Protection Association (NFPA) codes or local building codes. Typically, the standards we are referring to will be primarily the Occupational Safety and Health Standards contained within 29 C.F.R. § 1910 and 1926. Other types of standards that may be relevant include applicable insurance standards, such as those of Factory Mutual. References, such as the NFPA's Industrial Fire Hazards Handbook, many NFPA codes, and insurance publications can be used to describe fire hazards in major industries and special process hazards based on current technology and past loss experience.

The safety professional must not only have knowledge of these standards and codes but also of how they should be applied in particular situations. For example, with new construction, the safety professional may want to evaluate the site under consideration for any of the following (Schneid and Collins 2002, 1–4):

- Exposure to natural disasters such as floods or exposures from adjacent facilities and processes;
- Availability of adequate water supply for fire protection;
- Acceptability of local emergency support forces, such as a fire department;
- Presence of impediments to site access such as traffic, terrain, or other buildings;
- Incorporation of building design factors such as fire-resistant building materials, fire areas and segregation for high-value or high-hazards areas, alarm and automatic suppression systems, and sufficient exits.

QUANTIFICATION OF RISK

Once we have identified a fire hazard, we need to assess the risk. The type and level of risk assessment conducted will depend on cost and time limitations, the significance of the decision, and the complexity of the problem. For example, routine code-compliance problems many times can be handled by making simple choices. However, complex problems involving new technologies or high hazard operations will require application of more detailed risk assessment methods.

Basic risk assessment involves identifying the probability and severity of potential fire losses. With both factors, an element of uncertainty must be recognized. Several documents have emerged to support the application of fire risk analysis. Examples of these documents include (NFPA 2008, 3–136):

- SFPE Engineering Guide to Application of Risk Assessment in Fire Protection Design
- NFPA 551: Guide for the Evaluation of Fire Risk Assessments
- ISO/TS 16732 Fire Safety Engineering—Guidance on Fire Risk Assessment

When addressing the severity of a fire risk, it is important to consider both direct and indirect loss potentials. Direct losses include damage to buildings, equipment, and contents. Indirect losses include business interruption, liability for injury or death, environmental contamination, and damage to company image. For most quantification studies, loss potential is expressed in equivalent monetary terms.

Once we have the probability and severity of the fire risk, this data is used to decide about the acceptability of the risk. If the risk is acceptable, no immediate action may be necessary. However, we may still need to monitor the risk for changes that could render it no longer acceptable. If the risk is unacceptable, then decisions must be made about how to deal with it.

FIRE PROTECTION AND PREVENTION STRATEGIES

Safety professionals are responsible for recommending the appropriate fire prevention and fire protection strategies to an organization. There are two broad categories for these strategies: engineering and administrative controls. Engineering controls are typically the priority for a safety and health professional because these controls can eliminate the risk of a fire or explosion.

Examples of engineering controls may include any of the following: substitution of a nonflammable liquid for a flammable liquid, pressure relief devices, explosion-proof electrical wiring, and ventilation in a spray painting booth. These engineering controls eliminate the fire or explosion risk by eliminating the ignition source, preventing excessive pressure buildup, or reducing the concentration of flammable gases or vapors to below their flammable limits. It is worth noting that some engineering controls do not eliminate the risk but minimize damage after a fire starts. The most common example is an automatic sprinkler system. Such a system does not prevent a fire but minimizes the damage caused by fire. Related to this responsibility of recommending appropriate strategies is the need for the safety professional to coordinate inspection, testing, and maintenance of these fire suppression systems once they have been installed.

The fourth option for risk control discussed earlier was the development, implementation, and monitoring of fire risk management programs. The safety professional will be involved in the development of many of these programs, all of which should be in writing and define the program's specific purpose or objective and scope.

Employee training is an important part of many fire risk management programs. Employee training might include training for the use of fire extinguishers, emergency-response, and participation in a fire brigade if one is used at the establishment. The U.S. Fire Administration, which is part of the Federal Emergency Management Agency (FEMA), is an excellent source for fire safety training.

The last responsibility of a safety professional within fire program management is the development and evaluation of fire response strategies. Considering the events of September 11, 2001, these responsibilities have become increasingly more important and have expanded beyond simple fire response plans to include emergency-response plans.

Once a risk management decision has been made that involves loss control improvements, it may be necessary to conduct a cost–benefit analysis. Determining the cost of fire loss control alternatives, which includes design, installation, system maintenance, and training expenses, is usually a very straightforward process. However, evaluating the benefit or the amount of risk reduction is a much more difficult task. Determining this benefit involves assessing the reduced probability of fire occurrence, reducing the severity of losses, or both. Estimating the fire risk is not an exact science and requires considerable judgment. However, valuable information may be available about loss experience provided by insurance reports and the National Fire Incident Reporting System. In addition, fire protection engineering analysis can also help in estimating fire risk after control strategies have been implemented. Finally, another tool available to the safety professional for risk assessment is computer modeling, which integrates deterministic fire hazard modeling, probabilistic modeling, and risk profile (loss-data) information (NFPA 2000, 10–153).

MEASUREMENT OF THE EFFECTIVENESS OF FIRE STRATEGIES

In essence, this last step reevaluates the probability of the fire risk after the strategies have been implemented. If they have been implemented successfully, the probability will be reduced to the level identified in the cost–benefit study discussed earlier. This measurement of effectiveness is important with all strategies but especially those directed at fire risk management programs.

It cannot be overemphasized that a written program is only as effective as its implementation. The safety professional's responsibility to measure and evaluate a program is critical to ensuring that the program is properly implemented and effective in reducing fire risk.

In conclusion, a properly conducted risk assessment will provide management with an idea of the relative degree of risk that a facility may be vulnerable to, the facility's level of preparedness to handle a fire or other emergency, and its ability to survive the emergency situation and remain in business.

EMERGENCY-RESPONSE PLANS

An emergency-response plan is essentially a standard operating procedure for handling emergency situations. Twenty years ago, emergency-response plans focused on fire and natural emergencies. Such a narrow scope is no longer acceptable with today's emergency risks evolving substantially. Emergencies now encompass areas such as cyberterrorism, product tampering, biological attacks, and ecological terrorism, threats that were virtually unheard of fifteen years ago. The results of these risks can be just as devastating to an organization as a fire or natural emergency, and the preventative and proactive measures taken are substantially different. Appropriate planning and preparedness before emergencies happen is essential to minimizing the risks and resulting damages (Schneid and Collins 2002, i–ii).

Reaction after the emergency must be coordinated to minimize damage as well as to avoid further damage to remaining assets. This planning and preparedness through the development of an emergency-response plan is an important responsibility of the safety professional. To assist the safety professional in meeting this responsibility this section does the following:

- Provides an overview of federal regulations that have emergency-response requirements;
- Discusses preplanning activities necessary for developing a response plan; and
- Provides suggestions for the elements to include in a response plan.

Federal Regulations Related to Emergency Response

On a federal level, there are three primary governmental agencies with regulatory responsibilities related to emergency response. For the public sector, the primary governmental agency is FEMA; in the private sector, the two primary agencies are the Occupational Safety and Health Administration (OSHA) and the U.S. Environmental Protection Agency (EPA) (Schneid and Collins 2002, 29–30). The following is a summary of some of the FEMA, OSHA, and EPA regulations impacting emergency response within an organization. This list is not meant to be all inclusive and is only a summary of the regulations. Safety professionals must review the full text of the regulations to determine their application and specific requirements. Compliance with the applicable regulations is essential to avoid potential penalties, as well as to ensure a complete and thorough emergency-response plan.

FEMA

FEMA's responsibilities for national preparedness to emergencies include development of federal program policy guidance and plans to ensure that governments at all levels can cope with and recover from emergencies. This includes the development of concepts, plans, and systems for the management of resources for a variety of national emergencies. FEMA also supports state and local governments in fulfilling their emergency-response responsibilities by providing funding, technical assistance, services, supplies, and equipment.

OSHA

OSHA has established standards for the development of emergency action plans and fire prevention plans, as well as standards for emergency responders. The following is a summary of some of these standards:

OSHA EMPLOYEE EMERGENCY PLANS AND FIRE PREVENTION PLANS: 29 C.F.R. § 1910.38

This OSHA standard details requirements for emergency action plans required by any other section of OSHA standards. Emergency and fire prevention plans must be in writing, kept in the workplace, and available to employees for review. However, OSHA does permit an employer with ten or fewer employees to communicate the plan orally to them.

At a minimum, OSHA standards require that the emergency action plan address the following (USDOL 2013a, 29 C.F.R. § 1910.38):

- Procedures for reporting fires and other emergencies
- Procedures for emergency escape and assignment of emergency escape routes
- Procedures to be followed by employees who remain to operate critical plant operations before they evacuate
- Procedures to account for all employees after emergency evacuation has been completed
- Procedures to be followed by employees performing rescue or medical duties
- Identification of the names or regular job titles of persons and departments to contact for further information or explanation of duties under the plan
- Procedures for maintaining an alarm system that has a distinctive signal for each purpose and that complies with 29 C.F.R. § 1910.165
- Procedures for designating and training enough persons to assist in the safe and orderly emergency evacuation of employees
- Procedures for reviewing the emergency-action plan with each employee covered by the plan when the plan is developed, the employee is initially assigned to a job, the employee's responsibilities under the plan change, or the plan is changed.

OSHA standards also provide specific requirements for fire prevention plans. The purpose of the fire prevention plan is to prevent a fire from occurring in a workplace. As with the OSHA emergency plan, the fire prevention plan must be in writing, kept in the workplace, and made available to employees for review. However, OSHA does allow an employer with ten or fewer employees to communicate the plan orally to them. At a minimum, OSHA requires the fire prevention plan to include the following (USDOL 2013b, 29 C.F.R. § 1910.39):

- A list of all major fire hazards, proper handling and storage procedures for hazardous materials, identification of potential ignition sources and procedures for their control, and the type of fire protection equipment necessary to control each major hazard;
- Procedures to control accumulations of flammable and combustible waste materials;
- Procedures for regular maintenance of safeguards installed on heat-producing equipment to prevent the accidental ignition of combustible materials;
- Identification of the name or job title of employees responsible for maintaining equipment to prevent or control sources of ignition or fires;
- Identification of the name or job title of employees responsible for the control of fuel source hazards;
- Procedures for informing employees upon initial assignment to a job of the fire hazards to which they are exposed and of those parts of the fire prevention plan that are necessary for their self-protection.

Two definitions are important to understanding OSHA's requirements for fire prevention plans. An incipient fire is a fire in the initial stage that can be extinguished by portable fire extinguishers. Interior structural fires are fires that are beyond the incipient stage and therefore cannot be controlled by portable fire extinguishers (USDOL 2013b, 29 C.F.R. § 1910.155). Because of the costs associated with implementing a fire brigade, many organizations may develop fire strategies to address only incipient-stage fires and allow the municipality to handle interior structural fires. If such a strategy is implemented, all employees who are involved in fighting incipient-stage fires must be trained annually in the general principles of using portable fire extinguishers as well as the hazards involved in incipient-stage firefighting. It is important to note that the designated employees must also receive hands-on training in the use of portable fire extinguishers on an annual basis.

OSHA HAZARDOUS WASTE OPERATIONS AND EMERGENCY RESPONSE: 29 C.F.R. § 1910.120

This OSHA standard was initially developed in response to the EPA's Superfund Amendments and Reauthorization Act (SARA) Title III Emergency Planning and Community Right-to-Know Act, but it tends to have a much broader application within industry. Even though an organization may not be covered under this act, it may have to comply with this regulation if its employees are required to respond to a hazardous material release or spill. Therefore, it is important to review all this OSHA regulation, as well as the Emergency Planning and Community Right to Know Act discussed later in this section, before developing an emergency-response plan. The following general requirements of this OSHA standard are specifically related to emergency response (USDOL 2013d, 29 C.F.R. § 1910.120):

- Development of a safety and health program designed to identify, evaluate, and control safety and health hazards and provide for emergency response.
- A preliminary evaluation of the site's hazards prior to entry by a trained person to identify potential site hazards and to aid in the selection of appropriate employee-protection methods.
- Training of employees before they can engage in hazardous waste operations or emergency response that could expose them to safety and health hazards. Persons completing specific training for hazardous waste operations shall be certified.
- Medical surveillance at least annually and at the end of employment for all employees exposed to any hazardous substance at or above established exposure levels or those who wear approved respirators for thirty days or more on site. Such surveillance will also be conducted if a worker is exposed to unexpected or emergency releases.
- An emergency-response plan to handle possible on-site emergencies prior to beginning hazardous waste operations. Such plans must address personnel roles; lines of authority, training, and communications; emergency recognition and prevention; safe places of refuge; site security; evacuation routes and procedures; emergency medical treatment; and emergency alerting.
- An off-site emergency-response plan to better coordinate emergency action by the local services and to implement appropriate control action.

OSHA CHEMICAL PROCESS SAFETY MANAGEMENT: 29 C.F.R. § 1910.119

This OSHA standard was developed in response to Clean Air Act amendments and applies to companies in certain chemical-processing and -handling fields. The primary purpose of the Chemical Process Safety Management standard is to eliminate or minimize the consequences of catastrophic releases of toxic, reactive, flammable, or explosive chemicals. This standard is like the EPA's Risk Management Programs for Chemical Accident Release Prevention Standard, which focuses primarily on community safety rather than employee safety. Provisions of the chemical process safety standards that are related to emergency management include the following (USDOL 2013c, 29 C.F.R. § 1910.119):

- Development of a written emergency action plan for the entire facility that must be kept at the workplace and made available for employee review;
- Review of the emergency action plan with employees when it is developed, when duties or responsibilities change, or when the plan is changed;
- Audits of the emergency action plan at least once every three years.

OSHA FIRE BRIGADES: 29 C.F.R. § 1910.156

One of the first OSHA requirements for the establishment of a fire brigade is the development of an organizational statement. This statement is basically a required policy statement that addresses the existence of the brigade, the number of members, a description of the fire brigade's function, and the type, amount, and frequency of training (USDOL 2013e, 29 C.F.R. § 1910.156). A fire brigade may perform the following duties during a fire emergency:

- Sound the alarm and aid in employee evacuation
- Shut off machinery and utilities and ensure that fire-suppression systems are working properly, and fire doors are closed
- Move motor vehicles away from plant
- Direct firefighters to the scene of a fire
- Stand by at sprinkler valves
- Extinguish the fire and maintain a fire watch after the fire is extinguished
- Assist with salvage operations and put fire protection equipment back into service

FIRE PROGRAM MANAGEMENT

OSHA standards require that all brigade members be physically capable of performing the duties assigned to them. The OSHA standard identifies no specific testing or examination to determine if a person is physically capable. For this reason, during the examination, all the functions and duties associated with being a member of the fire brigade, including requirements for personal protective equipment (PPE), must be made clear to the physician. It should also be noted that the requirement for the brigade members to be physically capable of performing their duties only applies to those members performing interior structural firefighting (USDOL 2003, 29 C.F.R. § 1910.156).

Another major requirement of OSHA standards on fire brigades is the training of members. OSHA standards require that training of brigade members be provided before they perform fire brigade activities and at least annually thereafter. In addition to the annual training required by OSHA, members conducting structural firefighting must receive education or training quarterly. The quality of training is expected to be comparable to that offered by state fire academies and schools mentioned in the OSHA standards. Training content must be commensurate with the duties performed by brigade members and must be hands-on. Examples of training content include principles and practices of firefighting and the handling of other emergencies. Members should get experience with all firefighting equipment. A review of emergency plans and procedures, equipment operation, special fire hazards, fire drills, and coordination and communication with community emergency-response agencies are also recommended training topics (USDOL 2013e, 29 C.F.R. § 1910.156).

OSHA requires that the employer shall maintain and inspect, at least annually, all firefighting equipment to assure its safe operation. Portable fire extinguishers and respirators must be inspected at least monthly. Firefighting equipment that is damaged or unserviceable must be removed from service and replaced (USDOL 2003, 29 C.F.R. § 1910.156).

OSHA standards also require that the employer shall provide to all employees involved in the fire brigade, at no cost, the necessary PPE. It is also the responsibility of the employer to assure that all fire brigade members wear the PPE when engaged in interior structural firefighting. Specifically, the employer shall ensure that the protective clothing protects the head, body, and extremities and consists of at least the following components (USDOL 2013e, 29 C.F.R. § 1910.156):

- Foot and leg protection: Such protection can be achieved either through fully extended boots that provide protection for the legs or through protective shoes or boots worn in combination with protective trousers, both of which must meet requirements for Class 75 footwear, be water resistant for at least 5 in. from the bottom, and equipped with slip-resistant outer soles.
- Body protection: Such protection shall be coordinated with foot and leg protection to ensure full body protection for the wearer. This can be achieved through the wearing of a fire-resistant coat in combination with fully extended boots or protective trousers. In either case, the fire-resistant coat or trousers must meet the NFPA 1971 "Standard on Protective Clothing for Structural Firefighting."
- Hand protection: Such protection shall consist of protective gloves or glove systems that provide protection against cuts, punctures, and heat penetration. All gloves or glove systems shall be tested in accordance with the test method contained in the National Institute for Occupational Safety and Health's "The Development of Criteria for Firefighters Gloves."
- Head, eye, and face protection: Head protection shall consist of a protective head device with earflaps and chin straps that meet the performance construction testing requirements of the National Fire Safety and Research Office of the National Fire Prevention and Control Administration. Protective eye and face devices shall be used by fire brigade members when performing operations in which the hazards of flying or falling materials that may cause eye and face injuries are present. Protective eye and face devices provided as accessories to protective head devices are permitted when such devices meet the requirements of 29 C.F.R. § 1910.133.
- Respiratory protection: The employer shall provide at no cost to employees and ensure the use of respirators that comply with 29 C.F.R. § 1910.144. An approved, self-contained breathing apparatus with a full face piece or with approved helmet or hood configurations must be worn by fire brigade members involved in interior structural firefighting or when in confined spaces where toxic products of combustion or oxygen deficiency is present.

EPA
Superfund Amendments and Reauthorization Act Title III
Emergency Planning and Community Right-to-Know Act of 1986

The Emergency Planning and Community Right-to-Know Act was created to help communities plan for emergencies involving hazardous substances

(USEPA 2013b). The Act established planning and reporting requirements for hazardous and toxic chemicals. The Community Right-to-Know provisions help increase the public's knowledge and access to information on chemicals at individual facilities, their uses, and releases into the environment (USEPA 2013b). At the state level, this act requires all states to establish a state emergency response commission (SERC), which approves districts or areas where local emergency planning communities (LEPCs) will be formed. The LEPCs must develop a local emergency-response plan based on an evaluation of available resources for preparing for and responding to potential hazardous material incidents. The plan should include the identification of facilities, transportation routes, response, and notification procedures, evacuation plans, training programs, and designation of a community coordinator (Vulpitta and Larson 2011, 8–9).

The following general requirements of the Emergency Planning and Community Right-to-Know Act apply specifically to emergency response (USEPA 2004b):

- Both a written safety and health program and an emergency-response plan must be in place.
- A facility emergency coordinator must be appointed to work with the LEPC to ensure that the emergency-response plan is compatible with and integrated into a community emergency-response plan.
- Evaluation of the site characteristics by a trained person must include a list of extremely hazardous substances around which planning is carried out and provide a hazardous-chemical inventory to the state and local fire department. Facilities required to have a material safety data sheet (MSDS) available under OSHA must submit the MSDS or a list of MSDS hazardous materials to the state commission, local planning committee, and local fire department.
- Training and certification, as well as annual medical surveillance, must be provided for all involved personnel.

Another important aspect of this act is the creation of the EPA's Superfund Emergency Response program. This program provides quick response to the release, or threatened release, of hazardous substances wherever and whenever they occur. This is one of two major components of the Superfund

Response program designed to protect human health and the environment from the multiple threats posed by hazardous substances. The program has three main priorities (USEPA 2004b):

- Readiness to respond twenty-four hours per day to a release incident
- Response with whatever resources are required to eliminate immediate dangers to the public and the environment
- Community relations that can be used to inform the public about a release, response activities, and the substances involved

RESOURCE CONSERVATION AND RECOVERY ACT

The Resource Conservation and Recovery Act (RCRA) is primarily an environmental regulation that covers facilities having hazardous waste; however, this act does have requirements for written contingency plans that a safety professional may need to comply with. This Act along with 29 CFR 1910.120, OSHA's Hazardous Waste Operations and Emergency Response (HAZWOPR) standards establish requirements for employers engaged in hazardous waste handling, treatment, and disposal. As a rule, most planning and training developed to comply with the OSHA 29 CFR 1910.120 standards will be acceptable under RCRA. However, in some areas, the requirements are more stringent under RCRA, therefore, an understanding of both the OSHA standards and the EPA standards is a must. Examples of requirements under Section 264.16 of the EPA Standards for Owners and Operators of Hazardous Waste Treatment, Storage, and Disposal Facilities include (USEPA 2013a, 40 C.F.R. § 264.16):

- Personnel training related to emergency management must be provided by a person trained in hazardous waste management and must include instruction in hazardous waste procedures and contingency plan implementation. All personnel must be able to respond effectively to emergencies that require them to be familiar with emergency procedures, equipment, and other systems.
- Emergency equipment related to internal communications or alarm systems must be provided. This would include a device for summoning community emergency assistance, as well as extinguishment equipment to include, at a minimum, an adequate supply of water for firefighting, portable fire

extinguishers, fire control equipment, spill control equipment, and decontamination equipment.
- Planning with local responding agencies must be completed to ensure that all agencies are familiar with the facility layout, the properties of the hazardous waste handled and associated hazards, places where people will be working, and facility access and evacuation routes. This planning should also extend to state emergency-response teams, contractors, and equipment suppliers, as well as local hospitals.
- Contingency plans must be developed for each facility to minimize the hazards to human health or the environment from fires, explosions, or unplanned releases of hazardous wastes. These contingency plans should describe the actions plant personnel will take in response to emergencies. The plan should include a list of the names, addresses, and phone numbers of all qualified emergency coordinators, as well as a list of all emergency equipment at the facility, its location, a physical description of each item, and a brief outline of the equipment's capabilities.
- Emergency procedures must be developed, outlining the responsibilities of the emergency coordinator. Examples of responsibilities include activating the facility alarm; notifying appropriate state or local agencies; identifying the character, source, amount, and extent of the release of hazardous waste; assessing the health and environmental hazards from the release; taking all reasonable measures to control and stop the emergency; monitoring for leaks, pressure buildup, and other problems; and providing for post emergency treatment, storage, or disposal of recovered waste.

NFPA 1600: RECOMMENDED PRACTICE FOR EMERGENCY MANAGEMENT

This NFPA standard provides recommendations for the minimum criteria for emergency management planning for both private and public organizations. Specifically, this standard recommends the following planning areas (NFPA 2013, 6):

- Prevention: activities designed to prevent the event from occurring
- Mitigation: the types of hazards that make the organization vulnerable to emergencies and steps that are taken to prevent or reduce the effects of the emergency

- Preparedness: the activities, programs, and systems developed prior to an emergency that are used to support the facility's response program
- Response: the activities that will help to stabilize and control the emergency
- Continuity: the activities undertaken to keep the organization's operations running
- Recovery: the activities that will help to return the facility to a functional status

The National Response Team (NRT), which is chaired by the EPA, is made up of sixteen federal agencies, each with responsibilities and expertise in various aspects of emergency response to pollution incidents. Prior to an incident, the

NRT provides policy guidance and assistance. During an incident, the NRT provides technical advice and access to resources and equipment from its member agencies. The NRT also helps the private sector with prevention, preparedness, and response efforts by encouraging innovation and collaboration to increase the effectiveness and reduce the cost related to compliance with response regulations. This interagency planning and coordination framework is replicated at the regional, subregional, and local levels. There are thirteen regional response teams, one for each of the ten federal regions, to help ensure that appropriate federal and state assistance will reach an incident scene quickly and efficiently when needed (USNRT 2013).

In June 1996, the NRT published the Integrated Contingency Plan (ICP). Rather than being a regulatory initiative, the ICP document provides guidance. It presents a sample contingency-plan outline that addresses the requirements of the following federal regulations: the Clean Water Act; the EPA's Risk Management Program Regulation, Oil Pollution Prevention Regulation, and RCRA Contingency Planning Requirements; and OSHA's Emergency-Action Plan Regulation, Process Safety Management Standards, and Hazardous Waste Operations and Emergency-Response Regulation. The ICP has three primary objectives (USEPA 1996, 5–6):

- Provide a mechanism for consolidating multiple facility response plans into one plan that can be used during an emergency
- Improve coordination of planning and response activities within the facility and with public and commercial responders

- Minimize duplication of effort and unnecessary paperwork burdens and simplify plan development and maintenance

The ICP sample format is based on the Incident Command System (ICS), which allows the plan to dovetail with established response-management practices. The NRT intends to continue promoting the use of the ICP guidance by regulated industries and encourages federal and state agencies to rely on the ICP guidance when developing future regulations (USEPA 1996, 7).

Planning an Emergency-Response Strategy

The development of an effective emergency response starts with planning. Facility management is responsible for seeing that an emergency-response program is implemented and that it is frequently evaluated and updated (Vulpitta and Larson 2011, 5). The input and support of all employees, as well as the community, must be obtained to ensure an effective program. Therefore, an important part of the planning process starts with developing an emergency-response committee. This committee will be responsible for coordinating the emergency-response plan's development and its implementation, including training, emergency drills, equipment, and plan evaluation. Members of this committee should be involved with responding to the emergency and, at a minimum, include membership from community emergency-response agencies and facility personnel, including representatives from management, maintenance, engineering, transportation, safety, and human resources. The involvement of community emergency-response agencies is an absolute necessity in maintaining ongoing relationships and communications between the facility and the local community.

One of the first activities of the emergency-response committee is to identify the potential risks, assess their viability, evaluate the probability of their occurring, and appraise the potential damage. Such a survey will focus on the following:

- Facility operations, processes, and raw materials. The emergency-response planning committee must assess facility operations, processes, and materials to determine if they pose substantial potential emergency risks or if the facility is located near other operations or facilities posing such risks (Stringfield 2000, 18–20). Although the risks from emergencies will vary from operation to operation, risks are often classified as man-made or natural. Natural risks are inherent but are often overlooked when assessing potential risks. The emergency-response planning committee needs to determine if the facility is located near any natural risks, such as an earthquake fault, volcano, hurricane zone, heavy snow area, flood zone, or forest-fire area. Man-made risks include those associated with fire and explosions, hazardous-material incidents, aircraft crashes, shipwrecks, and railroad and truck accidents. The committee must also consider emerging risks such as workplace violence, terrorism, bioterrorism, and cyberterrorism. Workplace violence, for example, has become the leading cause of work-related deaths in some retail industries, opening an expanding area of potential liability against those employers who fail to safeguard their workers.

- Prevention and preparedness level of the facility. The emergency-response planning committee must assess the prevention activities, procedures, programs, and plans that currently exist to prevent the risk of emergencies (Stringfield 2000, 26–27). This includes having knowledge of the fire protection systems in a building, of how these systems operate, and of the actions necessary to supplement these systems. Preincident planning must determine not only whether these systems exist but whether they are adequate for the building occupancy. The committee will also have to evaluate the capability of the facility to respond to an emergency. Life safety considerations should be the top priority for preincident planning. A facility that has a good training program, instructing employees and conducting drills in emergency procedures, will mitigate life safety risks.

Based on this, the committee will need to review facility information regarding drawings, process flow diagrams, phone lists, rosters, material safety datasheets, evacuation plans, training records, community emergency plans, and so forth. This type of information should be available not only to emergency responders within the facility but also to community emergency-response agencies that may be involved.

It is critical that the emergency-response committee properly identify and assess each facility on an individual basis to identify the actual or potential risk of emergencies. Based on this assessment, a customized emergency-response program should be developed and implemented to address the specific risks for each facility.

Another important decision in planning a response strategy is determining the extent to which the organization is willing to commit time and money to developing and implementing an emergency-response plan. The emergency-response committee must know the time and resource commitments at their disposal so that it can plan accordingly. Examples of costs include PPE, fire and other emergency equipment and supplies, medical costs, and training costs. This training will include conducting drills regularly for all employees and community emergency-response agencies. This is both expensive and very time-consuming. Part of this commitment also includes taking on the increased risk of injury to employees involved in emergency-response activities. In some organizations, where this increased risk of injury may be unacceptable, employees' only emergency-response strategies may be simply to evacuate.

Once the emergency-response committee has completed the risk assessment and determined the organization's level of support, it is then time to evaluate emergency support at the local, regional, and state levels.

An evaluation of the support and capability of local community emergency-response agencies may take into account whether an agency is a paid or volunteer organization, the type of equipment it has available, its expected response times, and the training of its members. Since many facilities have hazards that community emergency-response agencies may not recognize, a list of all the hazards and emergency risks that the responders may meet is crucial for planning purposes. The emergency-response team may also want to draw on the expertise of the LEPC to assess the capabilities of community response agencies (Schneid and Collins 2002, 39–41).

Part of this planning step may also include developing formal agreements with nearby organizations, such as restaurants, hospitals, and schools, to provide food, shelter, and medical assistance in the event of a major emergency. The emergency-response plan will want to specifically identify the organization, a contact name, the phone number and address, and the travel distance. All of this planning with the community will allow the emergency-response committee to develop an inventory of key resources and equipment that will be needed during an emergency and to show the amount of equipment and number of resources that may also be available within the community. This might include communications equipment, food and drink, lighting, medical supplies, material-handling equipment, power tools, spill cleanup materials, rescue equipment, and traffic-control supplies.

The last aspect of the planning process that the emergency-response committee may want to consider is communication with the community and media during an emergency. At the very least, this committee needs to determine the best way to communicate with the community, employees, local and state emergency-response agencies, federal and state regulatory agencies, and the media. All employees involved will want to know what has happened, as will the local community, and it is critical that the facility determine the best way to communicate this information accurately.

In conclusion, when preincident planning by the emergency-response committee is successful, everyone benefits. Both internal and external emergency-response agencies can more effectively manage their response activities, which results in a safer response and minimizes property loss.

This preincident planning is a continuous process, and all facility changes must be evaluated to determine their effect on emergency risks and responses. Keeping up with changes in a facility requires a high level of commitment, communication, and cooperation among all organizations involved in emergency response.

Developing the Written Emergency-Response Plan

After completing the risk assessment and reviewing applicable regulations, the emergency-response committee is now ready to develop the written draft emergency-response plan. Various model emergency-response plans are available for the committee to use as a starting point; however, it is recommended that companies use the ICP as a benchmark. This plan was developed collaboratively by the EPA, the U.S. Coast Guard, the Minerals Management Service, the Research and Special Programs Administration, and OSHA. The ICP is intended to provide guidance to facility management in the preparation of a single emergency-response plan that will eliminate the need for the multiple emergency-response plans that facilities may have prepared in the past to comply with various regulations (USEPA 1998, 2). The structure of the ICP guidance is based on the National Interagency Incident Management System (NIIMS) ICS, a nationally recognized system that allows for effective interaction among response personnel and that is currently being used by numerous federal, state, and local organizations (USEPA 2004d, 8). The

ICP format is organized into three main sections: an introductory section, a core plan, and a series of supporting annexes. The core plan is intended to reflect the essential steps necessary to initiate, conduct, and terminate an emergency-response action: recognition, notification, and initial response, including assessment, mobilization, and implementation. The core plan should reflect a hierarchy of emergency-response levels. The use of response levels by an organization allows response personnel to match the emergency and its potential impacts with the appropriate resources, personnel, and emergency-response actions. The consideration and development of response levels should be consistent with similar efforts that may have been taken by the LEPC or mutual-aid organization. The concept of these response levels should be considered in developing checklists or flowcharts designed to serve as the basis for the core plan. The annexes are designed to provide key supporting information for conducting an emergency-response under the core plan, as well as document compliance with regulatory requirements not addressed elsewhere in the ICP. Annexes 1 to 3 are not meant to duplicate information that is already contained in the core plan but to provide more detailed, supporting information on response actions that are specific to the hazards encountered. The ICP encourages the use of checklists or flowcharts wherever possible to capture these emergency-response actions in a concise, easy-to-understand manner. Annexes 4 through 8 are dedicated to providing information that is noncritical at the time of a response, such as cross-references to demonstrate regulatory compliance and background planning information (USEPA 2004d, 7–12).

In conclusion, an emergency-response plan is the foundation for operations during an emergency. A good plan provides important information that assists the incident commander in implementing appropriate strategies and tactics for managing the incident. Use of the ICP model emergency plan is not required to comply with federal regulatory requirements, and facilities can continue to maintain multiple plans for compliance if they so desire. However, an emergency-response plan prepared in the ICP format is the federally preferred method of response planning. The ICP model will also minimize duplication in the preparation and use of emergency-response plans at the same facility and will improve economic efficiency for both the regulated and regulating communities. Facility expenditures for the preparation, maintenance, submission, and update of a single plan should be much lower than for multiple plans.

The use of a single emergency-response plan per facility will eliminate confusion on the part of facility first responders, who must often decide which plan is applicable to a particular emergency. Use of a single, integrated plan should also improve coordination between facility response personnel and local, state, and federal emergency-response personnel (USEPA 2004d, 4–7).

Emergency Medical Care

Emergency-response plans must include provisions for emergency medical care. During most emergencies, injuries and illnesses that require emergency medical care can reasonably be expected to occur; response time is a crucial factor in minimizing these injuries and illnesses. Emergency-response planning would start with a survey of local medical facilities regarding medical capabilities and response times; then, arrangements would be made to handle emergencies based on facilities' capabilities. Ambulance services need to be familiar with a facility's location and access routes in advance (Vulpitta and Larson 2011, 14).

OSHA standard 29 C.F.R. § 1910.151 sets forth three general requirements regarding emergency care:

- The employer should ensure the ready availability of medical personnel for advice and consultation on matters of plant health.
- Where the eyes or body of any person may be exposed to injurious corrosive materials, suitable facilities for quick drenching or flushing of the eyes and body should be provided within the work area for immediate emergency use.
- In the absence of an infirmary, clinic, or hospital in near proximity to the workplace, which is used for the treatment of all injured employees, a person or persons should be adequately trained to render first aid. Adequate first aid supplies should be readily available.

Three key terms in the preceding standard are near proximity, adequately trained, and adequate first aid supplies. OSHA Letters of Interpretation addresses each of these as follows (USDOL 2002):

- Near proximity: In areas where accidents resulting in suffocation, severe bleeding, or other life-threatening or permanently disabling injuries or illnesses can be expected, a three- to four-minute response time, from time of injury to time of administering first aid, is required. In other circumstances (i.e., where a life-threatening or permanently disabling injury is unlikely), a longer response time, such as fifteen minutes, is acceptable.

- Adequate training: OSHA's "Guidelines for Basic First Aid Training Programs" recommends the following topics for first aid training: teaching methods, responding to a health emergency, surveying the scene, basic adult cardiopulmonary resuscitation (CPR), basic first aid intervention, universal precautions, first aid supplies, trainee assessments, and program update. Refresher training to maintain the CPR and first aid certificates must be conducted annually for CPR and every three years for first aid. An OSHA directive also indicates that persons with a current training certificate in the

American Red Cross Basic, Standard or Advanced First Aid Course shall be considered adequately trained to render first aid (USDOL 1976).

- Adequate first aid supplies: OSHA has often referred employers to ANSI Z308.1, "Minimum Requirements for Workplace First Aid Kits," for guidance in the minimum requirements for first aid kits; however, it should be noted that OSHA has not adopted this ANSI standard, which is therefore not mandatory. OSHA cannot provide a list of exact requirements for first aid supplies that will apply to every workplace. Therefore, each workplace must be evaluated on a case-by-case basis, considering the types of injuries and illnesses that are likely to occur. To assist in this evaluation, a safety professional may want to consult with the local fire and rescue department or a licensed health-care provider.

Another aspect of medical services related to emergency-response teams is preplacement medical examinations and medical surveillance. Specifically, before assigning personnel to these teams, the employer must ensure that employees are physically capable of performing the duties that may be assigned to them.

Training

Training is absolutely critical to the effectiveness of an emergency plan. When addressing training, OSHA uses two distinct terms: education and training. OSHA defines education as the process of imparting knowledge or skill through systematic instruction. Training, on the other hand, is the process of making trainees proficient through instruction and hands-on practice in the operation of equipment that is to be used in the performance of assigned duties. Before implementing an emergency-action plan, enough people must be trained to assist in administrating the critical elements of the plan. All other employees need to be trained in how to respond to each type of emergency. In addition to the specialized training for key people, other employees should be trained on the following (Vulpitta and Larson 2011, 11):

1. Evacuation plans and routes;
2. Shutdown procedure responses for emergencies;
3. Alarm identification; and
4. Preferred means of reporting emergencies.

Emergency-response training needs to be completed initially when the plan is developed and for all new employees so that people know their roles and responsibilities in an emergency. Agencies, organizations, and individuals likely to be active in a response effort must learn the division of labor and tasks to be undertaken (Phillips, Neal and Webb 2012, 146).

A particularly important part of emergency-response training is the use of emergency drills, which also serve as a tool to measure planning effectiveness. Exercises and drills are used to determine the effectiveness of the emergency program, the people, the policies and implementing procedures, the organization, the equipment and the mutual-aid agreements (Vulpitta and Larson 2011, 89).

Personal Protective Equipment

PPE is a consideration in fire program management, especially in emergency-response planning. Effective personal protection is essential for anyone who may be exposed to potentially hazardous substances during an emergency incident (Vulpitta and Larson 2011, 12). In emergency situations, employees can be exposed to a variety of hazardous circumstances, including any of the following:

- Fire, smoke, and electrical hazards;
- Chemical splashes or contact with toxic materials;

- Explosion hazards such as flying particles; and
- Unknown atmospheres that may have inadequate oxygen levels or contain toxic gases, vapors, and mists.

It is extremely important that employees be adequately protected in these situations, and these employees must also receive medical clearance to wear PPE.

Media Control

As mentioned in the ICP model emergency-response plan, it is critical to plan for the communication of information to the outside world through the media during an emergency. All areas of possible information flow must be addressed to ensure that the organization is putting the best "spin" on an already bad situation. Some suggested measures for addressing the media as part of the overall emergency-response plan include the following (Schneid and Collins 2002, 84–86):

- Designate a safe area that is away from the flow of emergency traffic for all media vehicles and for all media communications.
- Maintain security in the designated media area and prohibit media representatives from accessing the emergency area.
- Identify a specific member of management to be the spokesperson for the company and allow no other employees to talk to the media. The person selected should have experience in public relations and dealing with the media.
- Send the media to appropriate areas to acquire video footage when necessary.
- Provide informational packets with company information to the media.
- Have legal counsel review all information prior to its presentation to the media and keep questions from the media to a minimum.

If the emergency involves environmental spills, a possible source of help is the EPA's Community Relations Program. This program has three primary objectives (USEPA 2013c):

- To provide information to the community on the health and environmental effects of the release and the response actions under consideration;

- To encourage citizens to provide information about the site and its surrounding areas and to express any concerns about the actions being undertaken; and
- To include citizen comments and concerns in the decision-making process at an emergency-response site.

An official EPA spokesperson is appointed for each emergency-response action to keep the public informed and to respond to any questions. Meeting with citizens in the community, responding to inquiries from the media, and providing local officials with site-status information are some of the activities that the EPA is likely to undertake.

Recovery After an Emergency

An effective plan for the recovery of your organization after an emergency incident must be a part of an emergency-response plan. Business recovery starts immediately after the emergency phase of the incident is over. Business recovery has expanded its focus over the past fifteen years to include recovery of the entire business, including all technology, people, and processes, and is intended to ensure the continuity of a business after an incident occurs. An emergency incident can cause production loss, invoices not going out, wages not being paid, suppliers not being paid, orders missed or lost, and loss of customer confidence. Therefore, failure to have a formal business recovery plan puts the organization at much higher risk of not recovering and consequently going out of business (Stringfield 2000, 43).

The first step in developing a business continuity plan is to perform a risk assessment. The risk assessment will assist in identifying the events that have a greater probability of occurring along with the potential financial impact they can have upon the organization. The risk assessment identifies the potential loss scenarios one should address in the business continuity plan (FEMA 2013).

The next step is to complete a business-impact assessment to quantify the operational and financial impact of an inoperable business function on the organization's ability to conduct its critical business processes. Part of this assessment process includes identifying the operational and financial impacts resulting from the disruption of business functions and processes. Impacts to consider include lost sales, delayed income, and increased expenses (FEMA 2013).

FIRE PROGRAM MANAGEMENT

The business continuity plan is developed using the results from the business-impact analysis. Planning includes identifying gaps in preparedness activities from where they currently are to where they should be to effectively recover from a disaster. Recovery strategy options should be examined, and best practices identified. Recovery strategies are planned, coordinated with outside agencies, and documented. To ensure the plan will function accordingly, plan testing and maintenance activities should be an ongoing part of the business continuity program (FEMA 2013).

INVESTIGATION OF EMERGENCY INCIDENTS

To prevent future loss incidents, it is critical to investigate every incident that results in, or could have reasonably resulted in, a major release, fire, or explosion accident. One of the most important reasons for investigating loss incidents is to determine the causes so that preventive measures can be taken to prevent future occurrences.

As a rule, incident investigations should be initiated as soon as possible after the emergency phase of the incident. All investigators should have plant and process knowledge, as well as training in loss incident investigation. The organization must develop a formal, written, loss incident investigation form that, at a minimum, includes the following information: date of incident, date of investigation, name(s) of investigators, description of incident, causal factors, and resulting recommendations. These loss incident reports should be reviewed by appropriate personnel within the facility.

Fire investigations have four primary objectives: review of structural damage, fire-ignition sequence, fire development, and fire casualties. The actual investigation of the fire typically will start with a review of the exterior of the structure to document any fire damage or other damage that may have been caused by firefighting. Next, the investigator will review the interior of the structure to document any fire damage or other damage caused by firefighting.

Once the review of the exterior and interior is complete, the investigator will attempt to reconstruct the fire. The focus here is to try to identify the fire's progression, starting with its origin and the location of combustibles or other fuels. The investigator will examine the ignition sequence by identifying the heat source, combustible material, and the act that brought the heat source in contact with the combustible material. In evaluating the progression of the fire, the contents of the building, such as ordinary combustibles, are examined because of their influence on both fire growth and smoke and gases produced. Structural features of the building, such as compartmentation, fire walls, interior finish, concealed spaces, and so forth, will also be investigated to determine their effect on fire development (Cote and Bugbee 2001, 114–16).

After the fire has been reconstructed, the investigator will then focus on interviewing witnesses and firefighters at the scene. The investigator seeks witnesses' observations about conditions, such as flame spread, heat generated, smoke volume and color, property damage, security issues, and unusual activities. Based on the reconstruction of the fire and the interviews, the investigator may need to conduct laboratory tests on the materials to evaluate such things as burn characteristics of materials, composition of residues, or possible failures of mechanical equipment like heaters. If failure of equipment is a possibility, the investigator may need to review the service and maintenance records of the suspect equipment.

Once all this information is collected, the investigator will analyze it to determine incident causes. Here, the focus should be on both surface and underlying causes. Surface causes are typically related to unsafe acts and or conditions, while underlying causes are related to management system deficiencies that allowed the unsafe acts and conditions to exist. All causes must be identified, and appropriate recommendations developed if we hope to prevent future incidents. It should also be pointed out that we need to investigate incidents so that accurate information and documentation can be provided for insurance and legal reasons. Recommendations will address the causes identified, focusing on engineering controls first, then on administrative controls. All good investigations include a follow-up on recommendations to make sure that they have been effective in reducing or eliminating the causes of fire. It should be noted that if fire casualties occur or if arson is suspected, fires are typically investigated by the local fire marshal or police department (Cote and Bugbee 2001, 116).

The remainder of this chapter will cover other common safety professional responsibilities in fire program management, specifically, the maintenance of fire protection systems, fire inspections, and hot work permit programs.

Maintenance of Fire Protection Systems

As discussed in chapter 9, fire protection systems play a vital role in the reduction of fire losses in industrial facilities. Unfortunately, large fire losses stemming from malfunctioning fire protection equipment continue to occur for a variety of reasons, such as closure of sprinkler water control valves and inoperative fire pumps, fire detectors, alarm systems, and sprinkler heads. The maintenance requirements for fire protection systems have been addressed in earlier chapters; therefore, this chapter focuses on the development of procedures to follow when any fire system is inoperable as a result of maintenance, renovation, equipment failure, or an emergency incident. Effective procedures during fire system shutdown are critical in reducing the possibility of fire losses. These procedures should be in writing and, at a minimum, do the following:

- Assign responsibility and authority to control the shutdown of fire systems to one individual, such as a maintenance supervisor or a plant engineer, who must be notified immediately of any shutdowns to fire safety systems
- Provide training and education for all those involved in the shutdown procedures
- Limit the area affected by the shutdown
- Supplement manual fire protection, such as portable fire extinguishers in the shutdown area, with automatic fire suppression systems
- Avoid hot work and other spark-producing operations within the shutdown area
- Verify, by testing, that the fire protection systems are operational
- Shut down or isolate any hazardous production operation in the area where the fire protection system is shut down
- "Lock out" or "tag out" impaired fire protection systems
- Notify the public fire department, central station, or alarm company of the fire system shutdown and the extent to which the system is out of service
- Complete work in a timely manner to limit the duration of the shutdown
- Restore any fire protection equipment, alarms, or detection devices that have been disconnected
- Verify, by testing, that the fire protection systems are operational, that portable extinguishers are in place and fully charged, and that hose lines have been returned
- Notify plant personnel and the public fire department, central station, or alarm company that fire protection systems have been restored

Fire Inspections

Just as important as the maintenance of fire protection systems is the implementation of an inspection program for the fire safety of plant operations. These fire inspections can be part of a larger safety inspection or specific to fire hazards or fire protection systems. Once an adequate program has been established and implemented, inspections must be conducted by individuals knowledgeable in the operation and testing of the equipment and trained in completing the inspections. The inspections should be completed, at a minimum, monthly.

To simplify facility safety inspections, written forms specific to the occupancy and equipment are recommended. These forms should be comprehensive enough that no element of prevention or protection is overlooked. It is critical that the inspections identify all deficiencies, provide appropriate recommendations, and assign responsibilities for correction and follow-up activities. Possible areas to address in a fire safety inspection include the following (Ladwig 1991, 338–39):

- Potential fuel sources, such as those associated with poor housekeeping, improper storage of ordinary combustibles, and improper use and storage of flammable or combustible liquids
- Potential ignition sources, such as those associated with smoking, electrical deficiencies, static charges, and heating appliances
- Life safety issues, such as blocked exits, aisle ways, exit signs, and fire doors
- Compliance with procedures, such as hot work procedures, shutdown of fire systems, maintenance of fire systems, and emergency evacuation
- Fire protection systems, such as alarm and detection systems, sprinkler systems, and portable fire extinguishers

Examples of fire inspection forms are available from a variety of sources, such as OSHA, FEMA, NFPA, the local fire department, and the company's property insurance carrier. The OSHA fire-inspection audit form is available on OSHA's website, where it is referred to as the Fire Safety Advisor. This Web-based fire safety program can be downloaded and used by the public to complete a fire safety inspection. NFPA publishes the NFPA Inspection Manual, which concentrates on the identification of unsafe conditions or deficiencies in buildings related to current fire codes and standards. This

Property Name: _____ Owner: _____
Address: _____ Phone Number: _____

OCCUPANCY
Change from Last Inspection: Yes ☐ No ☐
Occupant Load:
Egress Capacity: Any Renovations: Yes ☐ No ☐
General Industrial ☐   Special-Purpose Industrial ☐   High Hazard ☐
High Rise: Yes ☐ No ☐   Windowless: Yes ☐ No ☐   Underground: Yes ☐ No ☐

BUILDING SERVICES
Electricity ☐ Gas ☐ Water ☐ Other ☐
Are Utilities in Good Working Order: Yes ☐ No ☐
Elevators: Yes ☐ No ☐
Fire-Service Control: Yes ☐ No ☐
Elevator Recall: Yes ☐ No ☐
Heat Type: Gas ☐ Oil ☐ Electric ☐ Coal ☐ Other ☐
In Good Working Order: Yes ☐ No ☐
Emergency Generator: Yes ☐ No ☐
Size:
Last Date Tested:
Date of Last Full-Load Test:
In Automatic Position: Yes ☐ No ☐
Fire Pump: Yes ☐ No ☐ GPM:
Suction Pressure: System Pressure:
Date Last Tested:
Date of Last Flow Test:
In Automatic Position: Yes ☐ No ☐
Jockey Pump: Yes ☐ No ☐

EMERGENCY LIGHTS
Operable: Yes ☐ No ☐
Tested Monthly: Yes ☐ No ☐
Properly Illuminate Egress Paths: Yes ☐ No ☐
In Good Condition: Yes ☐ No ☐

EXIT SIGNS
Illuminated: Internally ☐ Externally ☐
Emergency Power: Yes ☐ No ☐
Readily Visible: Yes ☐ No ☐

FIRE ALARM
Yes ☐ No ☐ Location of Panel:
Coverage: Building ☐ Partial ☐

FIGURE 10.1A

Monitored: Yes ☐ No ☐
Method:
Type of Initiation Devices: Smoke ☐ Heat ☐ Manual ☐ Water Flow ☐
Special Systems ☐
Date of Last Test:
Date of Last Inspection:
Notification Signal Adequate: Yes ☐ No ☐
Fire Department Notification: Yes ☐ No ☐

FIRE EXTINGUISHERS
Proper Type for Hazard Protecting: Yes ☐ No ☐
Mounted Properly: Yes ☐ No ☐
Date of Last Inspection:
Adequate Number: Yes ☐ No ☐

FIRE-PROTECTION SYSTEMS
Type: Sprinkler ☐ Halon ☐ CO_2 ☐ Standpipe ☐ Water Spray ☐ Foam ☐
Dry Chemical ☐ Wet Chemical ☐ Other ☐
Coverage: Building ☐ Partial ☐
Date of Last Inspection:
Cylinder or Gauge Pressure(s): 1 psi., 2 psi., 3 psi., 4 psi., 5 psi.
Valves Supervised: Electrical ☐ Lock ☐ Seal ☐ Other ☐
Are Valves Accessible: Yes ☐ No ☐
System Operational: Yes ☐ No ☐
Sprinkler Heads 18" from Storage: Yes ☐ No ☐

FIRE RESISTIVE (FR) CONSTRUCTION
Stairway FR: Yes ☐ No ☐ Hourly Rating:
Corridors FR: Yes ☐ No ☐ Hourly Rating:
Elevator Shaft FR: Yes ☐ No ☐ Hourly Rating:
Major Structural Members FR: Yes ☐ No ☐ Hourly Rating:
Floor-Ceiling Assemblies FR: Yes ☐ No ☐ Hourly Rating:
All Openings Protected in FR Walls and Floor-Ceiling Assemblies: Yes ☐ No ☐

HAZARDOUS AREAS
Protected by: Fire-Rated Separation ☐ Extinguishing System ☐ Both ☐
Door Self-Closures: Yes ☐ No ☐
Hazardous Materials: Yes ☐ No ☐
Properly Stored and Handled: Yes ☐ No ☐
Properly Protected: Yes ☐ No ☐
Are Lift Trucks Properly Stored: Yes ☐ No ☐
Is the Fuel Properly Stored: Yes ☐ No ☐
Is Fueling Done Properly: Yes ☐ No ☐
Are Extinguishers Provided: Yes ☐ No ☐
Hazardous Processes: Yes ☐ No ☐
Properly Protected: Yes ☐ No ☐

FIGURE 10.1B

HOUSEKEEPING
Areas Free of Excessive Combustibles: Yes ☐ No ☐
Smoking Regulated: Yes ☐ No ☐
Is Stock Stored Properly: Yes ☐ No ☐
Are Incompatible Materials Separated: Yes ☐ No ☐
Is Trash Removed on Regular Basis: Yes ☐ No ☐

INTERIOR FINISH
Walls and Ceilings Proper Rating: Yes ☐ No ☐
Floor Finish Proper Rating: Yes ☐ No ☐

MEANS OF EGRESS
Readily Visible: Yes ☐ No ☐
Clear and Unobstructed: Yes ☐ No ☐
Two Remote Exits Available: Yes ☐ No ☐
Travel Distance within Limits: Yes ☐ No ☐
Common Path of Travel within Limits: Yes ☐ No ☐
Dead-Ends within Limits: Yes ☐ No ☐
50% Maximum through Level of Exit Discharge: Yes ☐ No ☐
Adequate Illumination: Yes ☐ No ☐
Proper Rating on All Components: Yes ☐ No ☐
All Exit Enclosures Free of Storage: Yes ☐ No ☐
Door Swing in the Direction of Egress Travel (when required): Yes ☐ No ☐
Panic/Fire Exit Hardware Operable: Yes ☐ No ☐
Doors Open Easily: Yes ☐ No ☐
Self-Closures Operable: Yes ☐ No ☐
Doors Closed or Held Open with Automatic Closures: Yes ☐ No ☐
Corridors and Aisles of Sufficient Size: Yes ☐ No ☐
Stairwell Reentry: Yes ☐ No ☐
Mezzanines: Yes ☐ No ☐
Proper Exits: Yes ☐ No ☐

VERTICAL OPENINGS
Properly Protected: Yes ☐ No ☐
Atrium: Yes ☐ No ☐
Properly Protected: Yes ☐ No ☐
Are Fire Doors in Good Working Order: Yes ☐ No ☐

OPERATING FEATURES
Fire Drills Held: Yes ☐ No ☐
Employees Trained in Emergency Procedures: Yes ☐ No ☐

FIGURE 10.1C

manual also provides occupancy-specific guidelines for fire inspections. Figure 10.1 provides an example from this manual of a fire-inspection form for an industrial occupancy (NFPA 1994, 345–347):

Once the inspections are completed, the written forms or reports should be forwarded to facility management for review and action. For those deficiencies that are severe and present an imminent danger, facility management should be notified immediately so that corrective action can be initiated. Fire safety inspections play a critical role in the overall safety program by providing a before-the-fact, proactive measure to identify and correct fire hazards before they result in loss incidents.

HOT WORK PERMIT PROGRAMS
A potential source of fire in most industrial operations involves the completion of hot work. Fire hazards may occur in the use of both gas and electric welding and in flame cutting because of the production of sparks. Sparks from cutting tend to be more hazardous because they are more numerous and are carried greater distances. Sparks, in the presence of flammable vapors, may start fires immediately. Smoldering fires, not apparent when the work is completed, may later burst into flame when no one is present. Work done by an outside contractor that requires cutting or welding should also be closely supervised. Approximately one of every three cutting and welding fires reported occurs while outside contractors are engaged in cutting and welding operations.

For this reason, OSHA developed standards (29 C.F.R. § 1910.252) requiring that a fire prevention and fire protection plan be developed when welding, cutting, or other hot work is completed outside of a designated hot work area. One specific requirement is that a fire watch person be present whenever hot work is performed in a location outside the designated hot work area where other than a minor fire might develop or any of the following conditions exist:

- Appreciable combustible material is closer than 35 ft. to the hot work or more than 35 ft. away when the materials can easily be ignited by sparks.
- Wall or floor openings within a 35-ft. radius.
- Exposed combustible materials in adjacent spaces, including combustible materials in concealed spaces, such as between walls and below floors.

- Combustible materials are adjacent to the opposite side of metal partitions, walls, ceilings, or roofs where hot work is performed and are likely to be ignited by conduction or radiation.

A hot work permit program requires that a permit be issued for all hot work (including permanent locations) conducted on or near a covered process. The permit should include the following information:

1. Date
2. Work to be performed
3. Location of job
4. Type of immediate firefighting equipment available, to include noncombustible covers
5. Fire watch assigned
6. Inspection of area before commencing work
7. Authorized signature of individual in charge of the permit system
8. Signature of the individual who authorized the job to indicate the area was inspected upon completion of work

The designated fire watch mentioned earlier must have fire-extinguishing equipment available and be trained in its use. He or she must also be familiar with the operation of the alarm in the event of a fire and shall watch for fires in all exposed areas. If a fire should occur, the fire watch should attempt to extinguish it only within the capacity of the available equipment or otherwise sound the alarm. The fire watch should be maintained for at least a half hour after completion of hot work operations to detect and extinguish possible smoldering fires. Other precautions to consider prior to commencing hot work include the following:

- Sparks or molten metal should not be permitted to pass through doorways or through cracks or holes in walls and floors.
- All exposed combustibles should be moved a minimum of 50 ft. (15.3 m) from cutting and welding operations. Noncombustible curtains must be used between the operation and combustible materials when adequate separation cannot be maintained.
- Floors should be swept clean and wood floors should be wetted down or covered with a noncombustible fire blanket.

This chapter has discussed the role of the safety professional in fire program management. It is critical to end this chapter with the understanding that once these fire programs are developed, it is line management's responsibility to implement these programs. The safety professional is frequently in a staff position that provides technical support to line management. Line management is responsible for the people and property within his or her area of supervision; common responsibilities include the following:

- To work closely with the safety professional and staff to ensure that fire safety preplanning is completed prior to occupancy changes or the installation of new equipment or processes
- To enforce fire safety procedures, such as maintaining exits free of all obstructions, not permitting smoking in fire hazard areas, and following hot work permit procedures
- To work with the safety professional in selecting employees for program assignments and to ensure that these employees have been properly trained
- To ensure that emergency drills are held regularly and to work with the safety professional to evaluate the effectiveness of the emergency programsTo work with the safety professional to establish priorities for inspection, testing, and maintenance of fire protection systems

The programs discussed in this chapter play a critical role in the overall management of fire risks in an organization. However, a written program is only as good as its implementation, and management plays a critical role in both program implementation and evaluation.

CHAPTER QUESTIONS
1. What are the four primary steps in the fire risk management process?
2. What is one of the safety professional's primary roles in the risk assessment step of hazard identification?
3. A basic risk assessment of a hazard involves identifying what?
4. When addressing the severity of a fire risk, it is important to consider both direct and indirect loss potentials. What are some examples of indirect losses?
5. What are some general options available for handling fire risk exposures?

FIRE PROGRAM MANAGEMENT 259

6. A properly conducted risk assessment can give management an idea of a facility's relative vulnerability to risk as well as what else?
7. Twenty years ago, emergency-response plans focused on fire and natural emergencies; however, such a narrow scope is no longer acceptable with today's emergency risks having evolved substantially to include what other risks?
8. On a federal level, what are the three primary governmental agencies with regulatory responsibilities related to emergency response?
9. Identify three areas that must be addressed in a written emergency-action plan according to OSHA standards.
10. What is the primary purpose of the OSHA Chemical Process Safety Management standard?
11. What are three examples of duties that a fire brigade may perform during a fire emergency?
12. One of the purposes of the EPA Emergency Planning and Community Right-to-Know Act is to establish requirements for federal, state, and local governmental agencies, as well as many business facilities, regarding emergency response to environmental emergencies. Another major purpose of this act is to do what?
13. What is one of the purposes of the ICP published by the NRT?
14. Another important decision in planning a response strategy is determining the extent to which the organization is willing to commit time and money to developing and implementing an emergency-response plan. What are some examples of costs associated with the development and implementation of an emergency-response team?
15. Using the NRT ICP, the core plan is intended to do what?
16. One of OSHA's requirements for first aid is that, in the absence of an infirmary, clinic, or hospital in near proximity to the workplace that can be used for the treatment of all injured employees, a person or persons shall be adequately trained to render first aid. According to OSHA interpretations, what criteria are used to evaluate "near proximity"?
17. Identify two of the suggested measures for addressing the media as part of the overall emergency-response plan.
18. What is the primary reason for investigating loss incidents involving fires?

REFERENCES

Cote, Arthur, and Bugbee, Percy. (2001). *Principles of Fire Protection.* Quincy, MA: NFPA.

Federal Emergency Management Agency (FEMA). (2013). Business Continuity Planning. http://www.ready.gov/planning (accessed October 30, 2013).

Ladwig, Thomas H. (1991). *Industrial Fire Prevention and Protection.* New York: Van Nostrand Reinhold.

National Fire Protection Association (NFPA). (1994). *NFPA Inspection Manual,* 7th ed. Quincy, MA: NFPA.

NFPA. (2008). *Fire Protection Handbook,* 20th ed. Quincy, MA: NFPA.

NFPA. (2013). *NFPA 1600: Recommended Practice for Emergency Management.* Quincy, MA: NFPA.

Phillips, Brenda, David Neal, and Gary Webb (2011). *Introduction to Emergency Response.* Boca Raton, FL: CRC Press.

Schneid, Thomas D., and Larry Collins. (2002). *Disaster Management and Preparedness.* Boca Raton, FL: Lewis Publishers.

Stringfield, William H. (2000). *Emergency Planning and Management.* Rockville, MD: Government Institutes.

U.S. Department of Labor (USDOL). (1976, January 27). OSHA, Standard Interpretations—Clarification of 1910.151 First Aid Training.

USDOL. (2002, April 18). OSHA, Standard Interpretations—Clarification of 1910.151 Medical Services and First Aid.

USDOL. (2013a). *Occupational Safety and Health Standards for General Industry, 29 C.F.R. § 1910.38: Emergency Action Plans.* Washington, DC: U.S. Government Printing Office.

USDOL. (2013b). *Occupational Safety and Health Standards for General Industry, 29 C.F.R. § 1910.39: Fire Prevention Plans.* Washington, DC: U.S. Government Printing Office.

USDOL. (2013c). *Occupational Safety and Health Standards for General Industry, 29 C.F.R. § 1910.119: Process Safety of Highly Hazardous Chemicals.* Washington, DC: U.S. Government Printing Office.

USDOL. (2013d). *Occupational Safety and Health Standards for General Industry, 29 C.F.R. § 1910.120: Hazardous Waste Operations and Emergency Response.* Washington, DC: U.S. Government Printing Office.

USDOL. (2013e). *Occupational Safety and Health Standards for General Industry, 29 C.F.R. § 1910.156: Fire Brigades.* Washington, DC: U.S. Government Printing Office.

United States Environmental Protection Agency (USEPA). (1998). Integrated Contingency Plan ("One Plan") Guidance. USEPA.

USEPA. (2013a). *40 CFR 264: Standards for Owners and Operators of Hazardous Waste Treatment, Storage, and Disposal Facilities.* USEPA.

USEPA. (2013b). Emergency Planning and Community Right-to-Know Act Overview. http://www.epa.gov/osweroe1/content/lawsregs/epcraover.htm (accessed October 31, 2013).

USEPA. (2013c). Emergency Response Program—Community Relations. http://www.epa.gov/osweroe1/content/community/commrel.htm (accessed October 31, 2013).

USEPA. (1996). The National Response Team's Integrated Contingency Plan Guidance. USEPA.

U.S. National Response Team (USNRT). (2013). Working Together to Protect against Threats to Our Land, Air, and Water. www.nrt.org/Production/NRT/NRTweb.nsf/homepage?openform (accessed October 30, 2013).

Vulpitta, Richard, and Dean Larson. (2011). *On-Site Emergency Response Planning Guide.* Itasca, IL: National Safety Council.

11

Creating a Comprehensive Emergency and Evacuation Plan

ACTION PLAN AND EVACUATION

As discussed in chapter 10, creating the emergency action plan is not a simple process. Risks must be evaluated—all risks, not just fire risks. Critical business functions must be understood. The liability of an emergency must be compared to the probability of it happening. Contingency plans must be developed in any case. Assignments must be given, and employees must be trained. At some point, all of this must be written down and the compilation of all this information becomes your plan.

The emergency action plan must provide valuable, up-to-date information that can quickly be referenced during an emergency event. The manual must contain all actionable information without padding or irrelevant information that may create confusion.

The emergency action plan is not static, it is a living document. It will change and evolve over time. The initial time spent developing the plan will not likely need to be repeated, but the document must be maintained and updated as times and circumstances change. The initial development of the document may be painful and time-consuming, but it is project well worth the time and expense.

A benefit of developing the emergency action plan is that it forces you and your team to look at every aspect of your organization's operations. In the process you may find redundancies or better ways to do things which can increase production or efficiency and may boost the bottom line.

This section was created to assist safety professionals and emergency planners with the development of such a plan. It is impossible to provide an emergency action plan that satisfies the needs of all organizations. A turnkey plan that can be purchased and simply placed on a shelf until an emergency occurs does not exist. Even with the help of this resource, you will be required to research and develop a plan that is specific to your business. With this in mind, let us discuss the plan, its purpose, and what it does in greater detail.

What Does It Do?

The emergency action plan exists to save lives and reduce injuries in the event of an emergency. It provides the means and methods to help you plan, survive, and recover from an emergency. Because lives, property, and business continuity are at stake, the emergency action plan must be considered an important part of your organization's business plan. It must be championed at the highest levels of the organization. It should educate employees and serve as a basis for ongoing employee training. Because people and processes do change, the emergency action plan must be reviewed on a regular basis and updated as required.

What Does It Contain?

Your emergency action plan should contain all pertinent information that will allow you to plan, survive, and recover from an emergency. This will require an evaluation of your building, type of business, and geographic location. The plan must contain names and phone numbers of employees who are able to help in the event of an emergency. In 29 C.F.R. 1910.38(a)–(b), Occupational Safety and Health Administration (OSHA) has developed regulations that identify the following minimum requirements of an emergency action plan:

- Emergency escape procedures;
- Description of potential hazards;
- Emergency escape routes;
- Procedures to be followed by employees who remain to operate or shut down critical building operations before they evacuate;

- Procedures to account for all employees after evacuation has been completed;
- Rescue and medical duties for those employees assigned to perform them;
- The preferred means of reporting fires and other emergencies; and
- Names or job titles of employees who can further explain the emergency action plan.

OSHA also requires that each facility must have a fire prevention plan in place. This plan is specifically tailored for the threat of fire. The employer is required to train employees in aspects of the plan that pertain to their exposure. Components of the plan include:

- A list of fire hazards that exist at the workplace;
- Names or titles of employees who are responsible for fire source hazards;
- Names or titles of employees who are responsible for fire alarms and fire-fighting equipment;
- Housekeeping procedures that limit fuel sources; and
- Employee training techniques and frequencies.

In addition to these minimum OSHA requirements, a good emergency action plan should include the following items:

- Statement of policy, also known as a mission statement;
- Floor plans with exits, fire extinguishers, pull stations, first aid stations, and other important features marked;
- Location of MSDS information;
- Names and numbers of external emergency service providers;
- Inventory lists of emergency supplies;
- Maintenance schedules for generators, fire sprinklers, alarm systems, and other related emergency equipment; and
- Chain of command established for emergency events.

OSHA does not require every employee to have a copy of the plan. Instead, OSHA requires that the plan be kept at the workplace and made available to all employees. A suggested best practice would be to provide all members of your emergency management team with a completed copy of the plan and simply make the plan available to all employees.

Many organizations have done this by putting the plan on the company's intranet website. Other organizations have provided employees with a scaled-down version of the emergency action plan, usually during new employee orientation. A summary version explains proper evacuation procedures, alarms, staging areas, and other basic issues. One important issue is the regular revision of the emergency action plan. You do not go through all the effort of creating a plan and then let it sit on a shelf. The plan should be reviewed at least annually and revised as needed. When changes are made to the plan, training should be provided to all employees affected by the changes.

In the event of an emergency, the action plan will be a vital reference. You will consult the plan on a regular basis, looking for names and number of key support personnel. You'll use the plan as a guide for making decisions.

The plan will ensure that all the necessary actions are taken to promote business continuity during and after the emergency. Bottom line is that the plan is of utmost importance! But think for a moment about the actual plan. Where do you keep it? Probably in your office. Would you be able to access the plan if the building burned down? What if an earthquake occurred and the building was unsafe to enter? What if everybody on the emergency management team kept their plans in their offices? Of course, you could go online to the company's intranet site and print it out.

For these reasons, it is imperative that you keep a copy of the emergency plan at an off-site location such as your vehicle or home. It makes sense to keep a copy on your laptop or a thumb drive attached to your keychain as well. Do not get caught in a situation where the plan is not available!

THE EMERGENCY MANAGEMENT TEAM

Every good plan is supported by a team of dedicated people. The plan is just a reference manual. Decisions still need to be made. Direction must be given. Human interaction and leadership are a necessity. The plan comes to life when somebody picks it up and uses it. Furthermore, the emergency management function is not a one-person job. Instead, a team of people will be involved, each with a specific purpose.

Creation of an emergency management team is essential early in the planning process. Expertise will be needed from all areas of the organization. If the emergency management team consists of employees with similar backgrounds, education, responsibilities, and job titles, the plan will be written in a vacuum. Valuable expertise from outside your department will not be tapped into, so make sure you include a diverse group of employees on your team. This team will likely grow as the plan becomes a reality. Floor captains, runners, evacuation partners, and command post leaders will need to be elected. For now, though, the team should be kept small but functional.

The length of time a member would sit on an emergency team is subjective. Some organizations like to rotate a percentage of team members on an annual basis. The belief is that by allowing a greater number of employees the chance to sit on the team, a greater number of employees will respect the plan and be able to react during crisis. This way, the bulk of the emergency team is left intact while employees with fresh ideas and viewpoints can be introduced. It is also believed that in the event a team member leaves the organization, the vacancy can be quickly filled by a past member.

Identifying Team Members and Responsibilities

When developing your emergency management team, be sure to include a diverse roster of employees. In addition to the obvious goal of saving lives and reducing injuries, your goal is to identify critical business functions that may be interrupted because of an emergency event. This is not usually accomplished through the eyes of a single individual. Once the diverse team of subject matter experts has been finalized, assignments and a hierarchy should be developed. During a crisis there must be a clear line of responsibility and command. Listed here are samples of some of the expertise needed on your team:

- Risk management specialist
- Logistics and procurement manager
- Media relations contact
- Human resources personnel
- Facilities maintenance manager
- Legal counsel:

Listed here are a few of the assignments that should be given:

- Emergency director
- Business continuity manager

- Command post leader
- Group leader
- Floor captains
- Fire brigade
- Evacuation assistants
- Runners

Internal and External Support

The risk analysis matrix includes a column for internal support and a column for external support. Although you will want to take advantage of your internal resources and employees, that may not always be possible. Key to surviving any emergency event will be the support of external resources and vendors. Early identification of these resources is essential, as is involving them in the planning process.

When identifying internal and external resources, consider how quickly the resource will be available. This plays an important role when discussing internal resources. You may have an excellent staff of highly trained maintenance technicians who are skilled at solving building-related problems, but the emergency event may cause them to be unavailable. In the event of labor unrest, your internal resources may not be willing to cross a picket line. When an emergency event occurs, your employees and their families may be directly affected and, therefore, unavailable, so make plans for a backup.

Insurance providers play a critical role in your emergency planning and recovery efforts. Unfortunately, most people do not know in detail what coverage they have in place. The time to educate yourself on these important issues is prior to an emergency, not after it. Many organizations believed they were covered by their insurance policies, only to find later that an amendment to the policies excluded a specific event. Since coverage terms, property valuation, property size, and other issues change on a regular basis, you should meet annually with your insurance advisor to discuss your policy. Typical questions might include the following:

- How will the property be valued (replacement/actual/other)?
- Am I covered for spoilage of goods?
- Do I have all the policies needed (fire/flood/storm/contents/liability)?

- Do I have a policy protecting me in the event of lost income?
- If I do have a lost income policy, and what is the duration?
- In what way does the provider require proof of loss?
- What additional loss documentation is needed and is it stored in a safe place?
- Does it make sense to self-insure?
- What are my deductibles and are they at proper levels?

Identifying Vendors and Contractors

Contractors come and go; it is, therefore, imperative that you review your list of contractors with active purchase orders annually at a minimum. Be certain that you have updated office and cell phone numbers. In heavily populated metropolitan areas, area codes tend to change on a regular basis, so make sure you have the latest area code as well. Identifying a good contractor is just a starting point. A contractor may agree to help after the emergency event, but because of circumstances out of their control, may not be available. Access roads could be blocked, their office could be heavily damaged, or some other logistical problem might exist. For this reason, you need to have alternate contractors available. Using the services of a contractor after an emergency event sounds like an easy process. Simply call the contractor, issue a work ticket, and tell them where to show up and what to do. In actuality, the process is not so simple. First, do you need permission from your insurance company before you can start clean-up and repairs? Following an emergency, the insurance company will often tell you to proceed but to document in writing and pictures the position immediately following the emergency and then to continue to keep similar records as the repair work progresses. It could be many days before an insurance company claims adjustor arrives on the scene.

Often you know the contractor's services are needed, but you're not sure of the scope of work. For instance, after a tornado, you contact your landscaping contractor and dispatch them to your corporate headquarters. Based on the reports you've heard, trees are down and blocking access roads and parking lots. Other trees are not down but pose a danger and require removal. There will probably be some irrigation repairs as well. Because the scope is unclear, you simply instruct the contractor to get the property back in order. Ambiguous requests like this are likely to cause over-billing. To account for these issues, detailed records must be kept. Depending on the relationship you have with the contractor, you may require a field inspector from your organization to give approval prior to commencing work. This may keep costs under control, but it really slows down the recovery process.

A better way to handle large projects that are discovered by the contractor in the field is to require them to document the damage with a camera or video camera prior to commencing the cleanup efforts. If your relationship is good with the contractor, this process will protect you and keep the contractor busy. You may also elect to place a cap on cleanup efforts. For instance, any job that exceeds $2,500 must first be approved by the facilities department. By blending a quick efficient cleanup with cost protection measures, you'll get the job done without a financial hangover the next morning. Prior to using the services of a contractor after an emergency event, you must iron out a few of the contractual details. The emergency creates unique circumstances. The contractor will bring in extra help. Many of the contractors will be helping around the clock. Specialized equipment may need to be rented. When you consider the possible issues, you'll begin to realize the importance of creating contractor specifications for emergency events. Examples of these issues include:

- Overtime rates (time and a half, double time);
- Hotel reimbursement;
- Mileage reimbursement;
- Expense and receipt requirements;
- Food reimbursement;
- Laundry reimbursement;
- Equipment rental responsibilities;
- Personal protection equipment requirements; and
- Additional insurance requirements.

Because much of the help you use will not be your primary contractors, you'll be tapping into companies that may not meet your usual company expectations. Specifically, insurance requirements come to mind. Do you feel comfortable contracting with companies that carry inferior amounts of liability insurance? It is a risk and you must weigh that risk against the restoration needs you face. Be careful with this decision. Remember, after an emergency unusual and dangerous conditions will likely exist. If this contractor does not have adequate insurance for normal working conditions, how can you feel comfortable using them in a dangerous environment? Instead of lowering your standards, spend more time locating quality alternate contractors who already have the required coverage in place.

The indemnity agreement should be a standard document signed by all contractors, regardless of their insurance amounts. This document protects your company from lawsuits and judgments because of the contracted assignment. As a last resort, you may elect to use contractors with inferior coverage amounts if they agree to sign an indemnity agreement. This document does not provide total protection, but at least it is a start. Of course, when signing contracts, always consult with your legal counsel beforehand. An example of an indemnity agreement coupled with a safety clause may look like this:

> Contractor hereby agrees to indemnify and hold harmless [company name] its agents, officers, successors, and assigns against any and all claims, suits, judgments, and damages of any kind resulting from this contract or the work supplied by the contractor, its employees, or subcontractors. Furthermore, contractor shall be solely responsible for supervising its employees, providing personal protection equipment, training its employees, record keeping, and adhering to all federal Occupational Safety and Health Act regulations or state equivalent.

EMERGENCY FLOOR PLANS

Several tools can assist emergency teams during search and rescue missions after an emergency event. Most of these tools are readily available to the planner, such as floor plans annotated with exits, fire extinguishers, and first aid kits. Other tools, such as the Urban Search and Rescue Grid, are created by highly specialized professionals and are particularly important for use in high-rise buildings.

When creating your emergency plan, start by acquiring an updated copy of the building floor plan. In a leased building you will need to speak with the property owner or its agents. If the building is company owned, you should speak to the facility management department. Over the years, buildings are renovated, so the floor plan that you have may not be an accurate representation of the building's configuration. Take time to ensure the accuracy of the floor plan. If you cannot locate any floor plan information, or if the plan is hopelessly outdated, it would be advisable to create a basic revised floor plan. And if you have the tools and expertise in-house, use them to create a basic floor plan.

It's not necessary to go into great detail on your floor plans; remember, the purpose is not to build the structure. The purpose of this floor plan is to identify windows, doors, emergency exits, alarm panels, fire pull stations, and other emergency-related items. A basic single-line floor plan minus the dimensions, details, and other nonessential items will be sufficient. When developing an emergency floor plan, remember to include important details such as:

Egress concerns

- Designated escape route
- Stairways
- Windows
- Doors
- Emergency exits
- Evacuation staging areas
- Fire safety
- Fire alarm pull stations
- Fire extinguishers
- Smoke detectors
- Water sprinkler testing valves
- Alarm panels

First aid

- First aid kits and stations
- Automated external defibrillators (AED)
- Eye wash stations

Specialty areas

- Mechanical rooms
- Telecommunications rooms
- HazMat storage locations

- MSDS locations
- Restricted areas

Utilities

- Gas mains and valves
- Utility shutoffs
- Water lines and valves
- Sewer lines

Power systems

- Generator and switchgear locations
- Generator powered circuits
- Uninterruptible power system location

Drainage

- Storm drains
- Roof drains

Critical Building Information

Critical building information must be made available during and after emergency situations. This is usually not a problem if company employees are available. Facility management or maintenance personnel are a ready source of critical building information, but what would happen if the emergency event occurred after business hours? For instance, what if a fire were to start in your building after hours? When the fire department arrives, will they have keys to your building? How will they locate a floor plan? Will they be able to locate your MSDS information? If the fire department does not have keys, they will enter your building forcefully, but considering the emergency that will be the least of your problems. The floor plan would be helpful but, again, not necessary. The MSDS information, however, is especially important to the firefighters. Without an idea of what chemicals are being stored on the premises, the firefighters are putting themselves in a dangerous situation. It would be unfair to ask any rescue personnel to enter your building without first identifying the hidden hazards.

The KNOX-BOX® Rapid Entry System provides a means for nondestructive emergency access to commercial and residential property. Key vaults are available in small sizes just large enough to hold a set of building keys. Larger vaults are available that can hold keys, MSDSs, floor plans, and other critical building information. These vaults are specially keyed to the requirements of your local rescue personnel, much like an elevator key. Check with your local fire department to see if KNOX-BOX systems are used in your area.

Fire Extinguishers and Means of Egress

Portable fire extinguishers provide the first line of defense when fighting a small fire. In many cases, quick response to a fire coupled with the proper extinguisher can prevent loss of life, damage to property, and reduced productivity. In fact, if the fire is contained and extinguished, an evacuation of building's occupants may not be necessary. At worst, a partial evacuation of those areas nearest to the fire may be all that is required.

Portable fire extinguishers are critical for extinguishing small fires in their early stages. Prior to using a portable fire extinguisher, the operator should be professionally trained. Three major of classes of fire exist, and if the emergency responder chooses the wrong fire extinguisher for a specific fire, he or she risks causing further harm. Proper training starts with identification of the fire source and proper selection of extinguisher. The three major classifications of fires are:

- Class A fires: Fueled by common combustible materials such as wood, cloth, paper, and many plastics. Use a water-based extinguisher with a green label and Class "A" identifier.
- Class B fires: Fueled by flammable liquids and gasses such as gasoline, kitchen grease, and paint. Use a carbon dioxide or aqueous film-forming foam (AFFF) extinguisher with a red label and Class "B" identifier.
- Class C fires: Fueled by live electrical components or circuits including appliances, tools, and wires. Use an extinguisher with a blue label and Class "C" identifier.

Fire Suppression Systems

In the event of a fire, quick response from local firefighting professionals is critical. The response is often not quick enough. The firefighters often arrive in time to extinguish the fire, aid the victims, and ensure that the fire does not spread to adjacent properties. Unfortunately, the damage to the subject property is often great. A sensible solution, especially in buildings that are unoccupied during the evening, is the installation of fire sprinklers. In many areas, fire sprinkler systems are mandatory on new construction and renovations. In fact, the National Fire Protection Association (NFPA, www.nfpa.org) has been a leader in developing standards for the fire protection industry. The NFPA has determined that installation of a fire sprinkler system coupled with smoke detectors could reduce deaths attributed to fire by 82 percent. Armed with this information, we can appreciate the importance of functional fire suppression systems.

There are several fire sprinkler options:

- Dry-pipe systems are used in areas where freezing conditions exist. Pressurized air is maintained in the sprinkler pipes. When a fire occurs, the heat ruptures a seal in the sprinkler head, the air escapes the pipes, and the water exits the sprinkler heads.
- Wet-pipe systems do not use pressurized air. When a fire occurs, the heat ruptures a seal in the sprinkler head and the water exits the sprinkler heads.
- Deluge systems are like dry-pipe systems in that they use pressurized air in the pipes. This system may be activated manually or automatically.

After the terrorist attacks of September 11, 2001, engineers determined that World Trade Center towers collapsed due to weakened steel members and not due to the violent collision of the two aircraft into the buildings. The aircraft were fully loaded with fuel, and as the fuel burned inside of the towers, temperatures reached more than 2,000 degrees. The World Trade Center towers contained fire sprinkler systems, but the impact of the aircraft ruptured the lines and rendered them useless. This should not lessen the credibility of fire sprinklers as an effective way to extinguish fires, especially considering the unusual nature of that emergency. On that same day, another terrorist attack occurred at the Pentagon. An airplane taken over by terrorists slammed into wedge one and wedge two of the Pentagon, penetrating four out of the five rings that make up the structure. Engineers noted that damage to the newly renovated wedge one was minimized thanks to the fire sprinkler system. The fire did not extend to areas beyond where the fuel was spilled. At wedge two, however, fire spread easily and rapidly because a fire sprinkler system had never been installed in this soon-to-be renovated area.

PREPLANNING AND THE URBAN SEARCH AND RESCUE GRID

When the fire department responds to a building after an emergency, critical building information is needed if they are to deploy their resources effectively and safely. Consider the fact that most rescue personnel will be unfamiliar with your building, its floor plan, its contents, and other important building issues. To assist fire rescue personnel with this dilemma, former fire officer Curtis Massey founded Massey Enterprises, Inc., in 1986 (now Massey Disaster Planning). Since that time Massey Disaster Plans have become an industry standard. These plans are taught as part of the curriculum at the National Fire Academy and are also the only preplanning system supported by every major fire department in North America. These plans address the emergency from the viewpoint of rescue personnel and include the not-so-obvious features that are critical to rescue personnel. Fire rescue, the Federal Emergency Management Agency, police departments, and other emergency personnel are assisted by Massey Disaster Plans. These plans are invaluable for all types of emergencies including fires, gas leaks, hazardous materials incidents, terrorism, biological/chemical attacks, mass shootings, hostage situations, technical rescues, and natural disasters.

Armed with critical building information on the plans, fire rescue personnel can best utilize the building systems to their greatest advantage. Detailed, easy-to-read graphics complement the text. Floor plans, site plans, and riser diagrams show the fire department where everything is located—both horizontally and vertically. The Massey Disaster Plans include two distinct features—the preplan and the urban search and rescue grid. The preplan features color-coded graphics that fire fighters understand and can quickly decipher. Items such as HVAC, shut-off valves, communications, alarms, and water supplies are noted on the preplan.

The second feature of the Massey Disaster Plan is the Urban Search and Rescue Grid. This document identifies support columns and other major features of the building that would enable firefighters to systematically search the building even if it experienced a collapse. Because of the detailed nature of these plans, their completion may take months and cost thousands of dollars; however, the effort is well worth the expense.

ALARM SYSTEMS

OSHA requires employers to devise a method for notifying employees of an emergency. These emergencies might include fire, chemical spill, bomb threat, or other related events. The alarm must be able to be heard above the noise level in the workplace. In some areas this may be difficult to do. In these cases, it is recommended that a secondary alarm warning system be used. For instance, an audible alarm coupled with a visual alarm such as a flashing strobe would provide much better protection.

The employer must also explain to the employees the preferred means of reporting emergencies. Methods may include manual pull box, public address systems, radio, or telephone. The Americans with Disabilities Accessibility Guidelines (ADAAG, 4.1.3 and 4.28) outline specifications for emergency alarms to ensure that they are accessible to persons with disabilities. ADAAG specifications for visual alarm devices specify intensity, flash rate, mounting location, and other characteristics. One major difference between the installation of audible alarms and visual alarms concerns location. It is always a requirement to install audible alarms less frequently than visual alarms. This requirement makes good sense because the reach of an audible alarm is much greater than a visual one. A visual alarm can be blocked by line of sight and unseen if the individual is in a room that does not have a visual alarm.

Understanding Your System

An alarm system is an integral part of your property's overall emergency management plan. The system must be operational and available during all hours of the day and night. It must be kept maintained, and you must test it on a regular basis. Of course, most organizations will contract this work out to others because they do not have the expertise in-house. If you elect to do that, it is not a problem; however, it would be advantageous to develop an in-depth understanding of your alarm system. Although this section does not pretend to be a technical resource that provides an overview of alarm and fire suppression systems, we will discuss a few of the basic regulations and guidelines concerning your alarm systems.

First, familiarize yourself with the alarm system currently employed at your property. Start by identifying the systems in place. Reference your manufacture's literature including the technical information. If you are subcontracting maintenance or repairs to an alarm company, tap into their expertise.

Ask them questions. Walk with them as they perform the regularly scheduled maintenance functions. In fact, invite them to a staff meeting and allow them the opportunity to conduct a training session on your alarm system. If you reach out and ask these contractors to assist you, you will begin to develop a partnership.

They will gladly provide training because an informed consumer is better equipped to understand the contractor and their advice. As mutual respect is gained between consumer and contractor, a longer and more lucrative relationship is likely to develop. In today's ultra-competitive market, contractors are looking for ways to prove their worth, and you should take advantage of this.

Regular Testing

Even the most expensive and best-designed alarm systems can be rendered useless if not properly maintained and tested. Time and time again, fire inspectors have discovered alarm systems that were adequate for the property where they were installed, but because of improper maintenance, the alarm system did not operate properly, and damage resulted. Businesses have been interrupted needlessly and in the worst scenarios, lives have been lost.

The cost of maintenance is low, but the risk of not performing it is extremely high. The problem with maintaining an alarm system usually stems from two areas. First, the building owners or managers do not understand the need for maintenance and testing, or the incremental cost to maintain the system is cut from the operating budgets. Those responsible for safety and emergency planning must ensure that neither of these two scenarios takes place in your organization.

When evaluating your alarm system, it is important to know the age and condition of your system. A system that is more than five years old should be examined for potential component breakdown. Normal wear and tear set in around this time, especially if the system is exposed to voltage fluctuations, humidity, dust, and temperature fluctuation. Periodic testing by a NICET-certified contractor should identify these problems before they break down. NICET technicians are certified by the National Institute for the Certification of Engineering Technologies (NICET) and specialize in life safety issues. Technicians who hold NICET certification have a thorough knowledge of system installation and life cycle inspection, testing, and maintenance protocols.

CREATING A COMPREHENSIVE EMERGENCY AND EVACUATION PLAN 279

If your system has not been inspected and maintained regularly, or if it has but only sporadically, beware. If the system is older than ten years, the equipment is susceptible to frequent and costly breakdown. Equipment installed more than twenty years ago should be evaluated and possibly scheduled for replacement. The technologies used in your system may be antiquated and many of the parts may be unavailable. Because the purchase of a new alarm system can be costly, evaluate the current system and develop a five-year plan for replacement if the condition warrants.

Most alarm system manufacturers recommend at least one full annual test and inspection; however, various agencies, organizations, and local authorities recommend—and in some cases mandate—more frequent testing intervals. Although not an all-inclusive list, the maintenance activities for fire alarm systems can be summed up in five basic areas:

1. Test and calibrate: Sensors such as smoke detectors must be tested and calibrated.
2. Inputs and annunciators: Place the system under test conditions and evaluate for problems.
3. Set sensitivity: Sensitivity must be adjusted based the system, location, and the potential emergencies. Specialized testing equipment is used for this task.
4. Coordination: Coordinate with the fire department and the monitoring company to test inputs to their system and communication.
5. Battery: The battery must be examined for corrosion and other visual indicators. Check expiration date and test the battery.

The NFPA provides the National Fire Alarm Code (NFPA 72). This standard covers the application, installation, performance, and maintenance of protective signaling systems and their components. Included in this code is information devoted to inspection, testing, and maintenance. The National Fire Alarm Code does not address whether a fire alarm is required in a building; it simply addresses the alarm post-installation. In addition to NFPA standards, OSHA has developed minimum requirements that often mirror NFPA code. These OSHA regulations can be found in 29 C.F.R. 1910.165. Furthermore, Appendix B to Subpart L of 1910 provides a cross-reference table that matches OSHA regulations to other applicable codes including

NFPA and ANSI. An important point to remember is that in virtually all cases, these standards represent minimum requirements. These standards may be superseded by local jurisdictions and even manufactures' recommendations. Even if you meet the strictest of standards, you may still not have adequate protection for your facility. Unique issues such as voltage fluctuations, weather, dust, humidity, saltwater exposure, and other environmental concerns can shorten the life of your system or otherwise affect the system in a negative way.

THE COMMAND CENTER

In the event of a major emergency, especially during the recovery stage, a command center must be established. This command center may be on site or off site if the emergency location is dangerous or if multiple locations are involved. Based on the emergency event and the location, the command center should be easily set up and transferred. If you are responsible for multiple structures it would be wise to have multiple potential command center locations so that if one is rendered unusable, the command center can be moved to another. This would serve to keep the command center from being damaged and would preserve utility of the center.

OFF-SITE DATA STORAGE

Now more than ever before, data security is paramount. Given the role it plays in business continuity, companies can ill afford to be without it. In many cases, off-site storage of data is a necessity, especially in the event of a major fire. To understand this, think about a disaster event like the World Trade Center attacks. Because the buildings were leveled, any data stored on site was lost. When considering backup options, solutions such as cloud storage, magnetic tape, CDROM, microfiche, and digital formats exist. These decisions should not be made without fully understanding the strengths and weaknesses of each. Those issues cannot be adequately covered here, so consult with information technology professionals who can evaluate your needs and your unique situation.

In a world so driven by digital technology, loss of computer functions can be devastating. Contingency strategies must be developed and implemented to protect the digital integrity of your organization. One of the greatest concerns for any contingency site is the need for seamless communications.

What systems are in place to ensure telecommunications will be available? Will computers operate as planned? Can backed-up data be loaded on the hot site computers in a quick fashion? How quickly? If business continuity is critical to your organization, consult with your information technology employees and discuss available contingencies. It will likely be necessary to bring in specialized assistance that can match your needs with the proper contingency strategy.

LACK OF ELECTRICITY DOES NOT MEAN "GO HOME"

Loss of utility is an ever-increasing concern. Although rarely considered a disaster (or even an emergency), loss of utility can negatively impact an organization's productivity. Even momentary blips in electrical service can result in loss of digital data and disconnection of telecommunications. Several solutions exist and can be realistically implemented to solve the problem of loss of electricity. There are pros and cons, of course, that must be addressed prior to making the investment. A good starting point is to identify the areas of your business that can least afford to be without power. These functions can be identified as "mission critical" functions.

Suppose your organization produces widgets. Your factory has an assembly line that produces 100 widgets per hour. Often, your line workers find themselves up against critical deadlines. For each hour that the assembly line is unable to produce widgets, your organization loses $10,000. It is very easy to identify the production floor as a mission-critical function. Other times, it is not so easy to identify the mission-critical functions of an organization. The problem usually occurs when every business unit believes it is equally important.

When an organization cannot prioritize the critical nature of its business, problems will exist. For this reason, the emergency planning effort needs to include members of executive management. They will be able to provide direction and clarification concerning the strategic direction of the organization. Too often, if the task is left to mid-level managers, critical functions become synonymous with whatever those managers are in charge of. We tend to place more importance on our jobs, perhaps not understanding what other departments do and how that affects the organization's business plan. In our example of the widget factory, peripheral tasks such as human resources and marketing would not be considered mission-critical tasks. This is not to say these tasks are not important. We are saying you should prioritize which business functions need to be restored first; human resources and marketing would come after production.

When evaluating business continuity and mission-critical functions, some thought should be given to options that are available in the event of loss of utility. For example, based on your organization's business, is it too expensive to purchase and maintain an emergency generator? Is the probability of loss of utility a small one? In the event of prolonged loss of utility, would it make better sense to move processing or manufacturing to an alternate site? Are the consequences of a loss of information technology resources sufficiently high to justify the cost of various recovery options? Performing a risk assessment will allow you to focus on areas in which it is not clear which strategy is the best.

EIGHT STEPS TO CREATING THE FINISHED EMERGENCY ACTION PLAN

Finally, in the development of your emergency action plan and collecting information, you need to actually create it. Here are eight suggested steps in plan development.

1. First draft
2. Review
3. Second draft
4. Tabletop exercises
5. Final draft
6. Print and distribute
7. Employee training
8. Annual review and revisions

The first three steps are often the most difficult. When developing the plan, understand that the first draft is not an exercise in perfection. It is a starting point, a launching point, if you will. It allows you to look at your business under a microscope and to identify all at-risk areas. When working on the plan you may well discover actions that can be taken immediately that could improve efficiency, cut costs, or reduce vulnerabilities. You might identify areas in the building where access should be restricted but is not—the computer server room, for instance—and you can take steps to ensure that from now on only authorized personnel have access to those areas. You may discover that the route taken each day to deliver the mail is not the most efficient and that orders going to the fulfilment department are delivered last rather than first so that they can be processed immediately. You may discover vulnerabilities in your chemical storage area or find that your HVAC system could be tampered with outside the building—again remedial steps can be taken immediately.

The second step is the review process. This step can be discouraging because you will be poking holes in step one. When conducting the review, ask participants to "bring me solutions, not problems." This encourages constructive ideas instead of just random complaining. It is okay to play devil's advocate. Ask tough questions. If you do not understand a process, say so. The second draft will be a better representation of the finished plan. It is not perfect, but it is getting close. The second draft is tested for validity through tabletop exercises. This is often known as the fire testing stage. It is like the beta launch of a software program—you want as many people as possible testing it to find all the problems so that they can be fixed. In tabletop exercises, scenarios are drawn randomly from a hat and the emergency management team walks through them using the plan as a guide. It is essential that these exercises are as spontaneous as possible. The team needs to get used to transitioning at moment's notice from normal routine to full emergency management mode. Any conflicts or gaps discovered during the exercises would be corrected in the final draft.

At this point, the plan has been battle tested and is ready to go. Printing and distributing the plan is a precursor to the most critical stage in this eight-step model—employee training. Employee training may consist of such diverse activities as electing floor captains, performing evacuation drills, and attaining CPR certification. A great emergency plan is worthless without employee involvement. Conversely, a good emergency plan supported by employee participation is worth its weight in gold.

Of course, writing the plan is an ongoing process. The emergency plan should be considered a living document, subject to growing pains and multiple revisions. A minimum annual review of the plan is a necessity and it should be looked at more often if your business is subject to frequent changes or other circumstances dictate review and revision.

REVIEWING THE EMERGENCY ACTION PLAN ANNUALLY

The emergency action plan should be evaluated on an annual basis at a minimum. No plan is ever perfect. Even if it were, people and processes change. A plan that sits on a shelf until an emergency occurs will be an outdated plan. The responsibility of auditing the emergency action plan should be assigned to a safety professional. One of the areas that frequently becomes outdated in most emergency action plans is important to support personnel data including names and phone numbers. This may include employees within the organization and contractors or governmental agencies outside of the organization. The auditor could confirm that individuals listed are still in the organization and still have the responsibilities that caused them to be included in the plan. Home, cellular, and work telephone numbers should be verified.

In addition to names and phone numbers of support personnel, an auditor should be able to answer the following questions:

- Does the organization have a formal plan for training its employees in emergency procedures?
- Is training being completed?
- Has the organization met with local community and government agencies and briefed them on its plan?
- Is the organization's insurance coverage adequate?
- Are there current photos or video of the facility and its assets?
- Does the plan reflect changes in the building layout?
- Does the plan reflect changes in policies and procedures?
- Does the plan reflect lessons learned from the drills and actual emergency events?

TRAINING, DRILLS, AND EVACUATIONS

Once the plan is finalized and all revisions have been made, the plan should be distributed to those playing a support role. These people might include emergency team members, floor captains, fire brigade members, evacuation assistants, runners, command post leaders, and anyone else who may be asked to play a key role in the emergency recovery efforts of the company. A good time to distribute the plan is at a training workshop for those employees who fill these support roles.

CREATING A COMPREHENSIVE EMERGENCY AND EVACUATION PLAN 285

To add credibility to the initial training workshops, open the meeting with a word from the company president or some other high-ranking management member or better still, get them to attend. This will impress upon those in attendance the importance of the roles they play. Lives are at stake. Business continuity and therefore jobs are at stake. The organization's reputation and productivity can be affected in a good or bad way. If this is true, back it up by a quick word of thanks and encouragement from management. It really does set a positive tone for the planning and recovery team.

The second reason for the training meeting is the opportunity for key support members to meet one another. Since the group will be a diverse one, many of the members may not know each other. As the security professional, your goal should be to cultivate an atmosphere of trust and reliance. This only comes when the members of the support team know and believe in one another. One of the best ways to cultivate this atmosphere is to conduct team exercises. There is no need to resort to corny team-building exercises, just have the members of the support group work on "what if" scenarios with members of their team. Of course, growth will not be experienced if everybody pairs off with people they already know, so suggest that they join a group of employees they do not know very well.

Training the Emergency Action Team

Because the emergency action team will be taking the lead in the event of an emergency, team members must be familiar with the plan and the subject property. Both considerations will take time and training. The training will not be a one-time event. It will be an ongoing event with a goal of preparing those taking the lead. Specific training formats include the following:

- Orientation training should be delivered to new members of the emergency action team. Each member should understand not only their responsibilities, but also the responsibilities of the other team members. At the orientation, team members should be assigned a copy of the emergency action plan book. The book will be their personal property, and if they leave the team or the company, they must return the emergency action plan. The team member will be responsible for updating the plan as revisions are distributed.
- Tabletop exercises are an excellent way to train members of the emergency action team without disrupting the normal day-to-day activities of the

organization. In a typical tabletop training exercise, members of the team sit in a conference room and pull emergency scenarios from a hat. Circling around the conference room table, each member would then explain their responsibilities to the group. A second pass around the table would allow team members to ask questions of other team members. If a weakness is discovered or an opportunity to improve is noticed, the entire team would work together to improve the emergency plan.

- Walk-through drills allow the emergency action team to perform their emergency responsibilities more accurately. During a walk-through drill, members of the emergency action team visit areas that are crucial to a successful plan implementation. Such areas might include shut-off valves, alarm panels, automated external defibrillators (AEDs), first aid stations, safe rooms, and exterior employee staging areas. The walk-through drills are essential for building familiarity and can be performed without impacting normal business productivity.
- Pre-evacuation drills are an extension of the walk-through drills. Involvement in this drill is limited to members of the emergency action team. Participants are instructed to evacuate the building, paying particular attention to the signage, exit routes, doorways, and staging areas. This is a perfect opportunity to evaluate the evacuation process without impacting the entire workforce. Participants should quickly evacuate as if an emergency is present. During the evacuation, notes should be taken, but that may force the participants to stop and write down their observations. A solution to this problem would be to provide participants with small tape recorders. As they evacuate the building, they would simply record their observations into the tape recorder without even stopping. Most smart phones also have voice recording apps. After the exercise is completed, the participants would play back their recorded observations and make adjustments to the plan as necessary.
- Full-scale exercises are useful in many organizations. The purpose of a full-scale exercise is to closely simulate an actual emergency event as possible. This necessitates the involvement of emergency response personnel, employees, and community response organizations. Because of the involved nature of the full-scale exercise, this option may not be easily implemented. You will need the cooperation of a number of different agencies, and they are not usually able to fulfill all requests. As an option, you may wish to have your

emergency management team participate in another organization's full-scale exercise. This option will not address your organization's specific needs, but it will allow you to "go live" and experience a staged emergency event. The experience gained can be of tremendous value and should not be overlooked.

The orientation training should be delivered to new team members at the time of their assignment. Thereafter, training should be provided to all team members at least annually. Other occasions when training, or at least an update, would be required include:

- When a process or responsibility is revised
- When policies affecting emergency planning and recovery change
- When a change to the layout of the physical building occurs
- When new equipment is introduced that may affect emergency planning (i.e., a generator or chemical storage)
- When key personnel changes
- When external circumstances change
- After each emergency

Revising the Emergency Action Plan

Revisions to the emergency action plan will occur. It is critical that you track these changes and ensure that all owners of the plan have the correct and latest information. You can accomplish this by inserting the page number, document title, and the revision date at the bottom of the page. This will allow you to quickly look at the emergency plan and determine if the information is up to date. As information is revised, simply print the newly revised pages, and distribute them to your team. Be sure to leave three or four pages intentionally blank at the end of each section, so you do not have to reprint the entire manual.

Training the Employees

Training is particularly important for effective employee response during emergencies. There will be no time to check a manual to determine correct procedures if there is a fire. Depending on the nature of the emergency, there may or may not be time to protect equipment and other assets. Practice is necessary to react correctly, especially when human safety is involved. Although most employees will not receive the depth of training that the emergency action team members receive, they still need some training including new employee orientation, tabletop exercises, announced evacuations, and unannounced evacuations.

When training adult learners, three factors are required:

- Awareness
- Education
- Involvement

The program starts with awareness. Next, formal education must take place. Finally, if any of this information is going to be retained, you must get involvement.

Emergency orientation training should be delivered to new employees as they accept assignments with your organization. This may include contractors such as janitorial personnel who spend substantial time on your property. Realistically, when a new employee starts work at your organization, they spend the first day or two filling out paperwork, reading through the company policies, asking questions about health insurance, and other related activities. The emergency action policies that you provide them will be their first exposure to the safety culture your organization deems important. You can help them realize the importance placed on employee safety and business continuity. Do not miss this opportunity! It is likely that you will need to work with your human resources department to ensure that your emergency policies become part of the standard new employee orientation process.

In most organizations, especially mid-sized to large companies, a complete copy of the emergency action plan will not be given to employees. The plan contains information that is not relevant to most employees, and if you overburden these employees with information, they will not look at it. Instead, provide employees with a scaled-down version of the plan, perhaps an executive summary with important phone numbers and a simplified floor plan.

Tabletop exercises are an excellent way to train the general employee population without disrupting the normal day-to-day activities of the organization. In these tabletop exercises, many excellent ideas and improvements have come from the employees. Typically, the employees will challenge the members of the emergency action team with questions and issues that were not discussed previously.

CREATING A COMPREHENSIVE EMERGENCY AND EVACUATION PLAN

Announced evacuations are a great way for the employees and the emergency management team to gain confidence in one another. The problem with most evacuation drills is the timing. People are busy and the evacuation is rarely considered a priority. When the evacuation drill does occur, people are confused, meetings are interrupted, deadlines may be impacted, and more time is spent performing the evacuation because of the unfamiliarity with the procedures. By announcing the evacuation, you allow the employees to prepare for the event. You also allow the employees to find the emergency exits prior to being evacuated. You have a chance to explain the evacuation routes, staging areas, and other related procedures. Eventually, as the employees become more familiar with the evacuation procedures, they will be ready for the unannounced evacuation.

Unannounced evacuations more accurately simulate an actual emergency event. When faced with an emergency we often forget the obvious. Where do we go? Where is the closest emergency exit? Where is the secondary exit? Who needs assistance in evacuation? Where is our staging area? Who performs the employee's roll call? Questions like these are easy to answer during a tabletop exercise, but when the unannounced evacuation occurs, we often get rattled. By conducting unannounced evacuations, you provide employees with the opportunity to work out the kinks. The unannounced evacuation should not be conducted until your employees are comfortable with the announced evacuations.

Regular evacuation drills are a critical component of a successful emergency action plan, but who can afford the impact of evacuating an entire building during a single drill? Here's the good news: a total evacuation is not necessary, nor is it recommended. Smaller drills have proven to be quite effective, and they provide the training necessary for the employees to gain confidence in the event of an actual emergency.

The NFPA has embraced an evacuation model that conducts a partial evacuation of the affected floors and the floors immediately above and below the affected floors. For most fires in most buildings, this course of action is sufficient.

Floor Captains and Fire Brigades

Floor captains can be of great help in the event of an emergency. Their purpose is to assist the employees and visitors to evacuate the building safely and efficiently. They may be trained in first aid and basic fire extinguishing techniques. Floor captains provide special attention to those with disabilities, and they ensure that everybody has left the building.

In some businesses, both floor captains and a fire brigade may be established. The responsibilities of a fire brigade are greater and require more advanced training. OSHA defines a fire brigade as "an organized group of employees who are knowledgeable, trained, and skilled in at least basic firefighting operations." Fire brigades are used when the nature of a business exposes its employees to unusual danger and the delay in response from the local fire department would cause further risk. For example, a nuclear power plant may find it prudent to establish a fire brigade to commence the firefighting process prior to the fire department's arrival. Fire brigades are not required by OSHA, but if you do elect to create a fire brigade, a few regulations do exist. In addition to the many regulations, OSHA suggests that prefire training should be conducted by the local fire department so the fire brigade will be familiar with the workplace and potential hazards. This meeting and training also fosters cooperation between the local fire department and the fire brigade so they will be able to work together in the event of an emergency. If you wish to create a fire brigade, consult with your local fire department, and review OSHA's regulations at CFR 29 1910.156.

Shut Down Procedures in an Evacuation

When evacuating a building, it is usually not a good idea for everybody to just leave. Certain equipment and supply lines could present a major problem if left operational. For instance, suppose you have a natural gas service line entering your building. This line fuels four pieces of equipment. If these lines are left open during an evacuation for a fire, the risk of the fire being further fueled by the presence of natural gas. A small fire could quickly get out of control and endanger the lives of the firefighters and other rescue personnel. To solve this problem, assign knowledgeable personnel the responsibility to shut down lines and equipment that might interfere with recovery efforts prior to evacuating the building. This will requires identification of potentially dangerous equipment, storage tanks, containers, supply lines, and valves.

The location of storage tanks and containers used to hold flammable or toxic chemicals must be communicated to the fire chief. This information will

CREATING A COMPREHENSIVE EMERGENCY AND EVACUATION PLAN

play a definite role in how the fire department uses its personnel and how it will fight the fire. Attempting to explain the hazards and their locations without a floor plan is folly. To make the floor plan user-friendly, the following information should be prominently identified:

- Shut-off valves
- Shut-off switches
- Storage tanks and containers
- Supply lines
- Breaker panels

Even when a floor plan is available, in the haste and confusion of the emergency, important features could be difficult to locate. To solve this problem, the visual floor plan must be cross-referenced with a written location system. The location key (see table 11.1) is a good way to accomplish that. The location key is created in a table format. The critical points are identified in the leftmost column. The middle column identifies the location of the items. The location is derived from an easy-to-implement grid system. The grid system can be superimposed over the electronic version of the floor plan in a CAD system, or you can simply draw a grid over a hard copy of the floor plan. The grid system works on an x and y axis. The horizontal axis (x) would be divided into segments and labeled with ascending letters of the alphabet. The vertical axis (y) would also be divided into segments and labeled with ascending numbers. Much like a map, you can use this grid system to quickly locate important items.

The location key is most helpful when the items are arranged alphabetically. The final column includes a space for comments. This location grid should be attached, manually or electronically, to the actual hard copy of the floor plan.

At the actual shut-off locations, make certain the valve is clearly identified. A sign with letters no less than two inches in height should identify the type

Table 11.1. Location Key

Item	Location	Comment
Breaker panel	A-4	
Chemical storage	K-2/P-17	See MSDS/Inventory log
Natural gas shut-off valve	H-12	
Water shut-off valve	H-13	Valve key above valve

of shut-off. Have you ever been confused by how to turn a valve off? Do you turn the valve clockwise, or is it counterclockwise? It is not always apparent, so be sure to label (at the valve location) the correct directional rotation for on and off. It may be advantageous to color code the sign depending on its use.

For instance, a red sign with white letters may be used for all flammable and combustible valve locations. A blue sign with white letters may represent water shut-off valves, and so on. Finally, always be aware of special keys or wrenches that may be required to turn the valve. Valuable time will be wasted if you do not know where the key is located.

Assisting Those with Special Needs in an Evacuation

The orderly evacuation of all occupants, including those with disabilities, can be a challenge. The Americans with Disabilities Act (ADA) offers many excellent suggestions for assisting those with disabilities during evacuation of a building. Those with hearing, vision, mental, or mobility issues need to be identified and their egress from the building must be addressed prior to an emergency event, not during it.

When considering specific evacuation procedures for those with special needs, you will see that typically two types of solutions are available—engineering and procedure-driven. Engineering solutions might include slings or controlled descent devices that allow the disabled person to exit via the emergency egress stairwell. Procedure-driven solutions might include the development of a partner system or the creation of areas of refuge for those with disabilities to wait for assistance. Often a mixture of both solutions will provide the best protection. Even though you attempt to cover all the possibilities, you cannot fully protect your employees unless you are aware of their limitations. People with permanent or long-term disabilities are relatively easy to identify and assist. The challenge occurs when dealing with people with temporary limitations such as pregnancy or those aided by crutches. Think also about the individual with asthma, who might be able to easily evacuate the building under normal circumstances but, because of the stress and presence of smoke, may have difficulty breathing. When developing your emergency plan and building evacuation procedures, consider these situations.

Use of Elevators for Exit

During an emergency, elevators are not considered a permissible means of egress from the building. In most buildings, the elevator has been programmed to immediately descend to the bottom floor (unless a fire is present at the bottom floor) upon being triggered by the alarm. At this point, the elevator is rendered inoperable unless emergency response personnel activate it using a special key. Many people have wondered why elevators cannot be used. Some studies have shown that if elevators were used during evacuations, building occupants could leave the building up to 50 percent more quickly. In particular, those with disabilities could safely and efficiently evacuate, without impeding the progress of others.

Using elevators during an evacuation sounds like a good idea until you understand why they are off limits. Firefighters must have a way to get to the fire. Since building occupants are exiting the building and fire rescue personnel are entering, there exists an opportunity for interference. Use of the elevator by fire rescue personnel allows them to get to where the emergency is in a quick fashion. Imagine if the firefighters, with their heavy rescue gear, had to climb seventy flights of stairs! It is much easier for building occupants to walk down seventy flights of stairs than it is for fire rescue to walk up seventy flights of stairs.

Another problem with elevators is the increased potential for smoke inhalation. Elevators cars are not very airtight. Doorways at each floor allow smoke to enter the elevator shaft. Smoke inhalation is not as likely for fire rescue personnel because they carry self-contained breathing apparatuses (SCBA). Building occupants attempting to use an elevator during evacuation might not make it out alive.

The fact remains though, that elevators could speed the evacuation, especially for those with disabilities. The National Institute of Standards and Technology (NIST) has studied the feasibility of elevator use during an evacuation, and they have developed a series of recommendations. First, the enclosed lobby area at each floor and the elevator shaft would need to be pressurized so that both remain smoke free. Second, dual power systems would need to be installed for reliability. Lastly, elevator components would need to be sufficiently waterproof to prevent failure due to flooding of the shaft during firefighting operations. Given these requirements, there has not been a strong movement to build elevators that can be used by building occupants during an evacuation.

It is also safe to imagine that if such elevators did exist, their use would not be limited to those with disabilities. A bottleneck could occur as occupants wait for the elevator to arrive.

Evacuation Assistance Devices

Those persons with mobility limitations will often find building evacuations to be a difficult and sometimes precarious event. Imagine someone attempting to navigate a wheelchair down a stairwell. Now, add hundreds of people to the scene and you can appreciate the difficulty with many evacuations. Even people with temporary disabilities will find evacuation difficult. To assist these individuals, several evacuation assistance devices have been developed. If your building is a multistory structure, you should seriously investigate the evacuation assistance devices that are on the market.

When considering different evacuation devices, consider the distance the person must travel, the person's confidence in the device, the impact on stairwell space, and cost. It would be advisable to ask wheelchair users for their advice and comfort level prior to equipment selection. If they are uncomfortable with the device or they do not know how to operate it properly, the device will likely go unused. When a decision is made on evacuation devices, be sure to train disabled persons and others how to use it properly.

Among the evacuation devices available are mechanical devices that allow a controlled descent in the stairwell. One such device allows the wheelchair user to roll onto a platform, secure himself or herself to the platform, and then descend to safety. The advantage of this device is the wheelchair user never has to be removed from the wheelchair. Other devices require the disabled person to be separated from the chair. This introduces an opportunity for injuries and may create an uncomfortable emotional separation for the person with mobility limitations. Another device requires the wheelchair-user to transfer from their wheelchair to a transportation platform which travels on specially installed tracks down the stairs. The descent speed can be controlled by a braking mechanism. Other nonmechanical devices are available to assist disabled persons with an evacuation. Various straps and harnesses are available that assist in the evacuation.

If you do not have any of these types of evacuation assistance devices and you are faced with manually assisting someone in a wheelchair, at least use a safe technique. For instance, two people can efficiently guide a wheelchair and its occupant. Start by positioning one person behind the wheelchair grasping the pushing grips. Next, tilt the chair backward until balance is achieved. While keeping the balance, move the wheelchair forward toward the steps. As the chair moves down the steps, always remain one step above the chair. Keep your center of gravity low and let the back wheels gradually lower to the next step. The wheelchair occupant should be leaning back. The occupant may try to resist this position by leaning forward but remind him or her not to lean forward. Kindly ask them to remain in the chair, lean back, and let assistants control the chair's descent. Be sure to have a second person assist by positioning themselves in front of the wheelchair to keep the wheelchair from gaining too much speed during the descent. Both assistants will attempt to keep the descent as smooth as possible, but they should not lift the chair off the ground. The weight of the disabled person and the wheelchair could easily injure the assistants. A third assistant may be needed at the rear of the wheelchair to share control of the push grips.

Regardless of the techniques utilized, do not forget to use proper lifting techniques. In past evacuations, people have hurt backs, twisted knees, and pulled muscles because they did not use proper lifting techniques. Usually the people who are hurt during these evacuations are the big and strong men who believe they can lift someone on brute strength alone. Meanwhile the smaller person who uses good techniques will come out uninjured.

Some of the basic lifting techniques include:

- Bend your knees
- Keep your back straight
- Use your leg muscles to lift
- Hold the person close before lifting
- Do not lift from an awkward position
- Ask for help if it is needed

If any evacuation devices are used, make sure that the location of the device is noted in the emergency plan and on the emergency action floor plan. When an evacuation drill is conducted, use the devices. The time to get comfortable with these devices is not during the excitement and confusion of a real emergency. The emergency action plan must also address training on proper evacuation device use, including safe lifting techniques.

Areas of Refuge

The ADA outlines specific requirements for fire-resistant spaces where persons unable to use stairs can call for and wait for evacuation assistance from emergency personnel. These areas are commonly known as areas of refuge and are often located at egress stair landings, but other areas that meet specific fire rating and smoke protection can be used. ADA regulations require that two-way communication devices be installed in these areas of refuge so those awaiting assistance can be identified.

The refuge area should only be used when the disability seriously impedes the evacuation of the individual. For instance, a wheelchair may present problems because of its weight and size, and you may elect to bring that person to an area of refuge. On the other hand, a person with a hearing or visual disability should be evacuated with the other building occupants, using the stairwell and emergency exits. An area of refuge is not a place to bring all people with any disability.

The Evacuation Staging Area

An orderly evacuation from a building during an emergency saves lives. Think for a moment about an evacuation. Upon notification, personnel will begin to file out of the building, following the appropriate exit signage and routes. Eventually, all of the building occupants, perhaps thousands of people, will be standing in the parking lot or some other location. Now imagine the fire truck or ambulance driving up to the building. Will they be able to access the site? Will people be endangered as they wander around the site? Will they interfere with the emergency responders? For these reasons and many more, staging areas outside of the building must be developed.

- Employees should be instructed to move away from the exit doors.
- Ample staging area space should be provided for employees.
- Have several numbered staging areas, if necessary, with people allocated to a specific area.
- Groups should be kept relatively small so everyone can be accounted for.

CREATING A COMPREHENSIVE EMERGENCY AND EVACUATION PLAN

- Exit doors should not empty to parking spaces or streets.
- Inclement weather such as rain, snow, or heat should be considered.

Keeping groups small is important. A large, congested group of people wandering the parking lot creates confusion, and it is difficult to account for who has exited the building and who remains. Small groups are easier to manage. So, keep the groups in the staging area small, twenty people or less per group, if possible.

One concern that must be addressed when evacuating the building is safety. Is it safer to stay inside? Is the exterior staging area more dangerous than remaining indoors? Inclement weather, biological attack, or nuclear accidents are examples that may make an evacuation dangerous. The decision is probably not one to make internally. Guidance from law enforcement agencies or rescue personnel should be sought. If you do elect to evacuate during inclement weather, is an alternate staging area available? If an earthquake or fire has occurred, is the staging area far enough from the building to keep the evacuees safe? These are the types of roleplay questions you'll need to address in your safety plan. It is important to account for everyone during an evacuation because for instance, in the event of a biological or chemical attack you need to make sure that no one leaves the scene and possibly spreads contamination.

Accounting for Employees and Visitors

When evacuating a building, employees should have a specific area that they go to and remain at until the decision is made to go back into the building or to leave the premises. This staging area serves several purposes. First, the staging area keeps the crowd of building occupants away from the building and out of the way of emergency responders such as police, paramedics, and firefighters. It should be sufficiently far from the building so there is no danger from falling debris or glass if windows blow out. Second, the staging area provides emergency team members with an organized way to account for employees and visitors. OSHA 29 C.F.R. 1910.38(a)(2)(iii) states procedures must be in place to account for all employees after an emergency evacuation has been completed.

Think about the process. Where will employees gather? Who will perform the roll call? If somebody is unaccounted for, do you have their cell phone or pager numbers? If somebody is missing, how will this information be related to the emergency rescue personnel? Who will give the "all clear" order? These are the types of questions that must be answered if you hope to conduct an orderly and safe evacuation.

It is important, too, that employees understand the reasons for having staging areas. If the building occupants simply exit the building and mingle in the parking lot, it becomes difficult to perform a roll call. When this happens, emergency rescue personnel are placed in harm's way because they are searching the building unnecessarily for people who are already outside of the building. Someone should be assigned the group leader whose primary responsibilities would include keeping the group together, performing roll call, and serving as spokesperson to the emergency command post.

The Chain of Command in an Evacuation

During an evacuation, order must be kept. If not done in an orderly fashion, building occupants can be injured, time can be wasted, emergency response personnel can be injured, and company liability will be increased. It is critical that a clear chain of command be established prior to any evacuations whether they be training drills or actual emergencies. Here are a few of the important roles that make up the chain command and which must be filled if an effective evacuation is to be conducted.

FIRE BRIGADE

The fire brigade will perform basic first aid and extinguish small fires in the event of an actual emergency. In an industrial environment they may fight more advanced fires if professionally trained and supplied. They may also assist the floor captains by directed building occupants to the available evacuation routes, but this is not their primary responsibility.

FLOOR CAPTAINS

The floor captains ensure that all building occupants have recognized the alarm and are quickly headed for the evacuation routes. Floor captains may also assist those with disabilities or those who do not speak English. Floor captains may be placed at stairway and elevator locations to monitor their use. They also check restrooms, loud locations, and private offices for persons who may not have heard the alarm or chose to ignore it.

GROUP LEADER

The group leader gathers employees at the preassigned location in the parking lot. They will take roll call and will call missing employees or visitors on cell phones or pagers. If contact cannot be made, they will relay this information to the runner.

RUNNER

The runner serves as a liaison between the group leader and the command post leader. During the evacuation, the runner (or runners if there is a large workforce) visits each individual group and gets a status report from the group leader. The runner will then relay this information to the command post leader.

COMMAND POST LEADER

The command post leader collects employee status from each of the groups and compiles a list of unaccounted personnel. The command post leader would pass this information on to emergency response personnel. If the evacuation is a drill, the command post leader would announce the "all clear" and direct all employees back into the building in an orderly fashion—for instance, staff could be allowed back in groups at five-minute intervals based on their staging area number. If an actual emergency occurred, the command post leader would consult emergency response personnel to make the decision to reenter or leave the premises.

Because the group leader will be conducting a roll call, they must have a current list of employees and visitors. The list of employees is easy to compile. If the group is kept small, the list will be easy to keep current. It should include the employees' names and cell phone, pager, and home phone numbers. The list should be kept on a small wallet-sized laminated card. If your group includes fire brigade members, floor captains, or employees responsible for equipment shut down, they likely will not be exiting the building until they have completed their assigned responsibilities and may not be in the staging area when the roll call is conducted.

One of the most difficult aspects of the roll call is accounting for visitors such as vendors and customers. Most of the time these visitors will be in the presence of an employee. If an evacuation occurs, the visitor should follow the employee out of the building. The employee should instruct visitors to follow them. If the employee and their visitor are separated during the evacuation, make this known to the group leader as they are taking roll call. The names of visitors who are unaccounted for will be passed on to the command post leader. Any group that finds an unknown visitor in its midst should verify the person's name and the name of the person they were visiting. You should also instruct that person to stay put. It may be helpful to retrieve visitor logs from the reception area and use those for accounting for non-employees.

If you plan to conduct an evacuation drill, someone should be assigned as the drill coordinator prior to the drill. This person plans, conducts, and evaluates the drill. A best practice is to rotate this assignment so various members of the emergency action team will understand the position. It is always advisable to have trained alternates, and this is a great way to ensure that will happen.

Those Who Refuse to Participate in Evacuation Drills

There is no question about it, evacuation drills take time. Since time is valuable and we have so little of it to spare, some employees may resist taking part in the emergency evacuation drills. How will you handle this difficult situation? After all, you cannot physically remove someone from the building, especially during a drill. This issue must be resolved and resolved quickly.

If certain employees are allowed to sit out the drill because they are too busy, soon other employees will do the same. Before long you will lose control of the drill. The best way to ensure a well-participated evacuation drill starts with the organization's leadership. Management should clearly state their endorsement of the evacuation drills and make it clear that all employees are expected to participate. Of course, this includes executive management. They must be visibly participating as well! Management should put some bite into this requirement as well. For instance, a process could be established whereby the name of any employee not participating in the drill will be forwarded to the vice president of human resources. When the organization's leadership takes a leadership role in emergency planning, the employees will follow their example.

Employers also have the right to require nonparticipating employees to sign a waiver, releasing the company from liability if they fail to evacuate. The legal credibility of such a waiver may not be ironclad, but at least the employee will understand the importance placed upon the drill. Additionally,

CREATING A COMPREHENSIVE EMERGENCY AND EVACUATION PLAN 301

the burden of the added paperwork may be more painful than participation in the drill. Some organizations have made the required paperwork for nonparticipation so long and involved that the employees spend less time in an evacuation than they would filling out the paperwork!

The Reentry Process after an Evacuation

After an evacuation drill has been completed, the building occupants will be required to reenter the building in an orderly fashion. This may seem like an obvious and simple process, but it may not be as easy as you would think. If your building is a large multistory structure, the reentry process can be especially difficult and frustrating. Most large buildings are not equipped to handle the entry of all occupants at the same time. For instance, the thousands of tenants in a large building such as the John Hancock Center in Chicago realize that if everybody started work at 8:00 a.m., it could take an hour or longer to get to your office.

The lines at the elevator would be so long you would need to show up at the building's lobby at 7:00 a.m. just to get to work on time. To compensate for the rush of workers starting each day, most employers stagger their arrival times. This serves to alleviate the crowds. By requiring some employees to start work at 8:00 a.m., others at 8:15 a.m., yet others at 8:30 a.m., you effectively lessen the impact of the mass of people entering the building.

The same concept applies to building entry after an evacuation drill. When the drill is completed, employees will break from their staging areas and walk directly to the elevators. Imagine the long lines, the chaos, and the frustration. The next time you attempt to practice an evacuation drill, you'll find many people unwilling to participate because of the time consumed in reentry and loss of productivity. To solve this problem, two good options exist.

First, you may elect to conduct an evacuation drill in smaller groups, perhaps two or three floors at a time. This makes the management of the groups and reentry into the building much easier. You will need to devise a way to alert the occupants of those floors being evacuated, without tripping the alarm and triggering an entire building evacuation. If your alarm system does not allow for a selective siren or message, you may decide to use your floor captains and a bullhorn. If you use this option, try to simulate the actual sound of the regular building alarm. A simple tape recording of the alarm played over the bullhorn may suffice. Another option is to purchase a small alarm horn or siren at an electronic store.

Your desire is to condition the building occupants to recognize the alarm and immediately begin evacuation. The second option is to conduct a complete building evacuation, but instead of allowing the building occupants to reenter the building all at once, you will release them in small groups. This is best done by releasing various staging areas in a staggered fashion. Groups would be released based on their location in the building and which elevator banks were available. All elevators, including cargo elevators should be utilized during the drill. If an elevator or escalator is out of service, you might be better served by postponing the evacuation drill until you have all systems operational.

Evacuation captains and assistants will need to have communication via two-way radio to ensure a smooth release of building occupants. Some additional planning will be necessary, but this option more closely represents an actual emergency.

When an actual emergency occurs that requires an evacuation, law enforcement personnel may want to interview some of the building occupants. Employees who were in an area of concern or who witnessed a problem should be identified and made available to the investigating agencies. A simple process can be created to ensure that this happens. For instance, the involved employees should have evacuated the building and assembled at their assigned staging area. Involved employees should check in with their staging area group leader and inform them of what they witnessed. The group leader would share this information with the runner, who would in turn deliver the information to the command post leader.

This structure will allow important information to be relayed to a central location in a quick and efficient manner.

After the Drill: Evaluating Performance

It is crucial to review the evacuation procedures after a drill is performed. If the emergency evacuation was a drill, let the group know how they did. Was the clocked evacuation time within the established benchmarks? Did they do better or worse than the last time the drill was practiced? If it was better, share that success with the employees. They need to know that the effort they have expended has been worthwhile. If the timing was substantially worse,

CREATING A COMPREHENSIVE EMERGENCY AND EVACUATION PLAN

ask the group why they think the performance was poor. They may be able to provide valuable insight into problems and the corrective measures that should be taken.

If you keep your building occupants apprised of the evacuation results, they will take an interest. They will want to improve their evacuation time the next time the drill is conducted. Most importantly, in the event of an actual emergency, the building occupants will have the confidence to evacuate without resorting to pushing, running, or other inconsiderate and dangerous behavior.

POST-EVENT RESTORATION

After an emergency has occurred, successful restoration and recovery actions are crucial. For business owners, this restoration phase can be the difference between business continuation and bankruptcy. The employer will be faced with many difficult decisions, and how quickly these decisions are made becomes critical to survival. Confident and quick communication with the media, suppliers, clients, and other groups must not be overlooked.

During the restoration phase employee comfort and safety must not be minimized. Employees will be concerned about their jobs. Many employees will be affected by the emergency at home, and the stress of the events can be overwhelming. Employees involved in the restoration phase will likely be working long hours, and in their zeal for recovery will not always take proper precautions or get proper sleep. These are some of the issues that must be addressed during the restoration phase.

Activating the Emergency Response Team

When members of your emergency response team are called to action, they must be prepared mentally and physically for what they may find. If your team is an early responder, a group that responds within an hour after the emergency event, it is unlikely that team members will be able to predict the severity of what lies ahead. Water service could be disrupted. Heating or cooling may be nonexistent. Floodwater may impact their efforts. Lack of electric service will hamper recovery efforts. If the area has been hit hard by the emergency, food and water will be scarce. Finally, transportation in and out of the area may be difficult. Many an emergency team member has found themselves stranded in an inhospitable environment without food, water, or a way to quickly get back to civilization.

The emotional and physical trauma can cause early responders to feel hopeless and powerless. In severe cases, the responders may even shut down, physically and emotionally. This phenomenon is magnified with a lack of food, water, or sleep. Finally, the experience reaches a crescendo after the initial adrenaline rush has passed and the realities of restoration become evident.

Understand the physical and emotional swings during a restoration effort and address them proactively. A list of provisions must be provided to emergency team members and responders prior to an emergency. The list should be part of the emergency action plan, and friendly reminders should be given to team members on an annual basis at minimum.

Members of the emergency response team or employees asked to assist in the recovery effort should be accounted for and tracked on a frequent basis. Essential information such as telephone numbers, home addresses, vehicles, location assignments, skills, and tools should be recorded.

Post-Event Access

Murphy's law dictates that often an emergency will occur after business hours or when emergency team members are otherwise away from the office. After the emergency, employees will have been evacuated, roads may be closed, and access to "ground zero" may be limited. If this occurs, how will you and other emergency team members be able to reach and access your buildings?

Have you met with local law enforcement and rescue personnel? Have you introduced your emergency response team? Have you discussed your emergency action plan? If you take these actions and develop a relationship with the local agencies, they will be more apt to grant access to your emergency response team.

Assessing the Damage

Once you have arrived on the scene, you should immediately take steps to assess the damage to your building and the surrounding property. Depending on the nature of the emergency and the resulting damage, you may elect to bring in specialized help for this assessment. If you have any doubts about the structural integrity of the building, do not enter it! Building audits provide an excellent starting point for post-emergency recovery; however, they should be revised to incorporate the specifics of your property and site.

Insurance and Salvage Decisions

When entering a building for an assessment, make certain you have the proper personal protective equipment (PPE) available and in use. This might include hardhat, steel-toed shoes, long-sleeved shirt, nuisance dust mask, respirator, and safety glasses. Employees assigned PPE must be professionally trained in its use, and with many items such as respirators, the gear must be fit tested.

As you walk through your property after the emergency event, you will need to document all property damage that may have occurred. If you have an insurance policy on the building and its contents, this documentation will be essential if you hope to be reimbursed for your loss. If you are not familiar with the insurance coverage, immediately contact your risk management department or insurance provider to make certain you are following correct procedures for documentation. Some type of visual recording device should be used when performing an assessment. If possible, a video camera should be used to record the damage. Continuous video gives a more accurate and complete description of what has occurred than photographs.

Keep salvageable property separated from "damaged-beyond-repair" property. The damaged property should be kept until an insurance adjuster has had the opportunity to evaluate it. If the damaged goods are a hindrance to recovery and exposure to the elements will not further harm them, establish a temporary staging area on site and move those items to that location. Prior to moving these items, document their condition and where they came from. If salvaged or damaged items are sold or otherwise removed from the property, be certain to collect a signed inventory list from the party removing the goods.

After a particularly devastating disaster, insurance response may not be immediate. When Hurricane Katrina struck New Orleans, many people who were affected said that it took weeks, and in some cases months, for the insurance companies to meet with them.

The Restoration Phase

At some point after victims have been aided and the damage has been assessed the actual restoration phase must begin. During this time, if unprepared, confusion, exhaustion, and often depression can set in. Having a detailed plan of attack will decrease the downtime caused by the emergency event. A few words of warning are in order, though. The task of rebuilding often seems monumental, even impossible, but the ability to recover from adversity gracefully is uniquely human.

Logistics

One of the most difficult aspects of emergency recovery is logistics—the movement of equipment, supplies, and personnel to locations where they are needed most. This movement becomes difficult when the access roads have been damaged. It becomes complicated to achieve when telecommunications are down. It is nearly impossible when your employees and contractors are literally digging themselves out of the ruins. Even with all these issues satisfied, the rush to get back to business puts extra pressure on the restoration team.

Things can't happen quickly enough. The problem with logistics is made more difficult when you have multiple damaged properties in your portfolio. You find yourself juggling conflicting priorities, schedules, and demands.

There must be a central location where the management of the recovery efforts can be performed without interruption. Even in an evacuation drill, a command center should be established as the focal point of communication. Whether you elect to locate the command center at the damaged property or off-site typically depends on several factors. For instance, after an actual emergency that requires restoration, the command center must be equipped with phones, fax machines, computers, and other communication resources. If electrical service has not been restored and emergency generator service is not available, the command center should be located elsewhere.

Having said that, in most cases the command center would better be served at the location where restoration is being done. In scenarios where you have multiple properties that have been damaged, a central command center should be established and then backed up by individual on-site command centers. The command center serves as the "brain center" of the recovery effort. Contractors, employees, media, government agencies, insurance providers, or anyone else with an interest in the property needs to have a place to go to ask questions and get answers. The command center must know everything that is going on to ensure that priorities are being met and that resources are not being wasted.

Do Not Forget about Safety

Most employers are required to adhere to OSHA regulations, including specific recording and reporting responsibilities. The possibility of a workplace injury is increased while recovering from an emergency event, and for this reason particular attention must be given to employee and contractor safety.

In the event of an accident that results in any injury, even a minor one, a post-accident report should be completed. The purpose of the post-accident report is to identify all injuries, even those that do not trigger an OSHA recordable. Ultimately, the post-accident report form will serve as tool for trending analysis. If employees keep getting injured while moving sandbags, a proactive manager would look for ways to eliminate or reduce the risk involved.

OSHA has created three specific forms that must be completed by employers when specific injuries or illnesses occur. Some industries and organizations employing ten people or fewer are exempt for OSHA's record keeping requirements unless selected by the Bureau of Labor Statistics to participate in an annual survey. The required OSHA forms include:

- OSHA Form 300 (Log of Work-related Injuries and Illnesses). This is used to record information on all illnesses and injuries that require medical attention beyond basic first aid.
- OSHA Form 301 (Injury and Illness Incident Report). This is the individual record of each work-related injury or illness recorded on the 300 Form. It includes additional data about how the injury or illness occurred. The 301 Form replaces the former OSHA 101 Form.
- OSHA Form 300A (Summary of Work-related Injuries and Illnesses). This replaces the summary portion of the former OSHA 200 Log and is now a separate form, updated to make it easier to calculate incidence rates.

Monitoring Restoration Progress

Because of the psychological trauma a major emergency event causes, the restoration progress can be drawn out. Often the work seems overwhelming. There appears to be no end in sight. If you speak with people who have witnessed great destruction, they will tell you that the damage is so great you sometimes do not even know where to start.

They may not be able to focus on the large restoration task at hand. They may be easily distracted, or they may be concerned with the welfare of their families. For this reason, be very selective about who you choose to monitor your restoration progress. You will need to have a leader on board who can remove himself or herself from the emotional trauma of the event. This person may be an employee from a non-affected area, or it may be an outside contractor.

Regardless, make sure the person is emotionally able to handle the task. Like any large project, the restoration phase must be kept on track as much as possible. If not managed, tasks will be overlooked, and valuable time lost without significant progress being made. This is especially critical in cases in which business continuity is at stake. If your organization is unable to conduct business, and for every day it is unable to do so, the company loses $250,000, the impact is significant. After the emergency, building assessments must be performed. Based on your findings, you may decide to bring in an architect or structural engineer. At some point, a scope of work should be decided upon and timelines should be developed. A project manager is critical if the required restoration is substantial. In fact, you may elect to bring in a project manager who specializes in recovery from disaster events. They have unique experiences dealing with restoration projects, and they can be instrumental in locating federal assistance that may be available to your organization.

Safety is another concern after an emergency event. Many people have been unnecessarily injured or even killed during the restoration phase because of a lack of safety measures. In our zeal to get back to a sense of normalcy, we take shortcuts and practice unsafe work habits. Our property may not be structurally sound, but we enter it. Water may create a breeding ground for mold and other indoor air quality issues, but we breathe the air without a respirator. Employees are eager to help but are not professionally trained or clothed. Broken glass is strewn about the floor, yet we walk on it without proper work shoes. Do not minimize the risk associated with the recovery efforts.

Keeping the Lines of Communication Open

The flow of factual and comforting information is important, especially following an emergency. People are confused, even scared. Rumors run rampant. The emergency planner will play an important role in disseminating reliable information. Whether you are communicating with employees, the media, shareholders, or other interested stakeholders concise and truthful information must be shared.

The media will seek statements from your organization. Concerning the media, the worst action you could take is inaction. If you fail to return calls or respond to questions, you run the risk of the media consorting with other sources such as competitors, employees, or neighbors. These sources may be biased, or uninformed, and false information could be delivered. So, rule number one is: Do not avoid the media. Rule number two is: Do not say more than you should. No matter how articulate you might be, no matter how comfortable you might be in front of an audience, you absolutely must plan and practice the press conference. Practice in front of a "friendly" audience first. An audience of trusted employees in your organization would think of all the hard questions that could possibly be asked. Discuss what should and should not be said. Draft out responses to the questions you know will be asked.

As part of your emergency planning you should devote a section of your plan to communications and develop a series of media holding statements covering many scenarios that could be issued almost immediately to keep the press informed until a more definitive press release can be issued.

Communication with Employees

Fear, anxiety, and, in some organizations, distrust are natural byproducts of an emergency event that requires extended recovery. Employers have a responsibility to communicate with employees when emergency-related events occur. Employees want to know when it is safe to return to work. They may need additional assistance if they were directly affected by the event. Quick and helpful information is an effective way for organizations to show that management cares for the well-being of the employees. Consider setting up a telephone hotline, creating a special information page on your company's website, or texting breaking news to employees.

After an emergency event, communication should not be limited to verbal announcements. Written communication should also be provided to the employees. Although not actually a press release, the formal communication is an invaluable tool for the employee/employer relationship. Without communication, employees will turn to rumors and innuendo. Issues such as work locations, return-to-work dates, and compensation during restoration should be addressed. Reminders on how to handle media inquiries and rumors should also be covered. Keep the communication positive and always thank the employees for their dedication.

Communication with Families

When emergency events occur and the ramifications are protracted, the families of employees can be affected. Anxiety and fear can paralyze those who do not know the condition or whereabouts of their family members. If the emergency event causes extended financial hardship on the organization, employees and their families may be concerned about job security. A delay in delivery of a paycheck could prove disastrous for many families. If layoffs are necessary, employees and their families will be interested in severance packages, health insurance continuation, and job placement assistance. These concerns must be addressed as soon as possible.

Communication with Insurance Providers

Although insurance providers will send a claims agent to review damages to your property, you may consider a proactive approach to communication to be wise. This is especially important when an emergency is regional in nature and the insurance provider is overburdened with claims. A formal communication describing the event and how you were affected is suggested. The correspondence should describe the damage incurred and identify the employer's contact information.

Communication with Government Agencies

An often-overlooked target audience is government agencies. These agencies might include FEMA, Red Cross, police, National Guard, FBI, fire rescue, and even OSHA. Formal written communication may be recommended based on the nature and impact of the emergency. For instance, an on-site oil spill and resulting fire would warrant formal communication with the Environmental Protection Agency. They will be interested in what happened, when it occurred, what is being done to address the spill, and what measures are being taken to ensure the incident is not repeated. In scenarios such as this, it is better (and often required) to contact them before they contact you.

Communication with Customers and Stakeholders

Contact with customers is often overlooked after an emergency event. If the emergency event is great and critical operations are impacted, customers and other stakeholders, such as shareholders, may be impacted as well. They will have concerns about your ability to provide services or products in a timely fashion. They may also be concerned with your organization's ability to pay for services rendered. Remember too, your customer may also be the shareholders in a corporation. They have some very real concerns about their investment that should be addressed. Because of these legitimate concerns, you must be proactive in communicating with them in a formal manner.

REFERENCE
Emergency Preparedness, Don Philpott, Bernan Press 2016

Appendix
Building Design Guidelines for Emergency Mitigation

IDENTIFYING THREATS

Before you can take mitigating action, you must identify what threats you face and where these threats might come from. This means identifying all internal and external threats and hazards and understanding why they might impact you. This is done in several ways. You conduct a design basis threat assessment and consult with your local police, FBI, and security consultants to create "what if" scenario-based assessments to cover every conceivable eventuality. When you have identified all possible threats, you can determine the degree of threat that each one poses. You are also better able to design and incorporate safeguards to protect your facility and its critical assets.

While the following section is based on protecting buildings and structures from the security professional's perspective, it also highlights the many vulnerabilities that could result in an arson attack or fire because of a terrorist or other malicious act. It also illustrates how building components can be strengthened to prevent or slow down a building's collapse during a catastrophic fire.

RISK ASSESSMENT

Having identified your critical assets, you now must determine how vulnerable those assets are.

The risk assessment uses threat, asset value, and vulnerability assessments to determine the level of risk for each critical asset against each applicable threat. This considers the likelihood of the threat occurring (probability) and the effects (consequences) of the occurrence if the critical asset is damaged or destroyed. The risk assessment provides security consultants, engineers, designers, and architects with a relative risk profile that defines which assets are at the greatest risk against specific threats. Any threat subjects an organization to risk. Therefore, when a threat is exhibited, a risk rating needs to be applied to understand how to manage the risk. A threat management team should have processes and procedures in place for measuring the probability and the severity of loss, considering a threat. The following primary risk types should be considered in determining risk exposure:

- Mission or function risks
- Asset risks
- Security risks

You must decide what risk each critical asset faces. Is there a low, medium, or high risk of that asset being attacked, damaged, or destroyed? This analysis gives you your threat rating and will determine your priorities when it comes to mitigation. Because it is not possible to completely eliminate risk, and every project has resource limitations, you must analyze how mitigation measures would affect risk and decide on the best and most cost-effective measures to achieve the desired level of protection (risk management). When considering your options remember that the goal is to put in place procedures and systems that protect your critical assets and eliminate weak spots and vulnerabilities.

BUILDING AND SITE MITIGATION

Long before any emergency affects you, you can take steps to mitigate the effects. The following section discusses many of the measures that can be taken. These measures should also be considered if you must repair or rebuild your building damaged during an emergency.

When discussing building and site mitigation, it makes sense to separate the space into six distinct areas, working from the inside outward:

- Interior: Limited access
- Interior: Public access

- Point of entry
- Exterior: Building perimeter
- Exterior: Parking and other off-site areas
- Exterior: Neighbors
- Interior: Limited access

Interior: Limited Access

The interior limited access areas include spaces such as storage rooms, telecommunications areas, computer server rooms, and mechanical rooms. Because of the threat from suspicious packages, mail rooms should also be considered limited access areas. A building can have several different layers of restricted access depending on its function and each would need separate access controls. These areas are off limits to most people, including visitors and employees. Access can be controlled using keys, proximity cards, maglock, card swipe system, or any other feasible manner. Start by performing an analysis of your building's space and determine what spaces need to be off limits due to security and safety concerns.

Interior: Public Access

In most buildings, especially those that offer public access, the amount of limited access interior space will be minimal. Even if this is true in your building, care must be taken to secure areas open to the public. For instance, can visitors to the building gain access to the roof? Are they able to access HVAC systems including intakes and discharges? Could a visitor with bad intentions force his or her way into limited access areas and start a fire? By performing an assessment of your building's public spaces, you will be able to identify areas of concern. You must know who uses the building and when, how they access and exit it, where they go when they are inside, and whether their movement is controlled. You must also know what procedures are in place to screen staff and to train them in the event of an emergency.

Point of Entry

Spending large amounts of money and time securing the inside of a building without addressing points of entry could be akin to changing the oil in an automobile after the engine has blown. It might be too little too late. What will you do to ensure that bad people and their tools of destruction do not gain access into your building in the first place? The obvious starting point is to have some way to police the entrances. This can be accomplished by employing security guards, card access, employee escorts, or several different solutions. For after business hours, an intrusion detection system should be installed. However, the danger lies in assessing your building's point of entry security from a doorway access point of view only. Think about other less obvious points of entry such as windows, loading docks, fresh air intakes, and other such penetrations. Have you identified and protected all utility inlets? Prepare a map showing all the different utilities used—water, gas, electricity, telecoms, and so on—how they come on to the property, their pipelines or cables while on property, and where they exit the property. It should show all manholes, inspection points, and sewer lines, and so on. It should also show where water is stored on property, the source of water for the fire suppression system, the location of all electric service points, and additional storage sites for other fuels.

Exterior: Building Perimeter

The immediate exterior building perimeter can also be an area of weakness. Perimeter protection is extremely important for urban buildings because in most cases the facility perimeter is defined by the external walls of the building. There is no buffer zone with wires, fences, and controlled access points to keep terrorists and others a safe distance away from your building. Walk around your perimeter to ensure the immediate surroundings are secure. Planters, overhangs, and other building features could be the perfect place to hide an explosive or incendiary device. When performing a security assessment, take these areas into consideration. Is the immediate perimeter lighting adequate? Can visitors with bad intentions gain access to or sabotage utilities or the emergency generator? Could a toxic substance be introduced into your building via a fresh air intake? Does the perimeter landscaping create a possible hiding place for criminals or explosives?

Exterior: Parking and Other Off-site Areas

A person with bad intentions may not even need access inside of your building to do damage. In the case of the Alfred P. Murrah Federal Building in Oklahoma City, the terrorist was able to bring down most of the building and kill 168 people without ever entering the facility. By parking a van full of explosives near the building and detonating them from a distance, incredible damage was inflicted. Studies have been performed that suggest standoff distances of 100 ft. or more will lessen the impact of a bomb. Since the Oklahoma City bombing, we realize the importance of not allowing unattended vehicles to park in front of buildings.

Keeping this 100-ft. standoff distance in mind, can you keep distance between your parking lot and the building? If you do, you may wish to make special arrangements for those with disabilities because the distance may create hardships. Parking garages create a dilemma. Since they are a building structure that could fail, an explosion could create tremendous damage. Consider limiting access to your parking garages and installing closed-circuit television (CCTV) to monitor movement. If you employ underground parking, it may be wise to limit it to registered vehicles, motorcycles, and perhaps small cars. Better yet, it may be prudent to eliminate that parking area altogether if feasible. Place parking as far as practical from the building. Off-site parking is recommended for high-risk facilities vulnerable to terrorist attack. If onsite surface parking or underground parking is provided, take precautions such as limiting access to these areas only to the building occupants or having all vehicles inspected in areas near the building. If an underground area is used for a high-risk building, the garage should be placed adjacent to the building under a plaza area rather than directly underneath the building. To the extent practical, limit the size of vehicles able to enter the garage by imposing physical barriers on vehicle height.

Access control refers to points of controlled access to the facility through the perimeter line. The controlled access check or inspection points for vehicles require architectural features or barriers to maintain the defensible perimeter. Architects and engineers can accommodate these security functions by providing adequate design for these areas, which make it difficult for a vehicle to crash onto the site. Remember that you might have to implement different security layers at different points around or close to the perimeter and these will have to be planned into your physical protection system. Deterrence and delay are major attributes of the perimeter security design that should be consistent with the landscaping objectives, such as emphasizing the open nature characterizing high-population buildings.

Because it is impossible to thwart all possible threats, the objective is to make it difficult to successfully execute the easiest attack scenarios, such as a car bomb detonated along the curb or a vehicle jumping the curb and crashing into the building prior to detonation. Site design considerations for bomb resistance might include:

- Bollards or planters
- Elevation of building level
- Curbs
- Security lighting, sensors, and cameras
- Gates
- Limiting the number of access points
- Discuss with utility companies and local authorities about padlocking manholes.

The purpose of protective bollards, planters, or even large pieces of exterior artwork is to keep a vehicle from gaining close access to a building and detonating explosives. Although bollards and retaining walls are effective, they can be aesthetically unattractive if not designed properly. The use of natural barriers such as trees with thick trunks can prove to be an efficient barrier that is aesthetically pleasing as well.

When reviewing your building and the desired level of security, always remember aesthetics and the role it plays in employee satisfaction and desirability of public interaction. You could build an intimidating fortress without windows which affords a great measure of security but is that desirable? Would employees enjoy working there? Might their productivity be reduced in such an environment? Would the public be too intimidated to visit and do business there? On the other hand, you could design a high-tech, functionally proper, aesthetically pleasing masterpiece. But at what cost? Anything can be accomplished if the price is right, and when it comes to a totally secure and safe structure, the price will likely be too high. The trick is to find an acceptable balance of risk and cost. Performing a risk analysis would be a prudent starting point.

Consulting with a design specialist who understands security and safety would also make sense. The science of designing viable buildings that are safe and secure is known as "crime prevention through environmental design" (CPTED). Designers and consultants specializing in CPTED bring an added dimension to your building design. Their intimate knowledge of security and safety is integrated with an architecturally attractive and functional building.

These consultants will identify opportunities based on your need and budget and typically are well worth the price.

Exterior: Neighbors

Review all the facilities in your neighborhood to see which, if any, pose a threat to you. Are there high-risk targets that might be attacked and, if so, what would the impact be on your facility and your ability to continue operating? Is there a distribution depot in the next street that houses toxic chemicals? Are there roads or rail tracks running near the building? Could people in parked vehicles use electronic devices to eavesdrop on what is going on or use weapons to fire at us or launch a grenade attack? Use fences or walls to screen the building from these vantage points. Is the road or railroad used to transport hazardous or nuclear materials? What would happen if a tanker crashed on the road close to your facility or a train derailed and exploded?

STRUCTURES
Windows and Doors: The Weakest Link

The windows and doors of a building often present the path of least resistance during an emergency. Whether the event is terror related or weather related, windows are usually the weakest link. Next to total building collapse, more people are injured from flying glass during an explosion in a building. In fact, most people who are injured by flying glass are within 15 ft. of the windows when the blast occurred. An example of this phenomenon is seen in the coordinated terrorist bombings of U.S. embassies in Nairobi, Kenya, and Dar es Salaam, Tanzania, on August 7, 1998. In the more lethal Nairobi attack, the first explosions were produced by grenades detonated outside of the embassy.

The grenades did little damage but startled the occupants of the embassy. As the occupants of the building began to walk toward the windows to investigate the loud noise, a much larger bomb planted in an automobile was detonated. The tremendous explosion shattered glass and turned the shards into razor sharp missiles, killing 224 people and wounding more than 4,000.

When examining the problems with windows, glass is just one of the issues. Many windows are not installed in a way that offers true blast protection. Often during an explosion or storm, the entire window becomes dislodged from its wall opening. Several safety features are available that can mitigate the damage caused by windows and glass. In some areas, such as south

Florida, window shutters are now required to protect people and buildings from the effects of storms. Protective window film applied over existing windows can also provide some protection from pressure experienced during a storm or explosion. The protection, however, is limited. The real purpose of window film is to keep glass from shattering and dangerous shards from becoming airborne. Protective film does resist impact. The glass will break; it just will not disperse as readily as it would if unprotected. If you are looking for protection from shards and impact resistance, then ballistic-grade glass would be an option worth exploring. The cost may be high, but the protection provided, especially in a high-risk building, may well be worth the investment.

When installing blast-resistant glass, it is important to understand construction methods for installing the windows. If installed improperly, the window may absorb an impact and the glass would remain unbroken, but the entire window frame would become dislodged from its opening. Installation of blast-resistant glass should include extra reinforcing around the window opening. Consult with the manufacturer and structural engineers when installing blast-resistant glass.

Doors, louvers, and other openings in the exterior envelope should be designed so that the anchorage into the supporting structure has a lateral capacity greater than that of the element. There are two general recommendations for doors:

- Doors should open outward so that they bear against the jamb during the positive-pressure phase of the air blast loading.
- Door jambs can be filled with concrete to improve their resistance.
- For louvers that provide air to sensitive equipment, some recommendations are:
- Provide a baffle in front of the louver so that the air blast does not have direct line-of-sight access through the louver.
- Provide a grid of steel bars properly anchored into the structure behind the louver to catch any debris generated by the louver or other flying fragments.

Exterior Frame

There are two primary considerations for the exterior frame. The first is to design the exterior columns to resist the direct effects of the specified threats. The second is to ensure that the exterior frame has sufficient structural integrity to accept localized failure without initiating progressive collapse. Because columns do not have much surface area, air blast loads on columns tend to be mitigated by "clear-time effects." This refers to the pressure wave washing around these slender tall members, and consequently the entire duration of the pressure wave does not act upon them. On the other hand, the critical threat is directly across from them, so they are loaded with the peak reflected pressure, which is typically several times larger than the incident or overpressure wave that is propagating through the air.

For columns subjected to a vehicle weapon threat on an adjacent street, buckling and shear are the primary effects to be considered in analysis. If an exceptionally large weapon is detonated close to a column, shattering of the concrete due to multiple tensile reflections within the concrete section can destroy its integrity. Buckling is a concern if lateral support is lost due to the failure of a supporting floor system. This is particularly important for buildings that are close to public streets. In this case, exterior columns should be capable of spanning two or more stories without buckling. Slender steel columns are at substantially greater risk than concrete columns. Confinement of concrete using columns with closely spaced closed ties or spiral reinforcing will improve shear capacity, improve the performance of lap splices in the event of loss of concrete cover and greatly enhance column ductility. The potential benefit from providing closely spaced closed ties in exterior concrete columns is extremely high relative to the cost of the added reinforcement.

For steel columns, splices should be placed as far above grade level as practical. It is recommended that splices at exterior columns that are not specifically designed to resist air blast loads employ complete penetration welded flanges. Welding details, materials, and procedures should be selected to ensure toughness.

For a package weapon, column breach is a major consideration. Some suggestions for mitigating this concern:

- Do not use exposed columns that are fully or partially accessible from the building exterior. Arcade columns should be avoided.
- Use an architectural covering that is at least 6 in. from the structural member. This will make it considerably more difficult to place a weapon directly

against the structure. Because explosive pressures decay so rapidly, every inch of distance will help to protect the column.

Load-bearing reinforced concrete wall construction can provide a considerable level of protection if adequate reinforcement is provided to achieve ductile behavior. This may be an appropriate solution for the parts of the building that are closest to the secured perimeter line (within 20 ft.). Masonry is a much more brittle material that can generate highly hazardous flying debris in the event of an explosion. Its use is generally discouraged for new construction. Spandrel beams of limited depth generally do well when subjected to air blast. In general, edge beams are very strongly encouraged at the perimeter of concrete slab construction to afford frame action for redistribution of vertical loads and to enhance the shear connection of floors to columns.

Roof System

The primary loading on the roof is the downward air-blast pressure. The exterior bay roof system on the side(s) facing an exterior threat is the most critical. The air-blast pressure on the interior bays is less intense, so the roof there may require less hardening. Secondary loads include upward pressure due to the air blast penetrating through openings and upward suction during the negative loading phase. The upward pressure may have an increased duration due to multiple reflections of the internal air blast wave. It is important to consider the downward and upward loads separately.

The preferred system is cast-in-place reinforced concrete with beams in two directions. If this system is used, beams should have continuous top and bottom reinforcement with tension lap splices. Stirrups to develop the bending capacity of the beams closely spaced along the entire span are recommended. Somewhat lower levels of protection are afforded by conventional steel beam construction with a steel deck and concrete fill slab. The performance of this system can be enhanced by use of normal-weight concrete fill instead of lightweight fill, increasing the gauge of welded wire fabric reinforcement and making the connection between the slab and beams with shear connector studs. Because it is anticipated that the slab capacity will exceed that of the supporting beams, beam end connections should be capable of developing the ultimate flexural capacity of the beams to avoid brittle failure.

Beam-to-column connections should be capable of resisting upward as well as downward forces.

Precast and pre-/post-tensioned systems are generally considered less desirable, unless members and connections can resist upward forces generated by rebound from the direct pressure or the suction from the negative pressure phase of the air blast.

Concrete flat slab/plate systems are also less desirable because of the potential of shear-failure at the columns. When flat slab/plate systems are used, they should include features to enhance their punching shear resistance. Continuous bottom reinforcement should be provided through columns in two directions to retain the slab if punching shear failure occurs. Edge beams should be provided at the building exterior.

Lightweight systems, such as untopped steel deck or wood frame construction, are considered to afford minimal resistance to air blast. These systems are prone to failure due to their low capacity for downward and uplift pressures.

Floor System

The floor system design should consider three possible scenarios: air blast loading, redistributing load in the event of loss of a column or wall support below, and the ability to arrest debris falling from the floor or roof above.

For structures in which the interior is secured against bombs of moderate size by package inspection, the primary concern is the exterior bay framing. For buildings that are separated from a public street only by a sidewalk, the uplift pressures from a vehicle weapon may be significant enough to cause possible failure of the exterior bay floors for several levels above ground. Special concern exists in the case of vertical irregularities in the architectural system, either where the exterior wall is set back from the floor above or where the structure steps back to form terraces.

Structural hardening of floor systems above unsecured areas of the building such as lobbies, loading docks, garages, mailrooms, and retail spaces should be considered. In general, critical or heavily occupied areas should not be placed underneath unsecured areas, because it is virtually impossible to prevent against localized breach in conventional construction for package weapons placed on the floor. Precast panels are problematic because of their tendency to fail at the connections. Pre-/post-tensioned systems tend to fail in a brittle manner if stressed much beyond their elastic limit. These systems are also not able to accept upward loads without additional reinforcement. If pre-/post-tensioned systems are used, continuous mild steel needs to be added to the top and the bottom faces to provide the ductility needed to resist explosion loads.

Flat slab/plate systems are also less desirable because of limited two-way action and the potential for shear failure at the columns. When flat slab/plate systems are employed, they should include features to enhance their punching shear resistance, and continuous bottom reinforcement should be provided across columns to resist progressive collapse. Edge beams should be provided at the building exterior.

Interior Columns

Interior columns in unsecured areas are subject to many of the same issues as exterior columns. If possible, columns should not be accessible within these areas. If they are accessible, then obscure their location or impose a stand-off to the structural component using cladding. Methods of hardening columns include using closely spaced ties, spiral reinforcement, and architectural covering at least six inches from the structural elements. Composite steel and concrete sections or steel plating of concrete columns can provide higher levels of protection. Columns in unsecured areas should be designed to span two or three stories without buckling if the floors below and possibly above the detonation area have failed, as previously discussed.

Interior Walls

Interior walls surrounding unsecured spaces are designed to contain the explosive effects within the unsecured areas. Ideally, unsecured areas are located adjacent to the building exterior so that the explosive pressure may be vented outward as well. Fully grouted concrete masonry unit (CMU) block walls that are well-reinforced vertically and horizontally and adequately supported laterally are a common solution. Anchorage at the top and bottom of walls should be capable of developing the full flexural capacity of the wall.

Lateral support at the top of the walls may be achieved using steel angles anchored into the floor system above. Care should be taken to terminate bars at the top of the wall with hooks or heads and to ensure that the upper course of block is filled solid with grout. The base of the wall may be anchored by reinforcing bar dowels. Interior walls can also be effective in resisting progressive collapse if they are designed properly with sufficient load-bearing capacity and are tied into the floor systems below and above.

This design for hardened interior wall construction is also recommended for primary exit routes to protect against explosions, fire, and other hazards trapping occupants.

Structural

Consider these points:

- Incorporate measures to prevent progressive collapse.
- Design floor systems for uplift in unsecured areas and in exterior bays that may pose a hazard to occupants.
- Limit column spacing.
- Avoid transfer girders.
- Use two-way floor and roof systems.
- Use fully grouted, heavily reinforced CMU block walls that are properly anchored to separate unsecured areas from critical functions and occupied secured areas.
- Use dynamic nonlinear analysis methods for design of critical structural components.

Exterior Wall/Cladding Design

The exterior walls provide the first line of defense against the intrusion of the air-blast pressure and hazardous debris into the building. They are subject to direct reflected pressures from an explosive threat located directly across from the wall along the secured perimeter line. If the building is more than four stories high, it may be advantageous to consider the reduction in pressure with height due to the increased distance and angle of incidence. The objective of design at a minimum is to ensure that these members fail in a ductile mode, such as flexure, rather than a brittle mode, such as shear. The walls also need to be able to resist the loads transmitted by the windows and doors. It is not uncommon, for instance, for bullet-resistant windows to have a higher ultimate capacity than the walls to which they are attached. Beyond ensuring a ductile failure mode, the exterior wall may be designed to resist the actual or reduced pressure levels of the defined threat. Note that special reinforcing and anchors should be provided around blast-resistant window and door frames.

Poured-in-place, reinforced concrete will provide the highest level of protection, but solutions like pre-cast concrete, CMU block, and metal stud systems may also be used to achieve lower levels of protection. For pre-cast panels, consider a minimum thickness of five inches exclusive of reveals, with two-way, closely spaced reinforcing bars to increase ductility and reduce the chance of flying concrete fragments. The objective is to reduce the loads transmitted into the connections, which need to be designed to resist the ultimate flexural resistance of the panels.

Also, connections into the structure should provide as straight a line of load transmittal as practical.

For CMU block walls, use eight-inch block walls, fully grouted with vertically centered heavy reinforcing bars and horizontal reinforcement placed at each layer. Connections into the structure should be designed to resist the ultimate lateral capacity of the wall. For infill walls, avoid transferring loads into the columns if they are primary load-carrying elements. The connection details may be difficult to construct. It will be difficult to have all the blocks fit over the bars near the top, and it will be difficult to provide the required lateral restraint at the top connection. A preferred system is to have a continuous exterior CMU wall that laterally bears against the floor system. For increased protection, consider using twelve-inch blocks with two layers of vertical reinforcement. For metal stud systems, use metal studs back-to-back and mechanically attached to minimize lateral torsional effects. To catch exterior cladding fragments, attach a wire mesh or steel sheet to the exterior side of the metal stud system. The supports of the wall should be designed to resist the ultimate lateral out-of-plane bending capacity load of the system.

Brick veneers and other nonstructural elements attached to the building exterior are to be avoided or have strengthened connections to limit flying debris and to improve emergency egress by ensuring that exits remain passable.

Mechanical and Electrical Systems

The key concepts for providing secure and effective mechanical and electrical systems in buildings are the same as for the other building systems: separation, hardening, and redundancy. Keeping critical mechanical and electrical functions as far from high-threat areas as possible (lobbies, loading docks, mail rooms, garages, and retail spaces) increases their ability to survive an event. Separation is perhaps the most cost-effective option. In addition, physical hardening or protection of these systems (including the conduits, pipes, and ducts associated with life safety systems) provides increased likelihood that they will be able to survive the direct effects of the event if they are close enough to be affected.

Finally, by providing redundant emergency systems that are adequately separated, there is a greater likelihood that emergency systems will remain operational after the event to assist rescuers in the evacuation of the building.

Architecturally, enhancements to mechanical and electrical systems will require additional space to accommodate additional equipment. Fortunately, there are many incremental improvements that can be made that require only a small change to the design. Additional space can be provided for future enhancements as funds or the risk justify implementation. Structurally, the walls and floor systems adjacent to the areas where critical equipment is located need to be protected by means of hardening. Other areas where hardening is recommended include primary egress routes, feeders for emergency power distribution, sprinkler systems mains and risers, fire alarm system trunk wiring, and ducts used for smoke-control systems.

Emergency Egress Routes

To facilitate evacuation, consider these measures:
- Provide positive pressurization of stairwells and vestibules.
- Provide battery packs for lighting fixtures and exit signs.
- Harden walls using reinforced CMU block properly anchored at supports.
- Use nonslip phosphorescent treads.
- Do not cluster egress routes in single shaft. Separate them as far as possible.
- Use double doors for mass evacuation.
- Do not use glass along primary egress routes or stairwells.

Emergency Power System

An emergency generator provides an alternate source of power should utility power become unavailable to critical life safety systems such as alarm systems, egress lighting fixtures, exit signs, emergency communications systems, smoke-control equipment, and fire pumps. Emergency generators typically require large louvers to allow for ventilation of the generator while running. Care should be taken to locate the generator so that these louvers are not vulnerable to attack. A remote radiator system could be used to reduce the louver size.

Redundant emergency generator systems remotely located from each other enable the supply of emergency power from either of two locations. Consider locating emergency power-distribution feeders in hardened enclosures, or encased in concrete, and configured in redundant routing paths to enhance reliability. Emergency distribution panels and automatic transfer switches should be in rooms separate from the normal power system (hardened rooms, where possible).

Emergency lighting fixtures and exit signs along the egress path could be provided with integral battery packs, which locate the power source directly at the load, to provide lighting instantly in the event of a utility power outage.

Fuel Storage

A nonexplosive fuel source, such as diesel fuel, is acceptable for standby use for emergency generators and diesel fire pumps. Fuel tanks should be located away from building access points, in fire-rated, hardened enclosures. Fuel piping within the building should be in hardened enclosures, and redundant piping systems could be provided to enhance the reliability of the fuel distribution system. Fuel filling stations should be located away from public access points and monitored by the CCTV system.

Transformers

Main power transformer(s) should be in the interior of the building if possible, away from locations accessible to the public. For larger buildings, multiple transformers, located remotely from each other, could enhance reliability should one transformer be damaged by an explosion.

Ventilation Systems

Air-intake locations should be located as high up in the building as is practical to limit access to the public. Systems that serve public access areas, such as mail receiving rooms, loading docks, lobbies, and elevators, should be isolated and provided with dedicated air handling systems capable of 100 percent exhaust mode. Tie air intake locations and fan rooms into the security surveillance and alarm system.

Building HVAC systems are typically controlled by a building automation system, which allows for quick response to shut down or selectively control air conditioning systems. This system is coordinated with the smoke-control and fire-alarm systems.

Fire Control Center

A fire control center should be provided to monitor alarms and life safety components, operate smoke-control systems, communicate with occupants, and control the firefighting/evacuation process. Consider providing redundant fire control centers remotely located from each other to allow system operation and control from alternate locations. The fire control center should be located near the point of firefighter access to the building. If the control center is adjacent to the lobby, separate it from the lobby using a corridor or other buffer area. Provide hardened construction for the fire control center.

Emergency Elevators

Elevators are not used as a means of egress from a building in the event of a life safety emergency event, as conventional elevators are not suitably protected from the penetration of smoke into the elevator shaft. An unwitting passenger could be endangered if an elevator door opens onto a smoke-filled lobby. Firefighters may elect to manually use an elevator for firefighting or rescue operation. A dedicated elevator, within its own hardened, smoke-proof enclosure, could enhance the firefighting and rescue operation after a blast/fire event. The dedicated elevator should be supplied from the emergency generator, fed by conduit/wire that is protected in hardened enclosures. This shaft/lobby assembly should be sealed and positively pressurized to prevent the penetration of smoke into the protected area.

Smoke, Fire Detection, and Alarm Systems

A combination of early warning smoke detectors, sprinkler-flow switches, manual pull stations, and audible and visual alarms provide quick response and notification of an event. The activation of any device will automatically start the sequence of operation of smoke control, egress, and communication systems to allow occupants to quickly go to a safe area. System designs should include redundancy such as looped infrastructure wiring and distributed intelligence such that the severing of the loop will not disable the system. Install a fire alarm system consisting of distributed, intelligent fire alarm panels connected in a peer-to-peer network, such that each panel can function independently and process alarms and initiate sequences within its respective zone.

Sprinkler/Standpipe System

Sprinklers will automatically suppress fire in the area upon sensing heat. Sprinkler activation will activate the fire alarm system. Standpipes have water available locally in large quantities for use by professional firefighters. Multiple sprinkler and standpipe risers limit the possibility of an event severing all water supply available to fight a fire. Redundant water services would increase the reliability of the source for sprinkler protection and fire suppression. Appropriate valving should be provided where services are combined. Redundant fire pumps could be provided in remote locations. These pumps could rely on different sources, for example, one electric pump supplied from the utility or emergency generator and a second diesel fuel source fire pump. Diverse and separate routing of standpipe and sprinkler risers within hardened areas will enhance the system's reliability (such as reinforced masonry walls at stair shafts containing standpipes).

Smoke-Control Systems

Appropriate smoke-control systems maintain smoke-free paths of egress for building occupants through a series of fans, ductwork, and fire smoke dampers. Stair pressurization systems maintain a clear path of egress for occupants to safe areas or to evacuate the building. Smoke control fans should be located higher in a building rather than at lower floors to limit exposure/access to external vents. Vestibules at stairways with separate pressurization provide an additional layer of smoke control.

Communication System

A voice communication system facilitates the orderly control of occupants and evacuation of the danger area or the entire building. The system is typically zoned by floor, by stairwell, and by elevator bank for selective communication to building occupants. Emergency communication can be enhanced by providing:

- Extra emergency phones separate from the telephone system, connected directly to a constantly supervised central station.
- In-building repeater system for police, fire, and Emergency Medical Services (EMS) radios.
- Redundant or wireless firefighter's communications in building.

Protective Bollards, Planters, and Green Space

As experts study the destruction caused by explosions, they have unanimously agreed that "standoff distance" is the best solution for decreasing the impact of the explosion. Simply put, the more green space between the building and the bomb, the less of an effect the explosive will have on the structure. In a practical sense, standoff distances can be increased by installing protective bollards or planters around the building. The bollards are reinforced concrete posts, usually cylindrical, ranging from 4 in. in diameter to 12 in diameter or greater. These bollards are successful in keeping vehicular traffic from entering a restricted area. On the negative side, bollards tend to look unattractive. They look as if they were placed to keep people out.

Urban designers and architects agree that protective barriers should blend into the environment. They should not be readily noticeable to the average person. There are barriers cleverly disguised as decorative walls, planters, benches, or artwork. Landscaping and fencing can be used to keep vehicles at bay. Forcing points of entry into a building through a long corridor minimizes the effects of an explosion as well.

Virtual Reality Software and Related Technology

Scientists and engineers have realized the importance of recreating events with a goal of improving processes or finding out what went wrong. This study of events can be applied to building failure due to an emergency event. Building simulation and the creation of virtual environments have assisted engineers in determining why a building structure failed. In true proactive fashion, engineers have used this data to better design buildings that can absorb emergency events including fire, earthquakes, terrorism, and other such events. Typically, computer-aided design (CAD) software is used to create a three-dimensional model of a building. This virtual building would then be placed in a digital environment that mirrors the subject property. At this point, programmers can subject the building to a host of virtual events while recording the outcomes of each. Modifications can be made to the 3-D building including ballistic glass, increased standoff distances, or other design features, and the effectiveness of each modification can be measured. This process allows the engineer to select the appropriate modifications for the safety level desired.

Although the availability of such software and expertise is currently limited to certain high-risk structures, the value of these studies is evident and growing in popularity. As technology evolves and expertise is developed, the cost of these studies will become more tolerable as well.

Post-failure Analysis

Immediately after a building collapse, specialized scientists and engineers will comb the premises to collect as much data as they can. To the untrained eye, the pile of rubble where the building once stood seems to contain very few clues of any value. To the highly trained eye, the site is a treasure trove of information. What caused the building failure? Where did the problem start? Could it have been prevented? Could better building materials have averted this disaster? Could better building techniques have helped? These are the types of questions that can be answered by performing a post-failure analysis.

Mitigation Assessment Team Program

The Mitigation Assessment Team (MAT) Program is an updated and enhanced version of FEMA's Building Performance Assessment Team (BPAT) program. MAT—drawing on the resources of the public and private sectors—evaluates building and related infrastructure performance after natural and non-natural emergency events. The goal of the program is to better understand what makes a building survive emergencies and how the negative effects of emergencies can be mitigated. After an emergency event, a MAT is dispatched to the emergency location to inspect the affected buildings. The assessment team will conduct engineering analysis to determine the cause of failure or the reasons for successful structural integrity. At the end of the team's analysis, they will recommend actions that could reduce future damages and protect lives during similar hazards. The positive results of the

MAT program include improved disaster-resistant building codes, design methods, and material selection.

The MAT consists of FEMA representatives, state and local officials, and public and private sector technical experts.

Although we cannot eliminate many emergencies, we can change how we prepare for them and how we react to them when they do occur. We can also go back after the emergency has occurred and examine the affected buildings. Did the building survive? If it did, why did it survive? By analyzing the effect the emergency event had on the building, we develop better building codes. We can specify better building materials. We can take the lessons that history has taught us and design better buildings. Doing so will advance business continuity, reduce insurance costs, and—most importantly—save lives.

Glossary

ACCELERANT: In chemistry, it is a catalyst (a compound that may speed up a chemical process). In fire investigation, the term is used to describe liquids and compounds utilized by arsonists in the setting of fires, such as acohol, gasoline, charcoal lighter fluid, and paint thinner.

ACETYLENE: A colorless hydrocarbon gas, C_2 and H_2, that is produced by the action of water upon certain carbon compounds and used principally in welding and cutting operations.

ACKNOWLEDGE: To confirm that a message or signal has been received, such as by the pressing of a button or the selection of a software command.

AERATED POWDER: Any powdered material used as a coating material which is fluidized within a container by passing air uniformly from below. It is common practice to fluidize such materials to form a fluidized powder bed and then dip the part to be coated into the bed in a manner similar to that used in liquid dipping.

AEROSOL: A material which is dispensed from its container as a mist, spray, or foam by a propellant under pressure.

ALARM: A warning of fire danger.

ALARM VERIFICATION FEATURE: A feature of automatic fire detection and alarm systems to reduce unwanted alarms wherein smoke detectors report alarm conditions for a minimum period of time, or confirm alarm conditions within a given time period after being reset, in order to be accepted as a valid alarm initiation signal.

ALERT: The action the canine makes to indicate to the handler that an accelerant has been located.

ALERT POINT: The point where a canine indicates that an accelerant is located.

ALIPHATIC: Aliphatic is especially used in reference to open-chain (noncyclic) hydrocarbons. Aliphatic hydrocarbons are major components of everyday materials such as turpentine, gasoline, and oil-based paints. Aliphatic hydrocarbons and their chemical derivatives are often quite flammable.

ALLIGATORING: Describes visual patterns showing the degree of the carbonizing process of wood when exposed to flame or sufficient heat. This results in structural deterioration of the wood, characterized by cracks and ridges resembling the back of an alligator.

AMPERAGE: The strength of a current measured in amps.

AMPERE (AMP): A unit of rate of flow of electricity. Flow of electricity depends on the pressure in volts and resistance, identified as impedance, in alternating current circuits.

ANNEAL: Removal of tension from spring steel through the application of heat. An example would be coil springs that have collapsed within themselves and will not spring back after being exposed to heat and fire.

APPROVED: Approved for a purpose means that the equipment is suitable for a particular application which is determined by a recognized testing laboratory, inspection agency or other organization concerned with product evaluation as part of its labeling or listing program.

ARC/ARCING: Either intentionally (by a switch) or accidentally (because of a loosened terminal), heating that results from high resistance when an electric current is interrupted. The degree of heat or intensity of the arc depends on the amount of current in the circuit. Electric arcs can ignite combustibles because of sparks being thrown in the general vicinity of the fault, whereas fault type heating alone, in an overloaded conductor, may result in the ignition of insulation and of combustibles in the immediate proximity of the conductor. Either of the said conditions can result without opening the circuit breaker or other circuit safety device.

AREA OF ORIGIN: The term used by fire investigators to describe the general area where a fire started. It is not limited to the specific point of origin.

The term may describe an entire building in a multibuilding fire or a room in a structure.

ARSON: To willfully and maliciously burn or cause to be burned, or aid, counsel or procure the burning of any structure, forest land or property. Does not include a person burning own property unless there is intent to defraud or there is injury to another person or damage to another person's property.

AUTO-IGNITION TEMPERATURE: The minimum temperature to which material must be heated to initiate or create self-sustained combustion, independent of any outside heat source. The term is also used in the discussion of spontaneous heating resulting from physical, chemical, and biological reactions without an external spark or flame.

BEADING: A condition found on electrical wiring when arcing has occurred. The electrical wire ends appear beaded or formed into a series of melted round balls.

BLASTING AGENT: The Occupational Safety and Health Administration defines a blasting agent as any material or mixture, consisting of a fuel and oxidizer, intended for blasting, not otherwise classified as an explosive and in which none of the ingredients are classified as an explosive, provided that the finished product, as mixed and packaged for use or shipment, cannot be detonated by means of a No. 8 test blasting cap when unconfined.

BOILING LIQUID EXPANDING VAPOR EXPLOSION (BLEVE): The result of a liquid within a container reaching a temperature well above its boiling point at atmospheric temperature, causing the vessel to rupture into two or more pieces

BRITISH THERMAL UNIT (BTU): The amount of heat necessary to raise the temperature of one pound of water by 1°F, measured at 60°F. One BTU equals 252 calories. It is important to recognize the difference between heat and temperature. Heat is quantity of measurement whereas temperature is the measure of intensity. Specific heat or thermal capacity of a substance is the number of BTUs required to raise the temperature of one pound of the substance 1°F or the number of calories to raise 1 g of the substance through 1°C.

BURNING POINT: The temperature which a substance must attain before it will ignite on the application of an ignition source or a flame. Preheated materials ignite more quickly when a flame is introduced because it may have already reached the burning point.

BURN PATTERN: Used in defining and illustrating the progress of a fire by means of visible charring, decomposition, and displacement. It may be used to define carbonization from luminous and non-luminous combustion, spontaneous ignition process, and pyrolysis. It may be used in describing major fire progress or limited progress inside walls or isolated areas and spaces. When used by experts, they should be required to define the concept behind its use.

BUSS FUSE: Device used for over current protection. May also be called plug fuse.

CALORIE: The amount of heat required to raise the temperature of 1 g of water by 1°C, measured at 15°C. One BTU equals 252 calories. As heat can be converted to energy, it may be said that one BTU equals 1.055 joules; one BTU per minute equals 0.0236 horsepower.

CARBON: A nonmetallic chemical element found in many inorganic and all organic compounds. Diamonds and graphite are pure carbon, and carbon is also present in substances such as coal, coke, charcoal, soot, and smoke. Wood and some fabrics commonly deteriorate to carbon and charcoal in the process of decomposition and are subject to spontaneous heating in piles.

CARBON BLACK: This may be formed by the incomplete combustion of natural gas and a liquid hydrocarbon or by a liquid hydrocarbon alone. It occludes oxygen and slow smoldering combustion may easily result if stored without proper cooling and ventilation. After thorough cooling and airing, carbon black will not heat spontaneously although heating may result from mixture with oxidizable oils. Dust explosion hazards may exist where carbon black is processed or stored.

CARBON DIOXIDE: A gas which may be a product of combustion.

CARBON MONOXIDE: A as which may be a product of combustion.

CATEGORY 1 FLAMMABLE LIQUID: Under the Globally Harmonize System, a category 1 liquid is one that has a flash point < 23°C and initial boiling point ≤ 35°C (95°F).

CATEGORY 2 FLAMMABLE LIQUID: Under the Globally Harmonize System, a category 2 liquid is one that has a flash point < 23°C and initial boiling point > 35°C (95°F).

CATEGORY 3 FLAMMABLE LIQUID: Under the Globally Harmonize System, a category 3 liquid is one that has a flash point $\geq 23°C$ and $\leq 60°C$ (140°F).

CATEGORY 4 FLAMMABLE LIQUID: Under the Globally Harmonize System, a category 4 liquid is one that has a flash point $\geq 60°C$ (140°F) and $\leq 93°C$ (200°F).

CAUSE: As used in fire investigation, identifying, and describing the igniting agent or heat source of a fire. For example, some causes could be "match ignited and thrown into combustibles," "friction from electric motor belt from thrown bearings," and "incendiary origin Molotov cocktail ignited and thrown into window of dwelling which then ignited draperies and carpet."

CENTIGRADE: One Centigrade degree is 1/100 the difference between the temperature of the freezing and boiling points of water at one atmosphere of pressure. 0 is the freezing point and 100 is the boiling point of water.

CENTRAL STATION FIRE ALARM SYSTEM: Receives fire alarm signals, supervisory and trouble signals from a protected premise. These stations are controlled and operated by a person, firm, or corporation whose business is the furnishings of such systems.

CHARCOAL: A black form of carbon produced by partial burning of oxidizing wood or other organic matter and compounds. Under certain conditions, charcoal reacts with air to heat spontaneously and ignites into self-sustained flaming combustion. The more finely the charcoal is divided, the greater the hazard of combustion. Spontaneous heating may result from lack of sufficient cooling; lack of ventilation; becoming wet, then drying without ventilation; friction; being finely divided without ventilation; or leaving the residue in a chemically unstable condition.

CHARRED: Used to define carbonization from luminous and non-luminous combustion, spontaneous ignition process or pyrolysis.

CINDER: A hot, but not flaming, piece of partly burned material.

CIRCUIT BREAKER: A type of electrical overcurrent device. There are circuit breakers and oil circuit breakers; adjustable and nonadjustable; and instantaneous and delay, with thermal tripping mechanisms. (REF: National Electrical Code) Overcurrent protection devices include thermal cutouts which are not intended to open short circuits or grounds, but may be used to protect motors and heavy branch circuits from overload, and plug fuses which are produced in several types including the onetime and the time delay (Type S), which is designed to discourage tampering or bridging. Overcurrent devices also include thermal cutouts which are not intended to open short circuits but may be utilized to protect motors and motor branch circuits from overload. Others include various types and classes of cartridge fuses which may be equipped with drop out links, onetime fuses, super lag renewable, as well as dual element fuses of the blade and ferrule type. (REF: National Electrical Code)

CIRCUMSTANTIAL EVIDENCE: Proof of circumstances surrounding the transaction; proof of certain facts and circumstances in each case from which a jury may infer other connected facts which usually and reasonably follow according to the common experience of mankind.

CITY GAS: This gas may be natural, manufactured, or liquified petroleum gas, or mixtures of any of these. Natural gas seems to be in most common use today and consists mainly of methane, a small amount of ethane, possibly some propane, butane, small amounts of carbon dioxide, and varying amounts of nitrogen. In compound the result is a lighter than air mixture at the meter.

CLASS 1 COMMODITIES: Noncombustible products that meet one of the following criteria: the noncombustible products are placed directly on wooden pallets or placed in a single layer corrugated cartons or shrink wrapped or paper wrapped as a unit load with or without pallets.

CLASS 11 COMMODITIES: Noncombustible products that are placed in slatted wood crates, solid wood boxes, multiple layer corrugated cartons or equivalent combustible packaging material.

CLASS 111 COMMODITIES: Products made of wood, paper, natural fibers, or Group C plastics with or without cartons, boxes or crates and with or without pallets.

CLASS 1V COMMODITIES: Defined as a product with or without pallets that meets any of the following criteria: constructed partially or totally of Group B plastics, consist of free-flowing Group A plastic materials, or contains within itself or its packaging an appreciable amount (5-15%) of Group A plastics.

CLASS 1 HAZARDOUS LOCATIONS: Class I locations are those in which flammable gases or vapors are or may be present in the air in quantities sufficient to produce explosive or ignitable mixtures.

CLASS 11 HAZARDOUS LOCATIONS: Those that are hazardous because of the presence of combustible dust.

CLASS 111 HAZARDOUS LOCATIONS: Those that are hazardous because of the presence of easily ignitable fibers or flyings, but in which such fibers or flyings are not likely to be in suspension in the air in quantities sufficient to produce ignitable mixtures.

CLASS A FIRES: Involve carbon-based products such as wood and paper.

CLASS B FIRES: Fires involving flammable gases and liquids.

CLASS C FIRES: Fires involving any combustible materials where electricity may be present.

CLASS D FIRES: Fires involving combustible metals, such as aluminum, magnesium, titanium, and zirconium.

CLASS K FIRES: Class K fires are fires that involve vegetable oils, animal oils, or fats in cooking appliances.

COMBUSTIBLE: A combustible material can be a solid or liquid. The U.S. Occupational Health and Safety Administration (OSHA) defines a combustible liquid as "any liquid having a flash point at or above 100°F (37.8°C), but below 200°F (93.3°C).

COMBUSTIBLE LIQUIDS: Typically, will require some external heat source to produce a sufficient concentration of vapors. Combustible liquids are any liquid having a flash point at or above 100°F and are divided into two classes: Class II and Class III. When a combustible liquid is heated to or above its flash point, it may have some of the hazards of a flammable liquid. (See Flammable Liquids)

COMBUSTIBLE SOLIDS: Those solids found in greatest abundance in properties which constitute the greatest bulk of property destroyed by fire. These include all materials which will ignite and burn or undergo substantial chemical change when subjected to heat or flame. The chemical makeup of most ordinary combustible solids is carbon, hydrogen, and oxygen with lesser percentages of nitrogen and other elements. Cellulose is the main component of wood by weight; paper is almost pure cellulose. Cotton consists of 90 percent cellulose. Animal fibers consist of protein molecules containing high percentages of nitrogen as well as carbon, hydrogen, and oxygen.

COMBUSTION: Can be defined as an exothermic chemical reaction between some substance and oxygen. Combustion consists of chain reactions involving free hydrogen atoms, H_2, hydroxyl free radicals, OH, and free oxygen molecules

COMBINATION DETECTOR: A device that either responds to more than one of the fire phenomena or employs more than one operating principle to sense one of these phenomena. Typical examples are a combination of a heat detector or a combination rate-of-rise and fixed-temperature heat detector.

COMPETENCY OF A WITNESS: A person is competent to testify if he has sufficient understanding to receive, remember, and narrate impressions and is sensible to and understands the obligation of another. Competency also entails an adequate ability to observe, mental capacity, and other relevant factors.

CONDUCTION: The process by which heat is communicated from one body to another by direct contact or through an intervening solid liquid or heat conducting medium. Some examples are; a steam pipe in contact with wood; a chimney flue in contact with attic or ceiling framing; a chimney flue in contact with attic or ceiling framing; a pair of pliers in contact with electrical conductors that are grounded through an electrode. The amount of heat transferred depends on the heat conductivity of the materials and the area of thickness of the conducting mass. The rate of transfer is in direct proportion to the temperature differential between the points of entrance and departure.

CONSUME: To destroy as by fire. To waste or burn away.

CONTRACT: An agreement by which a person undertakes to do or not to do something. Within the meaning of the federal contract clause of the Federal Constitution, the term includes not only contracts as the word is ordinarily understood, but all instruments, ordinances, and measures which embody the inherent qualities or purposes of contacts and carry reciprocal obligations of good faith. (REF: Ballentine's definition of contracts.)

CONTRACT MALAISE: This term includes all contracts of an immoral nature, iniquitous in themselves, and those opposed to sound public policy. Where both parties are in pari delicto (in equal fault), neither as a rule, will be accorded relief in a court of law.

CONTRACT OF INSURANCE: An agreement by which one party in a consideration promises to pay money or its equivalent or do some act of values to the assured upon the destruction or injury of something in which the other party has an interest.

CONVECTION: The process by which heat is moved by air differences, usually in a rising or circular pattern. For example, in the case of air circulation in a room, heat from a stove will move laterally if there is an obstruction to vertical movement such as a ceiling and vertically if there is an obstruction to horizontal movement such as a wall.

CONVERSION OF CENTIGRADE TO FAHRENHEIT: The formula for converting Centigrade to Fahrenheit is: $F = 9/5C + 32$.

Example: convert 10°C to °F

$F = 9/5C + 32 F = 9/5(10) + 32$

$F = 18 + 32 = 50 F$

The formula for converting Fahrenheit to Centigrade is: $C = 9/5F + 32$.

CORPUS DELICTI: The body of the crime, assuming that the specific crime charged has actually been committed by someone. The corpus delicti is made up of two elements; first that a certain result has been produced (i.e., a man has died, or a building has been burned); and second, that someone is criminally responsible.

(REF: Ballentine on corpus delicti and 7R.C.L.774) The corpus delicti specifically implies the body of the offense, or the substance of the crime.

In an arson case, the corpus delicti consists of two elements; first, that the fire occurred (the burning or fire must at least extend to a charring of the wood or fabric to satisfy the legal requirements of burning); and second, that a fire resulted from the willful and intentional ignition by a criminal agency. The first must be shown by affirmative evidence. The second may be shown by direct or circumstantial evidence. For example, if flammable liquids were present in the library of a dwelling and ignited by a trailer extending from the door, the corpus delicti may be established by providing the occurrence of fire and second, eliminating all reasonable, accidental sources of ignition such as electrical, mechanical, smoking, lightning, etc.

CRIME: Under interpretations in our system, crime is a wrong of public character because it possesses elements of evil which effect the pubic as a whole and not merely the person whose rights of person or property have been invaded. The term includes felonies and misdemeanors. Crime is also defined as an act committed or omitted in violation of a public law either forbidding or commanding it.

CRIMINAL CHARGE: A charge, strictly speaking, exists only when a formal written complaint has been made and a prosecution initiated against the accused. In the eye of the law, a person is charged with crime only when he is called upon in a legal proceeding to answer to such a charge.

CRIMINAL CONSPIRACY: A combination of two or more persons attempting to accomplish, by some concerted action, some criminal or unlawful purpose, or to accomplish some purpose not in itself criminal, by criminal or unlawful means.

CRIMINAL INFORMATION: A formal declaration of the charge or offense against a person, made by the prosecuting or district attorney, and filed in the court in which the person is to be tried.

CRIMINAL INTENT: That evil state of a person's mind, accompanying an unlawful act, in the absence of which no crime is committed. In some crimes, the law requires specific intent such as in some statutes making it a felony to file a proof of loss with the intent to defraud an insurance company. In other crimes, the intent may be general if the act is willful and intentional, such as the willful and malicious act of setting fire to a railroad bridge. In certain crimes, the intent may be implied from the obvious willfulness of the act in which case malice may be inferred. It is important for the investigator to distinguish clearly between intent and motive. Motive is the reason, real or fancied, that a person had for committing an act. Intent, express or implied, is required under the law. Proof of motive is not a requirement under the law.

CRIMINAL OFFENSE: As used in statutes permitting the introduction of evidence of commission of a criminal offense to affect the credibility of a witness, the term is generally held to include both felonies and misdemeanors, but not to include violation of a municipal ordinance.

CRIMINAL PROCEEDING: The prosecution of a person charged with a criminal offense. The subsequent contemplation of the conviction and the punishment for the offense of the person so prosecuted. A proceeding against a juvenile offender is generally not a criminal proceeding.

CRIMINAL PROSECUTION: A prosecution in a court of justice, in the name of the sovereignty of the jurisdiction involved, against one or more individuals accused of a crime. Although instituted by an individual, a criminal prosecution is not in any sense an action between the person instituting it and the prisoner.

CRIMINAL STATUTE: A statute which describes an act or the omission of an act as a criminal offense. A statute authorizing punishment for contempt is a criminal statute. Statutes in this category may be either felonies or misdemeanors.

CURRENT, ELECTRIC: Electrical current is of two kinds: direct current (DC) and alternating current (AC). DC is a continuous flow in the same direction, as in batteries; AC creates a flow which is periodically reversed and creates a system of waves. The number of cycles is the number of waves per second. As a rule, AC circuits are in single or three phases. Single phase is carried by a two-wire circuit although a third or neutral wire is often used which does not alter the principle. In a three-phase, three-wire, balanced (delta) circuit, the current flow in each conductor leads or lags the current in the other two conductors and the algebraic sum of the three is zero. Any of the two wires of a three-phase circuit is much greater than that required to operate it once it is in motion or at normal speed. The common voltages are 120/208 and 277/480. (REF: National Electrical Codes) The flow of electric current creates magnetic force and a change in magnetic field surrounding a system of wires. This produces an electric voltage which, in turn, creates current flow. These facts are the basis of the generator and the electric motor. Voltage is produced in the windings of the motors and generators. In a motor, voltage created by the rotor is known as back electromotive force. To some extent, this restricts the flow of current through the motor. The current required to start a motor is much greater than that required to operate it once it is in motion or at normal speed.

CURTAIN BOARDS: A noncombustible barrier suspended to bank heat along a ceiling to enhance the effectiveness of a sprinkler or extinguishing system.

CRYOGENICS: This is the science of dealing with the behavior of materials at temperatures close to absolute zero. It is utilized in the handling of gases such as argon, oxygen, hydrogen, and nitrogen, for purposes of liquification, transport, and storage. (REF: NFPA Cryogenic Properties of Gases)

DANGEROUS WHEN WET MATERIALS: Those materials that react with water to become spontaneously flammable or to give off flammable gas or toxic gas at a rate greater than 1 L/kg of the material, per hour.

DEAD LOAD: The weight of the building itself and any equipment permanently attached or built on the building.

DEBRIS: The ruins or rubble which results from the burning of materials.

DECOMPOSITION EXPLOSION: Certain compounds are capable of decomposing almost instantaneously. Examples are commercial explosives, ammonium nitrate, and carbon disulfide. In addition, certain other endothermic compounds are subject to explosion under certain conditions. Explosive decomposition is usually accompanied by the release of large quantities of hot gases and in general, it is correct to say that the speed of release of the gases is proportionate to the violence of the explosion. Explosions and detonations may result in the formation of pressures attaining several hundred pounds per square inch. A pressure wave may spread in all directions at velocities from a few thousand to more than 25,000 ft./s and which, in turn, can cause damage to buildings and fixed objects. Some authorities classify explosions as low order (explosions of gunpowder and similar explosives) and high order (in groups of dynamite, nitro compounds, etc.)

DECOMPOSITION, HEAT OF: Heat of decomposition is the heat released by the decomposition of endothermic compounds, that is, those requiring the addition of heat for their formation.

DEFLAGRATION: The burning of a gas or aerosol that is characterized by a combustion wave. The combustion wave moves through the gas and oxygen burning until all the fuel is used.

DENSITY: Density is the amount of something per unit volume. Most typically, one expresses the mass per unit volume for a solid or liquid. For example, 5.2 g/cm^3. For gases or dusts we might express this as g/m^3.

DETECTOR: A device suitable for connection to a circuit that has a sensor that responds to a physical stimulus such as heat or smoke.

DETONATION: The burning of a gas or aerosol that is characterized by a shock wave. With detonation, the shock wave is traveling at a speed greater than the speed of sound and the wave is characterized by extremely high pressure that is initiated by a very rapid release of energy.

DETONATORS: The term commonly used to describe the device employed in initiating the decomposition of explosive compounds such as dynamite and blasting gelatin. Detonators generally require two types of explosives; one being sensitive to heat or fire, such as the safety fuses, which will explode a second more powerful charge that detonates or boosts to the principal explosive. Lead azide is commonly used as the first detonator because of its high sensitivity. The detonators are usually designed to attach to the safety fuse or another remote triggering device and are contained inside a tubular copper or aluminum case, with an attachment suitable for crimping. Detonators are commonly designed for electrical activation.

DIFFUSE EXPLOSION: The instant decomposition of combustible gases in appropriate mixture with air. In a limited space, this phenomenon may displace walls, bulkheads, and fixed objects. The combustion produces a great amount of sudden heat with expansion dependent on volume. This is not a high velocity type explosion but is heat producing and capable of displacement damage as well as continuing fire.

DIP TANK: Examples of operations that can pose fire hazards include dip tank operations in which objects are painted dipped, electroplated, pickled, quenched, tanned, degreased, stripped, roll coated, flow coated, and curtain coated.

DIRECT EVIDENCE: Evidence given by witnesses who testify directly concerning their own knowledge of the main facts to be proven. Eye witnesses are a good example; "A" takes the witness stand and testifies that he saw "B" toss a burning object through the window of the dwelling and that immediately thereafter flames burst from the window and "B" ran from the scene.

DISTILLATION: As used in fire investigation a laboratory process utilized in the recovery and analysis of accelerants. In simple distillation, the evidence specimen is heated to vaporize the residue and the vapors are condensed to reform as a liquid. In the process of steam distillation, the residue recovered from the scene is flooded with water and heated to the boiling point. Steam and vapors are introduced into the process and the condensed results are then separated. The accelerant, if any, is then identified. Vacuum distillation is a process in which any accelerant present in the suspected specimen is removed by simultaneous application of suction and heat.

DUST: Fine, small particles of dry matter. Dusts can be generated by handling, crushing, grinding, rapid impact, detonation, and breakdown of certain organic or inorganic materials, such as rocks, ore, metal, coal, wood and grains.

DUST EXPLOSION: The rapid order combustion of finely divided particles of combustible solids in an atmospheric oxygen environment when an igniting source is present. The fine particles or dust may be from coal, plastics, grain, explosive powders, wood cotton, industrial waste, or light metal. The phenomenon may be accompanied by considerable heat but is not considered in the high explosive order although low order explosive damage and displacement may result.

DYNAMITE: A high explosive, principally consisting of nitroglycerin absorbed in inert solid materials to reduce sensitivity to shock. Straight dynamite is essentially nitroglycerin absorbed in a combustible absorbent such as wood pulp and extra oxidizing materials such as sodium nitrate with a small amount of calcium carbonate or zinc oxide. Ammonia dynamite consists of ammonium nitrate with nitroglycerin, in smaller percentage, and may also include sulfur and cereal products along with a neutralizing zinc oxide. Gelatin dynamite broadly covers a variety of compounds with the common factor consisting of the substitution of smokeless nitrocellulose powder for relatively inactive wood pulp. The compounds are gelatinous and plastic in mass and are sometimes called blasting gelatin. It is generally considered more potent than ordinary dynamite. Dynamites are considered far more stable to shock than nitroglycerin although the dynamite compounds will explode on overheating and exposure to heavy shock, such as impact by a bullet. The dynamites become unstable in conditions of improper storage, aging, and crystallization. Disposal should be restricted to explosives experts.

ELECTRIC BOILERS: A power boiler, heating boiler, high or low temperature water boiler in which the source of heat is electricity.

ELECTRICITY: In the context of flowing through conductors, electricity is composed of identical particles called electrons that are much lighter than atoms. Electrons weigh only 1/18 37th as much as an atom of the lightest chemical element, hydrogen. (In this context, electricity may be described as "an electric current; a stream of moving electrons, setting up a magnetic field of force through which it produces kinetic energy")

ELECTROSTATIC FLUIDIZED BED: A container that holds powder coating material that is aerated from below so as to form an air-supported expanded cloud of such material that is electrically charged with a charge opposite to the charge of the object to be coated; such object is transported, through the container, immediately above the charged and aerated materials in order to be coated.

ELECTROSTATIC SPRAY PAINTING: The spray material such as paint is negatively charged (while atomized or after having been atomized by the air or airless methods) through a connection of the spraying-gun to a generator. The material being painted is positively charged and a result, the difference in charge leads the particles toward the material to obtain the coating.

ENDOTHERMIC: An endothermic process is one in which heat must be supplied to the system from the surroundings.

EVACUATION SIGNAL: Distinctive signal intended to be recognized by the occupants as requiring evacuation of the building.

EVAPORATION RATE: An evaporation rate is the rate at which a material will vaporize (evaporate, change from liquid to vapor) compared to the rate of vaporization of a specific known material. This quantity is a ratio; therefore it is unit less.

EVIDENCE: That which makes clear or verifies the truth of the very fact or point in issue, whether from the defense or the prosecution. Those rules of law whereby we determine what testimony is to be admitted, what is rejected in each case, and what weight is to be given to the testimony so admitted.

EXIT ACCESS: Consists of the route one must take to portions of the building that are protected from fire through their design, such as is the case in a hallway that has a 2-h fire rating.

EXIT DISCHARGE: The portion of the exit that separates the exit from the public area.

EXOTHERMIC: An exothermic process is one that gives off heat. This heat is transferred to the surroundings.

EXPERIMENTAL EVIDENCE: This is proof of a fact or theory in the form of an experiment made during the trial and in the presence of the jury and usually in the courtroom. The admission of such evidence is entirely within the discretion of the court and neither party can demand it as a matter of right.

EXPERT OPINION EVIDENCE: The testimony of an expert in which he is permitted to state his opinion on a question of science, skill, or trade. Thus, the opinions of medical men are regularly admitted as to the cause of disease, death, consequences of wounds, sanity, and state of mind. In many Federal and State jurisdictions, witnesses who qualify to the satisfaction of the Court to be especially skilled in the chemistry and behavior of fire are permitted to form and express opinions to the jury as to the origin and cause of the fire. They are never permitted to testify as to who set the fire; that is the exclusive province of the jury. The jury may believe or disregard the opinions of conclusions of an expert witness or weigh his credibility in the same manner they can weigh the testimony of any other witness. In some jurisdictions, courts only allow an expert fire witness to testify as to his findings; for example, that he found evidence of a bomb, accelerant, or separate and distinct points of origin. Courts in those jurisdictions that are more conservative about the admission of expert testimony will not permit the expert to state his conclusion of opinion as to cause. They hold that this area is the excessive province of the jury or the triers of fact.

EXPERT WITNESS: An expert witness is one who has acquired such special knowledge of the subject matter about which he is to testify, either by study with recognized authorities on the subject or by practical experience, so that he can give the jury assistance and guidance in solving a problem of which their good judgment and average knowledge is inadequate.

EXPLOSION: An explosion is an instantaneous decomposition of a solid or liquid substance, extending throughout the entire mass and accompanied by considerable disengagement of heat. The substance is partly or wholly converted into gaseous decomposition products.

EXPLOSIVE LIMITS (Flammable limits): In the contact of gases and vapors which form flammable mixtures with atmospheric oxygen (oxygen and nitrogen), and pure oxygen there is a minimum concentration of vapor in air or oxygen below which flame propagation does not occur on contact with an ignition source. There is also a maximum proportion of vapor or gas in air above which flame, or explosion will not occur. These boundary line mixtures of vapor or gas with air are known as the lower and upper flammable or explosive limits and are usually expressed in terms of percentage by volume. Generally, it may be said that if a mixture is below the lower flammable, or explosive limit, it leans toward exploding or burning; above the upper flammable or explosive limit, the mixture is too rich to burn or explode. (For example, in pumping liquids such as hydrocarbon fuels from storage tanks, it is a common safety practice to replace the volume with inert gases, thereby eliminating the possibility of the introduction of an explosive mixture in the oxygen gas vapor spectrum. An inert gas will neither flame nor explode).

EXAMPLES OF EXPLOSIVE LIMITS: (In percent by volume of vapor–air mixtures)
Substance Low Limit Upper Limit
Acetone 2.15% 13.0%
Allyl Chloride 3.11% 11.1%
Butyl Alcohol 2.4% 8.0%
Gasoline 1.4% 7.6%
Hydrogen 4.0% 75.0%

EXPLOSIVES: United States Department of Transportation defines an explosive as any substance or article, including a device, which is designed to function by explosion (i.e., an extremely rapid release of gas and heat) or which, by chemical reaction within itself, is able to function in a similar manner even if not designed to function by explosion.

Solid or liquid explosives represent gases or vapors condensed into the smallest possible compass. The substance used in the blasting or explosive arts are generally termed explosives or blasting materials. Their detonation results in a sudden and enormous expansion of gases and vapors which are liberated from the previous condition of chemical combination. In addition to the commonly known and special industrial explosives, certain fertilizers and chemical industrial compounds are unstable and require special handling in packaging, transportation and storage and will explode violently under conditions of mishandling or misuse. These compounds are described in various industrial and public safety manuals. Some of the more common blasting explosives are ammounite, blasting powder, chlorate powder, carbonite, explosive gelatin, guhur dynamite, gelatin dynamite, wetterdynamite, nitroglycerin, and nitrocellulose.

EXTRA HAZARD OCCUPANCIES: Represent the potential for the most severe fire conditions and therefore present the most severe challenge to fire protection systems.

FAHRENHEIT TEMPERATURE: A temperature scale which sets the boiling point of water at 212° and the freezing point at 32° above the zero point on the scale.

FELONY: As a rule, all crime punishable by death or imprisonment in the state prison are felonies; all others are misdemeanors.

FIRE (Combustion): As used in the discussion of combustion, fire is the process whereby substances, or individual constituents of the same, combine with oxygen accompanied by the liberation of heat. Three factors are necessary to the process in general sense: the body or supporting factor (oxygen). The substance to be consumed (such as wood, coal, etc.) and the disseminating factor (heat).

FIRE ALARM CONTROL PANEL: A system component that receives inputs from the automatic and manual fire alarm devices and might supply power to detection devices and to a transponder(s) or off-premises transmitter(s). The control unit might also provide transfer of power to the notification appliances and transfer of condition to relays or devices connected to the control unit. The fire alarm control unit can be a local fire alarm control unit or a master control unit.

FIRE ALARM SIGNAL: A signal initiates by a fire alarm-initiating device such as a manual fire alarm box, automatic fire detector, water flow switch, or other device in which activation is indicative of the presence of a fire or fire signature.

FIRE ALARM SYSTEM: A system or portion of a combination system that consists of components and circuits arranged to monitor and annunciate the status of fire alarm or supervisory signal-initiating devices and to initiate the appropriate response to those signals.

FIRE BRIGADE: A fire brigade is an organized group of employees who are knowledgeable, trained, and skilled in at least basic firefighting operations. Even employees engaged only in incipient-stage firefighting will be considered a fire brigade if they are organized in that manner.

FIRE GASES: Fire gases generally refer to the gaseous products of combustion. Most combustibles contain carbons which, when burned form carbon dioxide when the oxygen supply is good and carbon monoxide when the air supply is poor. Other gaseous products of combustion may include hydrogen sulfide, ammonia, sulfur dioxide, hydrogen cyanide, nitric oxide, phosgene, and hydrogen chloride. The products of gases depend mainly on the chemical composition of the fuel as well as ventilation and combustion temperatures during the process.

FIRE LOAD: The fire load of a space or occupancy is the expected amount of combustible materials in the area or space under consideration. It can generally be expressed in the terms of the weight of the total combustibles per square foot of the occupancy in relation to their heat producing capability in the calories or BTUs. Most fuels, such as wood fibers petroleum products, books, papers and furniture may be rated in BTUs per pound and the fire load thereby generally computed. However, construction in or ventilation of the space may produce substantial variables. Predictions can only be general. American and British laboratories have produced standard time temperature curves for various classes of occupancies based on calculated fire loads which may be utilized for references.

FIRE PARTITION: A partition serving to restrict spread of fire but not related as a fire wall.

FIRE POINT: The lowest point in temperature at which a flammable liquid in an open container will produce vapors in continuing combustion. The fire point is usually a few degrees above the flash point except in certain unstable compounds.

FIRE PREVENTION: The elimination of the possibility of a fire being started.

FIRE PROTECTION: Basic tools of engineering and science to help protect people, property, and operations from fire and explosions (ASSE & BCSP, 2000, 23).

FIRE RESISTANCE: Uniform building codes define fire resistance as a relative term used with a numerical rating or modifying adjective to indicate the extent to which building materials, structures, or other materials resist the effect of fire; for example, resistance of 2 h as measured on the standard time temperature curve.

FIRE SUPPRESSION: Those activities (usually performed by fire department personnel) required to extinguish a fire. Those activities may include the application of water on the burning material, venting of a structure or removal of debris.

FIRE TETRAHEDRON: Minimum components needed to have a fire. The components are the fuel, oxygen and heat or some other type of energy source and a chemical chain reaction.

FIRE TRIANGLE: A term used in some fire training courses to illustrate the need for the three factors essential for fire: fuel, heat, and oxygen. Basically, combustion depends on the substance to be consumed, oxygen (the supporter of combustion), and the disseminating factor which is heat.

The disseminating factor may result from physical, chemical, or biological reactions.

FIRE WALL: Broadly defined as a wall erected to prevent the spread of fire. Effective, approved walls must have sufficient fire resistance to withstand the effects of the most severe fire that can be expected in the building and must provide a complete barrier, along with any openings suitably protected.

FIXED-TEMPERATURE DETECTOR: A device that responds when its operating element becomes heated to a predetermined level.

FLAME: It is generally conceded that burning materials, when supported by sufficient oxygen, produce luminosity called flame. It is generally accepted that flame is a distinct product of combustion. The sufficiency of the flame and its extension depends on the fuel, availability of oxygen, and quality of fuel or fuel load. The color of the flame may vary with the type of fuel and also with conditions of draft; it may also vary with degree of temperature For example, a natural gas flame, under poor conditions of draft and ventilation, may appear lazy and yellow or orange; under appropriate and well-adjusted ventilation effects, it may appear blue.

FLAME DETECTOR: Flame detectors are categorized as ultraviolet, single-wavelength infrared, ultraviolet infrared, or multiple-wavelength infrared.

FLAMESPREAD: Once ignition occurs, it is the propagation or movement of flame from layer to layer and/or across surface exposures without regard to the source. There are several accepted methods of testing materials and surfaces for flame spread and scaling possible rate of spread under fire conditions.

FLAMMABLE: Generally used in fire investigation and code standard to describe combustible material that ignites readily and burns intensely with rapid flame spread. The term is used in a general sense without a clear definition of ignition temperature, surface burning rate, or other defined guidelines. The term is preferred over inflammable to avoid possible confusion due to the prefix "in" which suggest the negative, such as in the term nonflammable.

FLAMMABLE AEROSOL: A flammable aerosol is defined by the U.S. Occupational Health and Safety Administration (OSHA) as "an aerosol which is required to be labeled flammable under the Federal Hazardous

Substances Labeling Act (15 U.S.C. 1261). For the purposes of paragraph (d) of this section, such aerosols are considered "Class IA liquids."

FLAMMABLE LIMITS: Flammable limits apply generally to vapors and are defined as the concentration range in which a flammable substance can produce a fire or explosion when an ignition source (such as a spark or open flame) is present. The concentration is generally expressed as percentage fuel by volume. Above the upper flammable limit (UFL) the mixture of substance and air is too rich in fuel (deficient in oxygen) to burn. This is sometimes called the upper explosive limit (UEL).

FLAMMABLE LIQUID: Fire protection authorities, including the National Fire Protection Association, have established an arbitrary division between liquids and gases which is defined in NFPA #30 of the Flammable Liquids Code. Liquids are those having a vapor pressure not exceeding 40 psi absolute at 100F approximately 25 psi gauge pressure. Flammable liquids are those having closed cup flash point below 200°F. and are divided into three classes in the Flammable Liquids Code. Combustible liquids are those with a flash point at or above 200°F. When a combustible liquid is heated to or above its flash point, it may have some of the hazards produced by flammable liquids.

FLAMMABLE (EXPLOSIVE) RANGE: This is the range of combustible vapor or so-called gas air mixture between the lower and upper limits commonly described as the flammable or explosive range. Flammable or explosive ranges of percent of vapors in air vary widely. For example, the explosive range of gasoline is 1.4% (lower) and 7.6% (upper) limit; hydrogen gas is 4.0% (lower) and 75% (upper); ethyl alcohol is 4.3% (lower) and 19% (upper).

FLAMMABLE SOLID: A flammable solid is defined by the U.S. Department of Transportation (DOT) quite extensively (see 49 CFR 173.124). Three broad classes are desensitized explosives such as those wetted with sufficient water, alcohol, or plasticizer to suppress explosive properties.

FLASHOVER: Initial progress of a fire may be slow or fast depending on the characteristics of initial ignition and basic fuel. Important factors are the nature of the fuel or fabric, its surface, finish, environment, and available oxygen. If the fire continues, it may eventually reach a stage where all exposed combustible surfaces simultaneously burst into flame. This is what is generally known as flashover and is significant in the communication of fire from space to space, even in supposedly separated areas. Flashover is a common phenomenon a product of fuel load, combustible surfaces, temperature, and available oxygen. The results of flashover may be mistaken by inexperienced fire investigators as separate and distinct areas of origin; this sometimes creates the mistaken illusion that a fire was incendiary when, in fact, it was not.

FLASH POINT: The flash point is the lowest temperature at which a substance gives off vapor sufficient to form an ignitable mixture with the air near the surface of the substance. In respect to its use in the terms of vapors and flammable liquids, the term ignitable mixture is used in the context of flammable range. When ignited between the upper and lower limits, it is capable of flame propagation away from the source of ignition. Combustion may not be continuous at the flash point, depending on fuel, extent, and environment. As indicated earlier, the term flash point most commonly applies to flammable liquids although there are many solids that may slowly deteriorate, sublime, or become chemically active at normal ambient temperatures. Many others will react in slightly higher temperatures or when exposed to direct rays of the sun, producing vapors which will flash while the substance retains its basically solid state.

FRAUD: Conduct which operates prejudicially on the rights of others and is so intended; a deception practiced inducing another to part with something of value or surrender some legal right and which accomplishes a desired end. As commonly used, the word implies deceit, deception, artifice, and trickery.

FUEL BREAK: An artificially created or natural separation between buildings and improvements and forest or flammable vegetation, or through wild lands to enable fire crews to combat and control advancing fire. The construction, permitted cover, and widths depend on code and official discretion in relation to the risks involved, terrain, and adjacent natural vegetation.

FUME: A cloud of fine particles suspended in a gas.

FUSIBLE LINK: Metal parts designed to separate at desired temperatures to release or otherwise activate self-closing fire doors or openings. Fire doors, for example, may be made self-closing by a system of weights suspended by wires, or pulleys restrained by fusible link or other fixed temperature release or heat responsive device.

GALVANISM: Electric current in energy resulting from a chemical reaction.

GAS: Gas is an aeriform fluid of matter lacking independent shape or volume, which tends to expand indefinitely. The properties of gases, and their behavior, can only be explained by assuming that they are composed of extremely minute particles in constant motion the higher the temperature, the more violent the motion. For fire protection purposes, gases are generally divided into two broad categories; flammable and nonflammable. The hazards of flammable gases are generally similar to flammable liquids and there is really no sharp line of demarcation between gases and gases of liquids in respect to fire hazard; further any actual gas, at a sufficiently low temperature and high pressure, becomes a liquid and any flammable liquid at a sufficiently elevated temperature becomes a gas.

GAS CHROMATOGRAPH: An instrument used by forensic experts to identify accelerants in fire debris. A specimen is analyzed and graphed. The graphs produced by this instrument are compared to accepted standard graphs of known accelerants and the specimen can then be identified. This instrument cannot detect accelerants in minute amounts and should not be considered conclusive evidence of the presence or absence of an accelerant.

GASOMETER: An instrument designed to measure gas, such as an explosion meter.

GLOBALLY HARMONIZED SYSYEM OF CLASSIFICATION AND LABELLING OF CHEMICALS (GHS): A system for standardizing and harmonizing the classification and labelling of chemicals. (Aspects of this harmonization system harmonized definitions for health, physical and environmental hazards of chemicals.)

GRAM: The basic unit of mass in the metric system.

GRAVITY: In physics, gravity is used in terms of weight in resistance to terrestrial gravitation. The rate of acceleration of gravity is approximately 32 ft./s. It is also used in terms of the weight of liquids and oils.

HALOGENS: The members of this group are chemically active and have similar chemical compositions. Some halogens include bromine, chlorine, fluorine, and iodine. They differ from each other only in the degree of activity. They are noncombustible but will support combustion of certain compounds. Turpentine, phosphorus, and some finely divided metals will ignite spontaneously in the presence of halogens. Halogen fumes are poisonous, corrosive, and extremely irritating to eyes and throat.

HARDWOOD: Hardwoods such as oak, ash, mahogany, and gum are commonly used in finishing and veneers, and in certain key construction members. Coniferous woods such as pine, cypress, and some of the cedars are softwoods used in framing. There is very little difference between hard and soft woods in reaction to fire except in some redwoods.

HEAT: As related to combustion, heat may be manifested in the form of rapid oxidation, with the evolution under certain circumstances, of flame, smoke, and light. A knowledge of the conditions that determine whether rapid oxidation of a substance with evolution of light will occur is essential to the principles of fire control and investigation. All substances are made up of atoms or groups of atoms (molecules). Each atom consists of units of energy in the form of electrons, protons, and neutrons in continuous motion ultimately expressing individual composition in the form of energy. Energy can be expressed in the form of heat. Chemical reactions are endothermic or exothermic. Exothermic reactions are those which produce products with less total energy than the reacting substance whereupon the energy has been released in the form of heat. Endothermic reactions produce products with more total energy than the reacting substances whereupon energy in the form of heat has been absorbed. Oxidation reactions involved in fires are exothermic in the sense that one of the products of the reaction is heat. While, by far, the most common oxidizing material is the oxygen of the atmosphere (one fifth oxygen and four fifths nitrogen), certain chemicals, such as sodium nitrate and potassium chlorate, can also readily release oxygen under favorable conditions thereby supplying the flame environment or even creating an explosive environment. Certain pyroxylin plastics contain sufficient oxygen combined in their molecules to produce combustion without an outside source; zirconium dust may be ignited with carbon dioxide. Other explosive and industrial compounds, as earlier stated, will produce oxidizing conditions sufficient for graduated or instantaneous decomposition accompanied by the disengagement of heat.

HEAT EXPLOSION: A term used by some firemen and investigators which they interpret synonymously with flashover. There is considerable doubt about the accurate usage of such a term since it has been utilized to categorize widely different situations of fire behavior under the same classification.

HEAT OF COMBUSTION: The amount of heat released during the process of complete oxidation, generally referred to as calorific value. Heat or combustion depends on the arrangement and numbers of atoms in molecules. Calorific values are commonly expressed in BTUs per pound or in 1gram calories per gram. (Sawdust produces approximately 8.490 BTU/lb.; cotton about 7,000 BTU/lb.; while cottonseed oil produces approximately 17,200 BTU/lb.; and crude oil approximately 18,000 BTU/lb.)

HIGH TEMPERATURE WATER BOILER: A water boiler intended for operations at pressures more than 160 psig or temperatures more than 250°F.

HOT WATER HEATING BOILER: A boiler in which no steam is generated, from which hot water is circulated for heating purposes and then returned to the boiler, and which operates at a pressure not exceeding 160 psig or a temperature of 250°F at the boiler outlet.

HUMIDITY: The amount or degree of moisture in the air. Relative humidity is the amount of moisture in the air as compared to the amount that the air could contain at the same temperature. While relative humidity has only a moderate effect on a structure fire once ignition occurs, it is a factor in the moisture content of wood and vegetable materials as well as fabrics.

HYDROCARBON: Any compound containing only hydrogen and carbon; benzene and methane are examples of hydrocarbons.

IGNITION: The heating of a compound or substance to the point of continuing combustion or chemical change.

IGNITION, SPONTANEOUS: In general terms, this is the process wherein substances or compounds ignite because of an increase in temperature of the substance or compound without an independent outside ignition source and without drawing heat from its surroundings. It is sometimes called spontaneous combustion by laymen although it may occur over a long-term period such as in unventilated, piled, green hay or other vegetable products; coal dust; charcoal; and oil paint rags. The rate of heat generation, air supply, and insulation properties of the suspect fuel and surrounding environment have definite relevance to whether or not spontaneous heating will occur. In most vegetable compounds, the process may be slow; on the other hand, with unstable or rapidly oxidizing compounds, heat generation may be rapid and even explosive. The process may be luminous or nonluminous. It may occur with or without smoke.

IGNITION TEMPERATURE: Ignition temperature is generally defined as that point of minimum temperature to which a solid, liquid, or gaseous substance must be heated to initiate or cause self-continuing combustion, independent of the heating or igniting source. Experience and tests have shown that ignition temperatures may vary with change in conditions and environment and listed ignition temperatures should be considered as only approximations. For example, ignition temperatures of flammable liquids and vapors may vary depending on vapor or air gas mixture, size and shape of the contained environment, and other chemical and physical factors. This is why there is some variance in reported ignition temperatures by different laboratories. The ignition temperature of combustible solids may be reasonably influenced by the rate of airflow, rate of heating, surface conditions, and size of the particular solid.

IMPEDANCE: In reference to alternating electric current, the apparent resistance corresponding to the true resistance in direct current; a combination of inductive reactance and resistance is referred to as impedance. The symbol Z and the quantity are measured in ohms. There are two methods of determining the impedance of a circuit that is made of both inductive reactance and resistance. These are the vector or impedance triangle method and the computation method. (Load and resistance have relevance to fire cause.)

INCENDIARY: In fire investigation, describes a fire set by human hands. This fire is usually started by applying some type of flame-producing device (match or lighter) to flammable materials. The flame-producing device is usually removed after igniting the materials. Fire investigators may determine a fire to be incendiary after all accidental causes have been eliminated.

INERT: The term indicates a physically neutral or inactive substance such as nitrogen, helium, and other gases which present no fire hazard and can neutralize atmospheres which would otherwise be explosive. For example, inerting systems are used to pump inert gases into contained explosive atmospheres as when oil tanks are pumped out during discharge of tanker cargoes. Inerting systems may also be utilized to extinguish fires.

INITIATING DEVICE: Includes all types of sensors, ranging from manually operated fire alarm boxes to switches, that detect the operation of a fire suppression system.

GLOSSARY

INSURANCE: A contract whereby one undertakes to indemnify another against loss, damage, or liability arising from an unknown or contingent event.

INSURANCE ADJUSTER: An agent of an insurance company whose general duty it is to adjust and report losses to the principal officers of the insurance company. Under relatively recent usage, adjusters may be direct employees of the insurance carrier or agents by retainer or special assignment. Insurance adjusters may also be retained directly by the insured to represent the interest of the insured in his negotiation with the insurance carrier.

INSURANCE AGENT: One who negotiates policies of insurance for a commission paid to him by the insurance company.

INSURANCE BROKER: One who acts as a middleman between the insured and the insurer and who solicits insurance from the public under no employment from any special company, but having secured an order, places the insurance with a company selected by him. A broker is the agent for the insured, though at the same time, for some purposes, he may be the agent for the insurance carrier. Any acts and representations within the scope of his authority as such agent are binding on the insured.

INSURANCE CARRIER (Insurance Company): Within the general context, a company engaged in the business of making contracts by which it agrees to indemnify the other parties (the insured or designated) from a loss or damage which they may suffer from a specified peril.

INSURANCE CONTRACT: A contract whereby one person undertakes to indemnify another against loss, damage, or liability arising from an unknown or contingent event.

INSURANCE MONEY: Money recoverable on a policy of insurance by the insured against the insurance company which issued the policy.

INSURANCE POLICY: An agreement by which one person, for a consideration, promises to pay money, an equivalent, or do some act of value to another upon the destruction or injury of something by specified peril.

INSURE: To contract to indemnify a person against loss from stated perils; to enter into a contract of insurance as insurer.

INSURED: When used as a noun in fire policies, it is the person whose property and legal representatives is insured, but it should not be extended to include any other persons, such as the mortgagee of the property.

INSURER: A person who, by contract of insurance, agrees to indemnify another person, called the insured, against loss from certain specified perils.

INTENT: In the legal sense, intent is distinct from motive and may be defined as a deliberate act to affect a certain result. Motive is the reason which leads the mind to desire that result. In a trail for arson, the prosecution must show that the act of setting the fire was intentional as opposed to accidental.

INTERIOR FINISH: Class A materials have a flame spread rating from 0 to 25 and they are the best in terms of flame spread, in other words the flame does not propagate as far with these materials as they would with the other classes of materials. Class B materials have a flame spread index from 26 to 75 and Class C has a rating from 76 to 200 (Cote & Bugbee 1991, 152).

INTRINSICALLY SAFE: When equipment has been designed specifically for a hazardous environment, it is "intrinsically safe." Intrinsic safety is a protection concept employed in potentially explosive atmospheres.

INVESTIGATIVE LEAD: The continuing inquiry resulting from basic information. For example, stated facts must be verified by a disinterested third party before the information can be accepted as fact. The investigative leads are indispensable in factual investigation. The pursuit of investigative leads is all too often neglected.

IONIZATION SMOKE DETECTION: The principle of using a small amount of radioactive material to ionize the air between two differently charged electrodes to sense the presence of smoke particles. Smoke particles entering the ionization volume decrease the conductance of the air by reducing ion mobility. The reduced conductance signal is processed and used to convey an alarm condition when it meets present criteria.

KINDLING: Some fire authorities classify the ordinary fire fuels into three categories: tinder, kindling, and bulk fuel. Tinder is described as material ignitable by a common domestic match and will continue to burn. Kindling is defined as material that will ignite and burn if associated with sufficient tinder but in which a match will not produce a continuing fire. Bulk fuel includes the heavy timbers.

KINETIC: Movement resulting from motion.

KINETICS: The science that deals with the motion of masses in relation to the forces acting on them (as in explosions).

KINETIC THEORY: The theory that the minuet particles of all matter are in constant motion and that the temperature of a substance is dependent on the velocity of this motion. Increased motion is accompanied by increased temperature. According to the kinetic theory of gases, elasticity, diffusion, pressure, and other physical properties of gasses are due to the rapid motion (in straight lines) of the molecules; to their impact against each other and the walls of the container; to weak cohesive forces between the molecules, etc.

LATENT HEAT: In general, the additional heat required to change the state of a solid to liquid at its melting point or from liquid to gas at its boiling point after the temperature of the substance has reached either of these points.

LATENT HEAT CHANGE: The change in heat content of a substance when it undergoes a phase change only, no temperature change. A latent heat where liquid is converted to a gas is heat of vaporization while latent heat where a solid is converted to a liquid is heat of fusion.

LATENT HEAT OF VAPORIZATION: Latent heat of vaporization is the heat which is absorbed when one gram of liquid is transformed into vapor at the boiling point under one atmosphere of pressure. It is expressed in calories per gram or BTUs per pound.

LIGHT HAZARD OCCUPANCIES: Occupancies or portions of other occupancies where the quantity and/or combustibility of contents is low and fires with relatively low rates of heat release would be anticipated.

LISTED: Equipment, materials, or services included in a list published by an organization that is acceptable to the authority having jurisdiction and concerned with the evaluation of products or services, that maintains periodic inspection of production of listed equipment or materials or periodic evaluation of services, and whose listing states that either the equipment, material, or service meets appropriate designated standards or has been tested and found suitable for a specified purpose.

LOWER EXPLOSIVE LIMIT: See Lower Flammable Limit.

LOWER FLAMMABLE LIMIT (LFL): If the vapor concentration in air is too low, there will not be enough vapors to ignite (also referred to as the lower explosive limit (LEL). People commonly refer to the vapors as being too lean.

MAGAZINE: Magazines are used for the storage of explosives and are classified as either Class I magazines or Class II magazines.

MALICE: In legal sense, a condition of the mind which shows a heart devoid of social duty and fatally bent on mischief, the existence of which is inferred from the acts committed or words spoken. Malice means an intentionally wrongful act without just cause or excuse, or as the result of ill will; it does not necessarily signify ill will toward an individual. Hence the law implies malice where one deliberately injures another in an unlawful manner.

MANUAL FIRE ALARM BOX: A manually operated device used to initiate an alarm signal.

MEANS OF EGRESS: Consists of three parts: the exit access, the exit, and the exit discharge.

MELTING (failing) POINTS: The following are approximate melting points in degrees Fahrenheit. (REF: NFPA)
Aluminum: 150–1500
Brass: 575–1800
Copper: 1981 (average)
Glass: 400–1600
Iron (pure): 2795
Iron (gray cast): 2246
Iron (white cast): 2102
Iron (wrought): 2750
Iron (cast): 2000–3800
Steel: 2350–2675
Tin: 425–475
Zinc: 750–775

Unprotected steel columns fail in 15 minutes in fire tests. Unprotected cast iron columns fail in 30 minutes. Three-inch plan floor or four-inch laminated floors fail in 45 minutes.

MISDEMEANOR: Any crime which is neither punishable by death nor by imprisonment in a state prison.

MOTIVE: In criminal law, motive is that which leads or tempts the mind to indulge in a criminal act. It is not necessary to establish motive for a conviction, though it explains the actions of the parties or connects the defendant with the offense charged.

NATIONAL FIRE PROTECTION ASSOCIATION: The National Fire Protection Association (NFPA), a private nonprofit organization, is the leading authoritative source of technical background, data, and consumer advice on fire protection, problems, and prevention.

NEGLIGENCE: Failure to take the care to be expected from a reasonably prudent person.

NONCOMBUSTIBLE MATERIAL: A material in the form in which it is used and under the conditions anticipated will not ignite, burn, support combustion, or release flammable vapors when subjected to fire or heat.

NONWAIVER AGREEMENT: An agreement between an insurer and insured that the investigation activities of the insurer in connection with the claim shall not waive or invalidate any of the provisions of the insurance contract or any of the rights of either of the parties.

OCCUPANT-FIRE FIGHTING BEHAVIOR: Occurs with individuals that have economic or emotional ties to the building.

ODOMETER: An instrument for measurement of radio or electronic wavelength.

OHM, OHM'S LAW: An ohm is the unit of electric resistance. The relationship between volts, amps, and ohms is stated as follows:

I equals current in amps

E equals electromotive force in volts

R equals resistance in ohms

$I = E/R$ $R = E/I$ $E = IR$

ORDINARY HAZARD OCCUPANCIES: Subdivided into two groups. The first type occupancies, Ordinary Hazard Group 1, are occupancies where combustibility is low, quantity of combustibles is moderate, stockpiles of combustibles do not exceed 8 ft., and fires with a moderate rate of heat release would be expected. Examples of Ordinary Hazard Occupancies Group 1 would include bakeries and restaurant service areas. The second subgroup is Ordinary Hazard Group 2 which are occupancies where the quantity and combustibility of contents is moderate to high, stockpiles do not exceed 12 ft., and fires with moderate to high rates of heat release would be anticipated.

OUTSIDE STEM AND YOKE VALVE (OS&Y): These valves allow the building owner to shut off the water to the sprinkler system inside the property.

OXIDATION: Generally, the union of a substance with oxygen: the process of increasing the positive valence or decreasing the negative valence of an element or ion; the process by which electrons are removed from atoms or ions.

OXIDE: A binary compound of oxygen and some other element.

OXIDES: Oxides of metals and nonmetals react with water to form alkalines and acids respectively; this reaction may occur violently with sodium oxide. Calcium oxide, more commonly identified as quicklime, also reacts with water accompanied by the evolution of sufficient heat to ignite combustible materials such as cloth, paper, and wood.

OXIDIZING AGENT: A general definition of an oxidizing agent is a chemical substance in which one of the elements tends to gain electrons.

OXYGEN: A colorless, odorless, tasteless, gaseous chemical element that is the most abundant of all elements. It occurs free in our atmosphere of which it forms one-fifth, in volume, along with nitrogen. Oxygen is continually active and is able to combine with most other elements. It is essential to the life and growth processes, as well as to combustion. In an atmosphere of pure oxygen, combustion is more intense than in an environment of atmospheric oxygen which is approximately only one-fifth pure oxygen and four-fifths nitrogen. Oxygen may be manufactured by liquefying air and separating the oxygen from the nitrogen by a process based on their differences in boiling points called fractionating; oxygen may also be produced by the electrolytic decomposition of water separating it into its constituents; hydrogen and oxygen. Hydrogen gas is lighter than air, extremely explosive, and with a wide explosive range. Liquid oxygen is used widely in industry. Liquid oxygen, reduced to the liquid state under extremely low temperatures and container stored, weighs about 71 lbs. per square foot and, at atmospheric pressure, its temperature is 297 degrees F. Carbonaceous material, mixed with liquid oxygen, is used as an explosive.

OZONE: An allotropic form of oxygen usually resulting from or formed by electrical discharge in the air; also used as an oxidizing, bleaching, or deodorizing agent in the purification of water.

OZONIC: A solution of ethylic ether, hydrogen peroxide, and alcohol.

PANIC BEHAVIOR: Occurs when the occupants experience a sudden and excessive feeling of alarm or fear affecting persons in a fire leading to extravagant efforts to secure safety.

PATTERNS OF SPREAD: The term used by fire investigators to describe the remaining physical evidence of the progress of the fire from origin to control. A great deal of fire terminology is utilized by investigators to explain these patterns such as depth of char, smoke staining, and V-patterns. There is no known scientific authentication and no evidence that any one combination of the terms or methods is precise in any given fire. Because of fire load, weather conditions, fire suppression activities, and other various conditions, patterns of spread are unique to each fire.

PHOTOELECTRIC LIGHT-SCATTERING SMOKE DETECTION: Photoelectric light-scattering smoke detection is more responsive to visible particles (larger than 1 micron in size) produced by most smoldering fires. It is somewhat less responsive to black smoke than to lighter colored smoke. Smoke detectors that use the light-scattering principle are usually of the spot type.

PILOT LIGHT: In general, a gas-fueled burner, utilized to rekindle a principal burner (main burner) and commonly used in domestic and industrial gas fueled appliances. The more modern devices are utilized and regulated with automatic safety controls.

PLASTICS: Plastics may be generally described as compounds consisting of small molecules known as monomers that link together into long chains to form polymers and copolymers. In respect to their reaction to fire, plastics may generally be classified as either thermoplastic compounds which soften when heated or thermosetting plastics, once set, cannot be re-softened. Plastics reaction to heat and fire as well as toxicity of their combustion products show considerable variation.

PLUG FUSES: Devices used for overcurrent protection. There are several types including the ordinary one-time type, the time delay type, and the S type which does not have to be time delay. The S type may be designed so that the 16 to 30-amp classification cannot be used in fuse holders of the two initially described. Plug fuses are sometimes called buss fuses.

POINT OF ORIGIN: In the context of fire investigation, point of origin means the precise location where the initial ignition of the substance took place; usually, the location of the heat source is indicated. Point of origin is more specific than area of origin.

PRESSURE, EXPLOSIVE: Explosions may be characterized by energy release apparent in effect but not cause. Depending on their causes, explosions are of four principal kinds; energy release generated by rapid oxidation of gasoline vapor; release of energy generated by rapid decomposition such as in a dynamite explosion; release of energy caused by excessive pressure such as in a boiler explosion; and energy release created by nuclear fission of fusion, such as a hydrogen bomb explosion. A practical method of differentiating between fire and explosion is that an explosion develops forces which may cause violent displacement of structures or other objects and indicates the presence of pressures developed by the explosion of vapors and other materials.

PROCESS STEAM GENERATOR: A vessel or system of vessels comprised of one or more drums and one or more heat exchange surfaces as used in waste heat or heat recovery type steam boilers.

PRODUCTS OF COMBUSTION: The term is usually used in the context of physical and visible results of a fire in the form of combustion gases, flame, and smoke. Most combustible materials contain carbon, which burns to form carbon dioxide when the air supply is ample. When the air supply is limited, carbon monoxide may be apparent. Some of the other gases may be formed when the materials burn and these may include ammonia, hydrogen cyanide, hydrogen chloride, nitric acid, phosgene, hydrogen supplied, sulfur dioxide, and other gases, depending on the subject of combustion and temperature rise in the combustibles.

PROPRIETARY SUPERVISING STATION: A location to which alarm or supervisory signaling devices on propriety firm alarm systems are connected and where personnel attend all times to supervise operation and investigate signals.

PYROLYSIS: Generally defined, pyrolysis is physical and chemical decomposition by heat. Since the organic nature and structure of commonly combustible fuels vary, the process varies. When heated, organic compounds, such as wood, are subject to complex deterioration to simpler compounds which progressively may become more volatile and therefore more flammable than the previous structure. The heating need not occur in the form of apparent flaming combustion; it may be accomplished by exposure to steam pipes, electrical heat-producing sources, or inert gases. Pyrolysis has occurred in wood framing in sealed walls where adjacent heat-producing steam pipes have gradually carbonized (pyrolyzed) the wood framing over the years. The process will ultimately produce flaming combustion. At the other end of the time spectrum, the phenomenon may occur in a relatively short period of time.

PYROPHORIC MATERIALS: These are compounds that may ignite spontaneously. They may be called pyrophor and include materials such as titanium; hafnium; uranium; and thorium as well as phosphorated hydrogen; ethyl, methyl, and propyl compounds; and potassium sulfide.

RADIATION: As used in fire investigation, radiation is one of the three methods or media by which heat is transferred. The three methods being conduction, convection, and radiation. Radiation is the heat transfer, from one body to another, by heat rays, through intervening space, in much the same manner as light is transferred by light rays. Two examples are heat from the sun and heat from an electric heater adjacent to a wall. Radiated heat passes freely through a vacuum and through gases, like light, heat is reflected from glass or another bright surface.

REENTRY BEHAVIOR: The occupants successfully exited the building, but for various reasons, reenter. Typically, persons that reenter the building do so looking for loved ones, to assist others in exiting and to assist with firefighting.

REMOTE SUPERVISING STATION FIRE ALARM SYSTEM: A system installed in accordance with this code to transmit alarm, supervisory, and trouble signals from one or more protected premises to a remote location where appropriate action is taken.

SABOTAGE: The damaging of property or hindering of production in an underhanded effort to defeat or harm persons or property.

SAFETY FACTOR: The ratio of the strength of a material prior to failure to the safe working stress. For example, a safety factor of 10 would occur if the design load is only a tenth of the tested strength.

SEARCH PATTERN: The method used by the Accelerant Detection Canine Team to cover a complete area when searching for accelerants.

SELF-INCRIMINATION: The giving of testimony as a witness against one's will. The United States Constitution provides the Fifth Amendment guaranteeing that no one shall be compelled to testify against himself in the Federal courts and nearly every state has embodied the same provisions in their respective constitutions. The guarantee extends to self-incrimination situations during interviews by peace officers and certain governmental representatives.

SENSIBLE HEAT EXCHANGE: The change in heat content of a material due to a temperature change only, no phase change.

SMOKE: The vaporous matter resulting from some forms of combustion and made visible by minute particles of carbon suspended in the vapor. In chemistry, suspension of solid particles in gas. Under the usual conditions of insufficient oxygen for complete combustion, there may also be present methane, methanol, formaldehyde, and formic and acetic acids. The combination of the combustible, rate of heating, oxygen concentration at or near the combustible surface, along with temperatures, usually determines the composition of fire gases and the resulting smoke, if any. The color of the smoke may vary with the composition of the combustibles and by ventilation or application of water. Smoke particles may cool to the extent where water vapor, acids, and residues of the combustibles involved may be identified if recovered from areas such as windows or other surfaces. It is, of course, well established that moisture-laden particles, if inhaled, may carry highly poisonous or irritating compounds into the respiratory tract and eyes. If allowed to accumulate in a building, hot unburned products of combustion will ignite explosively when a supply of oxygen is suddenly made available. (See Flashover) Some firemen and investigators call this phenomenon a smoke explosion. This is one of the prime reasons why firemen vent fires, usually by opening windows, doors, or by cutting holes in the roof.

SMOKE DETECTOR: A device that detects visible or invisible particles of combustion.

SMOKE STAINING: Smoke particles will adhere to unburned surfaces such as walls and windows within a burned structure. Smoke staining assists the investigator in determining whether the fire spread was rapid or slow and whether accelerants were burned.

SPECIFIC GRAVITY: The ratio of the weight or mass of the given volume of a substance to that of an equal volume of another substance used as a standard. (Water for liquids and solids, air, or hydrogen for gases) The specific gravity of gasoline and other petroleum products, for example, is commonly measured in Degrees API (American Petroleum Institute.) Because specific gravity is a ratio, it is a unitless quantity. For example, the specific gravity of water at 4°C is 1.0 while its density is 1.0 g/cm^3.

SPECIFIC HEAT: The ratio of the amount of heat required to raise the temperature of a unit mass of a substance one degree to the amount of heat required to raise the temperature of the same mass of water one degree; the number of calories required to raise the temperature of 1 g of a given substance 1°C.

SPONTANEOUS COMBUSTION: Self-heating materials, those that exhibit spontaneous ignition or heat themselves to a temperature of 200°C (392°F) during a 24-h test period. (This behavior is called spontaneous combustion)

SPRAY BOOTH: A power-ventilated structure provided to enclose or accommodate a spraying operation.

STATIC ELECTRICITY: Designed as stationary electrical charges or discharges in the atmosphere; frictional electricity such as produced in movement of bodies or fabrics on surfaces. It is relevant to fire investigation in that electrical charges, without proper bonding or grounding, may introduce a spark or arc into an explosive or combustible environment. Some examples are improperly bonded or grounded fuel lines, tanks, and pumping facilities.

SUBROGATION: As used in civil actions, the substitution of one for another so that the new party succeeds to the former's rights or legal claims. It is frequently referred to as the doctrine of substitution. It is a device commonly used in insurance litigation, adopted, or invented by equity to compel the ultimate discharge of a debt or obligation. It is the mechanism by which the equity of one man (or party) is worked through the rights of another. The right of an insurer on payment of a loss to be subrogated pro tanto to any right of action which the insured may have had against any third person whose wrongful act or neglect caused the loss insured against by the insurer is one example.

TEMPERATURE: The degree of hotness or coldness of any substance or body, usually measured in terms of Kelvin or Rankine temperature.

THERMOSTAT: Thermostats are widely used, fixed temperature heat detectors used in signaling systems. Probably the most common type is the bimetallic type which utilizes two metals with different expansion coefficients resulting in movement of the strip and closing or opening of the contacts regulated to the temperature function.

TORT: An injury or wrong committed either with or without force, to a person or person's property. Such injury may arise out of nonfeasance, malfeasance, or misfeasance.

TRAILER: A term used to describe the means utilized to extend an ignition point from a location outside an occupancy into the space where the plant or booster may be located. Trailers often connect various "plants" or "sets" inside a space to insure complete involvement within a short period of time. Flammable liquids such as kerosene and charcoal lighter fluid have been frequently used. Gasoline is seldom used as a trailer because of the rapid evaporation characteristics that makes its use unpredictable. Various types of window cord, rope and newspaper have been commonly used, saturated in kerosene or other medium flash point fuels and distributed through the premises. Blasting cord has been used, particularly where explosions are desired.

TRANSFORMER: Generally, an apparatus or device for transforming or converting the voltage of an electric current. There are two types; a step-down transformer which changes high voltage to lower voltage; and a step-up transformer which changes lower voltage to higher voltage. Transformers vary in size and capacity from the liquid and mechanically cooled types found in substations and primary facilities right on down to the smaller units in primary service entrances and secondary areas.

TYPE 1 CONSTRUCTION: Buildings, commonly called fire resistive have structural members such as the frame, walls, floors, and roof that are all noncombustible with a minimum specified fire resistive rating.

TYPE 11 CONSTRUCTION: A construction type in which the structural elements are entirely of noncombustible or limited combustible materials, hence the common name of noncombustible.

TYPE 111 CONSTRUCTION: Commonly called ordinary construction, this is a construction type where the exterior walls are noncombustible with a minimum 2-h fire resistance, but the interior is constructed of combustible materials.

TYPE 1V CONSTRUCTION: Structural members are basically of unprotected wood with large cross-sectional areas and hence the common name of plank, timber, or mill construction. Bearing walls, bearing portions of walls, and exterior walls must be noncombustible and have at least a 2-h rating.

TYPE V CONSTRUCTION: A construction type where exterior walls and structural members are primarily made of wood or other combustible materials.

UNFIRED STEAM BOILER: A vessel or system of vessels intended for operation at a pressure more than 15 psig for the purpose of producing and controlling an output of thermal energy.

UPPER EXPLOSIVE LIMIT: See Upper Flammable Limit.

UPPER FLAMMABLE LIMIT (UFL): If the vapor concentrations in air are above the upper flammable limit (UFL) (also referred to as the upper explosive limit (UEL), the vapors will not ignite. This is commonly referred to as the vapors being too rich.

VAPOR: A vapor refers to a gas-phase material that normally exists as a liquid or solid under a given set of conditions. If the temperature is below a certain point (the critical temperature; this varies for each substance), the vapor can be condensed into a liquid or solid with the application of pressure.

VAPOR DENSITY: Vapor density is the relative density of a gas or vapor (minus air) as compared to air. A figure of less than 1 indicates that a vapor is lighter than air; a figure greater than 1 indicates that the vapor is heavier than air. The vapor density of a compound equals the molecular weight of the compound divided by 29.

In the formula, 29 is the composite of the molecular weight of air.

VAPOR PRESSURE: The pressure of a confined vapor that has accumulated above its liquid. It is determined by the nature of the liquid and the temperature. Vapor pressure figures for many substances may be found in chemical handbooks; vapor pressures of petroleum products are usually determined by the Reid method as recommended by the American Society for Testing Materials. ASTM Standard D 323. If we were to place a substance in an evacuated, closed container, some of it would vaporize. The pressure in the space above the liquid would increase from zero and eventually stabilize at a constant value, the vapor pressure.

VENTING: Firemen usually vent fires by opening doors or windows or cutting holes in the roof. The reason for this is to allow gases and hot unburned products of combustion to escape. If this is not done, smoke explosions or flashover may occur.

VOIR DIRE: To speak the truth. An oath is so called when it is administered to a prospective juryman or a witness as a preliminary step to examining his qualifications as a juror or witness.

VOLATILE: Those that may be readily vaporized. The hazards depend on their flash point and explosive range. Examples of volatile liquids are acetone, allyl alcohol, ethyl alcohol, gasoline, and kerosene.

VOLT (V): Volt is a unit of electrical pressure, the force which causes electricity to flow through a conductor. Voltage is not necessarily a measure of fire hazard. For example, 1.5 volts through a fine wire can cause that wire to become red hot. The heat of the wire, if in contact with combustibles, will result in fire.

WATER HEATER BOILER: A closed vessel in which water is heated by combustion of fuels, electricity or any other source and withdrawn for use external to the system at pressure not exceeding 160 psig and should include all controls and devices necessary to prevent water temperatures from exceeding 210°F.

WATT: The watt is a unit of electrical power. A current of one ampere flowing under pressure of one volt equals one watt. For example, a 100-watt light bulb rated at 110 v takes a current of 0.9 amp and has a resistance of 130 ohms. It is the demand for power when using items such as a light bulb, heaters, stoves, and motors that places the load on the conductors. If the demand is more than the size or capacity the conductors can carry (are rated for), heating of the conductors will follow. A fire may follow the heating or arcing growing out of the overload since circuit safety devices, such as fuses or circuit breakers may not open the circuit before ignition of combustibles occurs.

ZONE: A defined area within the protected premises. A zone can define an area from which a signal can be received, an area to which a signal can be sent, or an area in which a form of control can be executed.

Solutions to Chapter Questions

CHAPTER 1

1. Compare the fire death rates in the United States to the fire death rates in other industrialized countries.

    The U.S. fire deaths, based on one million per population, are almost twice the average fire death rates for other industrialized countries.

2. Effective fire prevention requires what three components?

    Effective fire prevention requires vigilance, action, and cooperation.

3. Describe the opportunities one has to intervene in a fire.

    Prevent the fire entirely

    Slow the initial growth of the fire

    Detect fire early, permitting effective intervention before the fire becomes too severe

    Opportunities for automatic or manual suppression

    Opportunities to confine the fire in a space

    Move the occupants to a safe location

4. Differentiate between fire protection and fire prevention.

    Fire prevention is the elimination of the possibility of a fire being started while fire protection strategies are those activities that are designed to minimize the extent of the fire.

5. Describe some of the job activities a fire protection engineer may be engaged in.

Examples of fire engineering safety job aspects include evaluating buildings to determine fire risks, designing fire detection and suppression systems, and research on materials and consumer products.

6. Describe the trends in the fire death experience in the United States since World War I.

    Overall, fire deaths in the United States have fallen nearly two-thirds since their peak levels around World War I.

7. What impact do fires have on occupational fatalities and injuries in the United States?

    Data involving fires and explosions from 1992 to 2000 indicated that fires and explosions account for approximately 3 percent of all nonfatal occupational injuries and illnesses involving days away from work. Between 1992 and 2000, fires and explosions accounted for approximately 1,760 deaths in the workplace.

8. What are some major sources of ignition in the workplace based on reported fires?

    The most common sources of ignition resulting in fires in the workplace are equipment, electrical courses, and open flames.

9. Describe some aspects of fire safety the Occupational Safety and Health Administration regulates in the workplace.

    OSHA regulates a variety of fire prevention activities in the workplace including fire extinguishers, life safety, fire prevention plans, the control of ignition sources, properly handling of hazardous materials, etc.

10. Describe OSHA's three key elements of fire safety.

    OSHA's three key elements of fire safety in the workplace are fire prevention, safe evacuation of the workplace in the event of fire, and protection of workers who fight fires or who work around fire suppression equipment.

CHAPTER 2

1. What is combustion?

    An exothermic chemical reaction between some substance and oxygen.

2. Specific heat is the amount of heat required to raise the temperature of a substance by.

    1 kg.; 1°K or 1 lb.; 1°F

3. With heat transfer by conduction, as the temperature increases the thermal conductivity of metals will do what?

   Decrease as the temperature increases.

4. In the NFPA 704 Hazard Rating System, the blue-, red-, and yellow-colored diamonds represent what?

   Blue indicates the health hazards; red indicates fire hazards; yellow indicates instability hazards.

5. When assessing the effect of heat on the body from a fire, what are the two main factors that determine severity?

   Length of exposure and temperature

6. What are the four elements of the fire tetrahedron?

   Fuel, oxygen, ignition/energy source, and chemical chain reaction

7. What is a hazardous material as defined by USDOT?

   A substance or material capable of posing an unreasonable risk to health, safety, and property when transported in commerce

8. Class K fires are fires that involve what type of fuel?

   Combustible cooking media

9. What are some examples of control methods for explosions?

   Containment, venting, suppression

10. The primary purpose of which reference guidebook is to aid first-aid responders in identifying the specific or generic classification of hazardous materials and to protect themselves and the public during the response to the hazardous materials incident?

    DOT Emergency Response Guidebook

CHAPTER 3

1. Describe some of the common electrical hazards that can result in fire ignition sources.

   Misuse of electric cords: Running the cords under rugs, over nails or through high traffic areas and the use of extension cords as permanent wiring.

   Poor maintenance: Lack of a preventive maintenance program designed to identify and correct potential problems before they occur.

   Failure of ground: Failure to maintain a continuous path to ground can expose entire electrical systems to damage and can expose the workers using unprotected equipment to electrical hazards.

Damaged insulation: Insulation protecting current-carrying wires can become damaged over time resulting in exposed wires. If the exposed hot and neutral wires touch, they can create a short circuit and an ignition source for fires.

Sparking: Friction sparking is a form of mechanic heat created when two hard surfaces, at least one which is metal, impact.

Overloaded circuit: A circuit becomes overloaded when there are more appliances on the circuit than it can safely handle. When overloaded, the wiring overheats and the fuse blows or the circuit breaker trips.

Short circuit: A short circuit occurs when a bare hot wire touches a bare neutral wire or a bare grounded wire (or some other ground). The flow of extra current blows a fuse or trips a circuit breaker.

2. Differentiate between the various NFPA classes of flammable and combustible liquids.

Flammable liquids are any liquids having a flash point below 100°0B0F, except any mixture having components with flash points of °00°F or higher, the total of which make up 99 percent or more of the total volume of the mixture.

Combustible liquids typically will require some external heating to produce a sufficient concentration of vapors.

Combustible liquids are any liquid having a flash point at or above °00°F and are divided into two classes, Class II and Class III.

3. What impact does the various classes and groups of flammable and combustible liquids have for the safety manager?

The various classes of flammable and combustible liquids are used as factors for determining the amounts of liquids that can be stored in a safety container and flammable liquid storage cabinets. The classes of liquids are also used to determine safe transferring methods to use.

4. Differentiate between the various NFPA hazardous environment classes.

Class I locations deal with the presence of flammable vapors, Class II locations deal with the presence of combustible dusts, and Class III hazardous locations deal with ignitable fibers and flyings.

5. What process should be followed when transferring a Category 1 flammable liquid from a 55-gallon drum to a 5-gallon safety container?

SOLUTIONS TO CHAPTER QUESTIONS

The flammable liquid should be transferred into the container by means of a device drawing through the top, or from a container or portable tanks by gravity through an approved self-closing valve. Adequate precautions shall be taken to prevent the ignition of flammable vapors. Category I liquids should not be dispensed into containers unless the nozzle and container are electrically interconnected through the use of a bonding wire. An alternative to using a bonding wire is the use of a metallic floorplate on which the container stands while filling is electrically connected to the fill stem. Transferring Category I liquids should be done inside buildings unless adequate ventilation is provided.

6. What type of fire hazard does oxygen present in the workplace?

    Oxygen is an oxidizer, serving as an oxygen source for other materials that are consumed as a fuel in a fire. Introducing pure oxygen to greases and oils can result in spontaneous combustion.

7. What fire prevention and spill control features would one expect to find on an aboveground storage tank that holds flammable liquids?

    Methods for controlling and preventing fires involving outside, aboveground tanks include the separation of storage tanks, diking and drainage, and venting.

8. Describe the safety features found in a flammable liquid storage room.

    Safety features found in a flammable liquid storage room include spill containment through the use of diking or drains, adequate ventilation, and appropriate electrical installations for hazardous environments. Fire protection features include fire extinguishers and/or sprinkler system, rated fire doors and rated fire construction of the walls.

9. Describe the safety features for an approved safety container.

    Approved safety cans have a maximum capacity of 5 gallons with a springclosing lid and spout cover. Safety cans are designed so that when they are subjected to heating, the can will relieve the internal pressure.

CHAPTER 4

1. Describe the mechanics of a BLEVE:

    The phenomenon known as a boiling-liquid expanding-vapor explosion (BLEVE) is the result of a liquid within a container reaching a temperature well above its boiling point at atmospheric temperature, causing the vessel to rupture into two or more pieces.

2. What is an oxidizer?

   A general definition of an oxidizing agent is a chemical substance in which one of the elements has a tendency to gain electrons.

3. What are four things one can do to prevent boiler explosions?
   A. Proper maintenance
   B. Inspections
   C. Proper installation
   D. Repairs made in accordance with codes

4. What is an explosion?

   The term explosion is defined as a rapid release of high-pressure gas into the environment.

5. Describe some of the safety precautions one must take when transporting explosives.

   A competent driver should be with the vehicle at all times.

   The driver or other attendant is not to leave the vehicle unattended for any reason.

   Properly maintained vehicles shall be used.

6. Define the various classes of oxidizers:
   - Class 1 oxidizers: Slightly increase the burning rate of combustible materials.
   - Do not cause spontaneous ignition when they come in contact with them.
   - Class 2 oxidizers:
   - Increase the burning rate of combustible materials moderately with which they come in contact.
   - May cause spontaneous ignition when in contact with a combustible material.
   - Class 3 oxidizers:
   - Severely increase the burning rate of combustible materials with which they come in contact.
   - Will cause sustained and vigorous decomposition if contaminated with a combustible material or if exposed to sufficient heat
   - Class 4 oxidizers:
   - Can explode when in contact with certain contaminants.

- Can explode if exposed to slight heat, shock, or friction.
- Will increase the burning rate of combustibles.
- Can cause combustibles to ignite spontaneously.

7. Describe some of the safety precautions one must take handling ammonium nitrate.
   A. Store in segregated areas
   B. Store in acceptable bins, storage containers, etc.
   C. Do not mix ammonium nitrate with other materials such as flammable liquids such as gasoline, kerosene, solvents, and light fuel oils, sulfur and finely divided metals and explosives and blasting agents.

8. Describe some of the more common safety devices found on boilers.

   Rupture disk device: A nonreclosing pressure-relief device actuated by inlet static pressure and designed to function by the bursting of a pressure-containing disk.

   Safety relief valve: An automatic pressure-relieving device actuated by a static pressure upstream of the valve which opens further with the increase in pressure over the opening pressure.

   Temperature limit control: Ensures boiler is operating within acceptable temperature ranges.

   Low water cutoffs: Indicate when water levels in the boiler get to a low level. Shut down heat source to the boiler.

   Flame supervisory unit (igniter): Checks to ensure gas is burning. Prevents accumulation of gas in the room.

   High and low gas pressure switches: Monitors gas pressure going into boiler.

   Trial for ignition limiting timer (15 seconds): Checks to ensure gas has been ignited. Prevents accumulation of gas in the room.

9. Describe some of the safety precautions one must take when using explosives.

   Blasting should only be performed by qualified individuals.

   Procedures for loading explosives in blast holes, initiating the explosive charges, and dealing with misfires should adhere to applicable safety standards.

   To control the potential hazards at the blasting site associated with explosives, sources of ignition such as matches, open light or other fire or flame should be prohibited.

Because accidental discharge of electric blasting caps can occur from current induced by radar, radio transmitters, lightning, adjacent powerlines, dust storms, or other sources of extraneous electricity, all blasting operations should be suspended and all persons removed from the blasting area during the approach and progress of an electric storm.

CHAPTER 5

1. What is fire resistance?

    Fire resistance refers to the ability of the material or assembly to resist the effects of the heat and flame from the fire. Therefore, fire-resistant construction would reduce the ease of ignition and flame spread of the building structure.
2. What are the three separate elements involved in a building fire?
3. The structural element of the building. A structural element is a member that if removed will affect the structural stability of the building
4. The contents of the building
5. The nonstructural building elements. Surface finishes, windows, interior vertical openings, decorative surfaces, air conditioning systems, etc.
6. What does a Type 222 classification for a building signify?

    In a Type 222 subclassification, the exterior bearing walls, interior bearing walls, columns, and beams have a 2-h fire resistance.
4. What are three characteristics of steel that can reduce its performance to withstanding a fire?
5. Steel conducts heat thereby aiding heat transfer.
6. Steel has a high coefficient of expansion at elevated temperatures. This high coefficient of expansion affects the steel structure because the ends of the structural member axially restrained and the attempted expansion due to the heat causes thermal stresses to be induced to the member. This stress combined with those of normal loading can cause a more rapid collapse.
7. Steel will lose its strength when subjected to high temperatures.
8. Why does gypsum have excellent fire resistance characteristics?

    Gypsum products have a high portion of chemically combined water and when exposed to a fire the evaporation of this water requires a great deal of heat energy.

6. Why is it important to consider the interior finish when evaluating the fire risk associated with a building?

The interior finish materials have a tremendous influence on smoke and toxic gas generation, as well as the speed at which the interior of the building can become involved once ignition has taken place, also known as flame spread.

7. Why do high-rise buildings pose a unique fire risk?

High-rise buildings affect fire department accessibility to the fire. Limitations exist in fire apparatus in reaching the upper floors of the exterior of the building.

The height of the fire and the number of fire service personnel required to deliver adequate types and amounts of equipment to the fire present additional challenges. More time and energy is required to deploy forces and equipment to the fire which could be exhausted before firefighting forces can mount an attack.

Delays in deploying equipment and fire fighters can indirectly affect fire growth, resulting in a fire of greater magnitude.

Due to their design, high-rise buildings significantly increase the occupant, equipment, and material load in a given building.

Due to building height, egress and people-movement systems within a high-rise building are limited.

8. What is fire load and why is it important when assessing the fire risk of a building?

The fire load in a building is the total amount of potential heat energy that might evolve during a fire in the building. The fire load can predict the fire severity for various occupancies and is also used to determine the fire resistance required of structural components.

9. What are some examples of ordinary hazard occupancies?

Examples of Ordinary Hazard Occupancies Group 1 would include bakeries and restaurant service areas. Examples of Ordinary Hazard Group 2 Occupancies would include libraries, dry cleaners, and wood working facilities.

10. Products that are primarily made up to Group B plastics on wooden pallets would be classified as what class of commodity?

Class IV Commodity

CHAPTER 6

1. Identify four factors that can influence how an individual will react during a fire emergency.
   A. Actions of others,
   B. Building characteristics such as marking of exits, perceived threat,
   C. Individual characteristics such as personality, and
   D. Previous experience/education
2. What is an occupancy under the "Life Safety Codes?"
   An occupancy is the principal use for the structure.
3. What three things comprise a means of egress?
   The means of egress has three components: the exit access, the exit, and the exit discharge.
4. What are the illumination requirements for emergency lighting?
   Emergency lighting should be of sufficient illumination levels and maintained for at least PA hours in the event of failure of normal lighting. Emergency lighting illumination levels should be not less than an average of 1-ft. candle and, at any point, not less than 0.1-ft. candle, measured along the path of egress at floor level.
5. When does OSHA require a fire prevention plan?
   A fire prevention plan is required when there is a potential for fire hazards in the workplace.
6. What does the term "protected pathway" imply for an exit?
   It implies a specific fire resistance.
7. What criteria are used to determine the maximum capacity for a means of egress?
   The clear width is divided by the capacity factors, which are based on occupancy classification and the surface (level or stairways).
8. What additional research was recommended following the Station Nightclub fire?
   Complete additional research to gain a better understanding of human behavior in emergency situations and to predict the impact of building design on safe egress in emergencies.
9. Exhibition halls would fall under what type of occupancy according to NFPA 101?
   Assembly Occupancy
10. In general, the number of exits in a building is influenced by what?

SOLUTIONS TO CHAPTER QUESTIONS 385

The type of occupancy, number of occupants, and the maximum travel distance for exit access.

CHAPTER 7

1. How does electrostatic spray painting work?

    In electrostatic spray painting, the spray material such as paint is negatively charged (while atomized or after having been atomized by the air or airless methods) through a connection of the spraying-gun to a generator. The material being painted is positively charged and as a result, the difference in charge leads the particles toward the material to obtain the coating.

2. What are acceptable methods of overflow protection on dip tanks?

    To prevent the overflow of burning liquid from the dipping or coating tank if a fire in the tank actuates automatic sprinklers, one or more of the following shall be done:

A. Drainboards shall be arranged so that sprinkler discharge will not flow into the tank.
B. Tanks shall be equipped with automatically closing covers.
C. Tanks shall be equipped with overflow pipes.

3. Why is liquid level control important in dip tank operations?

    Accidental overfilling of the tank or release of the liquid from the tank could result in rapidly spreading fire.

4. How does an aerated powder coating operation work?

    An aerated powder is any powdered material used as a coating material which shall be fluidized within a container by passing air uniformly from below.

5. What is the maximum quantity of flammable or combustible liquids that can be stored in a coating operation area?

    The maximum quantity of liquid located in the vicinity of the dipping or coating process area shall not exceed a supply for one day or 25 gallons of Class IA liquids in containers, plus 120 gallons of Class IB, IC, Class II, or Class III liquids in containers, plus two portable tanks each not exceeding 2500 660 gallons of Class IB, IC, Class II, or Class IIIA liquids, plus 20 portable tanks each not exceeding 660 gallons of Class IIIB liquids.

6. When are dipping and coating operations considered hazardous environments?

   Dipping and coating process areas where Class I liquids are used, or where Class II or Class III liquids are used at temperatures at or above their flash points are considered hazardous locations and as a result, electrical equipment used in these areas must meet applicable codes. The hazardous environment classifications can include the dip tank area, drain area and drying areas.
7. Describe the type of training workers in dipping and coating operations should receive.

   Personnel involved in dipping or coating processes should receive documented training on the safety and health hazards associated with dip tanks; the operational, maintenance, and emergency procedures required; and the importance of constant operator awareness.

CHAPTER 8

1. List and describe the three functions of an alarm system.

   (1) It provides an indication and warning of abnormal fire conditions; (2) it alerts building occupants and summons appropriate assistance in adequate time to allow for the occupants to travel to a safe place and for rescue operations to occur; and (3) it is an integral part of an overall life safety plan that also includes a combination of prevention, protection, egress, and other features unique to that occupancy.
2. Explain what factors need to be considered when selecting an alarm system.

   The two most important factors are the need for speed and accuracy of response to a fire with minimal chances of false alarms.
3. What is the primary purpose of signal annunciation?

   To enable responding personnel to identify the location of a fire quickly and accurately and to indicate the status of emergency equipment of fire safety functions that might affect the safety of the occupant in a fire situation.
4. Maintenance, inspection, and testing records must be retained until the next test and for one year thereafter. What information must be provided in these records?

SOLUTIONS TO CHAPTER QUESTIONS

Date, test frequency, name and address of property, name of person performing inspection, name/address/representative of approving agencies, designation of the detectors tested, functional test detectors, and functional test of required sequence of operations must be provided.

5. What is a zone, and what are its criteria?

   A zone is a defined area within the protected premises. A zone can define an area from which a signal can be received, an area to which a signal can be sent, or an area in which a form of control can be executed. Each floor of the building must be considered as a separate zone. If a floor area exceeds 20,000 ftA2 additional zoning should be provided. Length should not exceed 300 ft. in any direction. If the system serves more than one building, each building shall be indicated separately.

6. What are the three basic principles used for detecting heat from a fire?

   The three basic principles that heat detectors utilize are fixed temperature, rate compensation, and rate of rise.

7. What characteristic of smoke influences the density of a smoke detector in responding to a fire?

   The percentage of obscuration of the smoke required to produce a signal.

8. Which NFPA standard establishes the requirements for fire alarm notification systems?

   NFPA 72: National Fire Alarm and Signaling Code

9. What is the difference between a proprietary system and a central station monitoring system?

   A proprietary system is a monitoring system located on the protected property and is owned and operated by the property being protected. A central station system is controlled and monitored by an outside company whose business is to provide monitoring services.

CHAPTER 9

1. Why is a 2-in. main drain test completed on a wet-pipe sprinkler system?

   The 2-in. main drain test will alert the inspector of potential water supply problems such as inadequate water pressure or flow and the presence of debris or obstructions in the water supply line.

2. What is the purpose of completing a trip test on a wet-pipe sprinkler system?

   The trip test replicates the opening of the most remote sprinkler head in the system. This test is performed to test the flow and pressure at the most remote location on the system and to test the water flow alarm.

3. What is a Class B fire and what type of extinguishing medium is appropriate for using on such fires?

   Class B fires involve flammable and combustible liquids, oils, and greases. Appropriate extinguishing mediums include carbon dioxide, aqueous film-forming foam, multipurpose dry chemical (ammonium phosphate), and halogenated agents.

4. Explain how a bulb style sprinkler head activates?

   With bulb sprinklers, there is a glass bulb with a liquid inside. There is also an air bubble. As the liquid heats up it expands, the air bubble disappears and then the glass bubble shatters releasing the cap that was holding back the water in the branch line.

5. In what type of situations are dry-pipe sprinkler systems the most suitable?

   Dry-pipe systems are used in facilities where the sprinkler system could be subject to freezing temperature, below 40°F. In a dry system, air is under pressure in the system above the water flow alarm clapper valve.

6. Describe the three different design classes of standpipe hose systems?

   Class I systems have 2 1/2 in hose connections at designated locations in buildings for full-scale firefighting intended for use by fire department personnel. Class II systems have 1 1/2 inch hose connections. They are intended to be used as first-aid measures by trained fire brigades to battle afire before the fire department gets to the scene. With Class II systems, a hose, nozzle, and a rack are typically installed on each hose connection. Class III systems are provided for both first-aid response and full-scale firefighting with both 1 V2 and 2 V2 in hose connections available.

7. Supervisory initiating devices are used to supervise the operation of critical operating features of a sprinkler system such as?

   The closing of sprinkler water control valve, air pressure in a dry-pipe system, level of water in a storage tank, and integrity of a fire pump.

8. What are the two major types of fire hydrants used in the United States?

There are two major types of fire hydrants used in the United States: the dry barrel and the wet barrel.
9. What is the purpose of the water flow check valve in a sprinkler system?

The water flow check valve serves two purposes. First, if there is water pressure loss, below the check valve, the clapper will remain seated and keep the water in the sprinkler system piping above the clapper valve in a wet-pipe system. The second purpose of the check valve is to activate a water flow alarm in some systems.

CHAPTER 10

1. What are the four primary steps in the Fire Risk Management Process?
   - Identification of the fire and emergency hazards/events that could lead to significant loss.
   - Quantification of the risk: probability of a fire or emergency event occurrences and loss consequences.
   - Development and evaluation of alternative prevention and protection strategies to reduce the fire and emergency risk.
   - Measurement to determine the effectiveness of the strategies in reducing the fire and emergency risk associated with the implemented alternatives.
2. What is one of the primary roles of safety professionals in the risk assessment step of hazard identification?

   The safety professional will be responsible for providing the technical knowledge related to the fire codes and standards that may be used to identify the actual/potential fire hazards.
3. A basic risk assessment of a hazard involves identifying what?

   Identifying the probability and severity of potential fire losses.
4. When addressing the severity of a fire risk it is important to consider both direct and indirect loss potentials. What are some examples of indirect losses?

   Indirect losses include business interruption, liability for injury or death, environmental contamination, and damage to company image.
5. What are some general options available for handling fire risk exposures?
   - Avoiding the risk by not completing the activity,

- Risk transfer by purchasing insurance to cover potential losses, or other alternate risk transfer arrangements such as self-insurance,
- Providing loss control improvements, or
- Developing a risk management program that includes a combination of the bove.

6. A properly conducted risk assessment can provide management with an idea of the relative degree of risk that a facility may be vulnerable to as well as what else?

    The facility's level of preparedness to handle a fire or other emergency and its ability to survive the emergency situation and remain in business.

7. Twenty years ago emergency response plans focused on fire and natural emergencies. However, such a narrow scope is no longer acceptable with today's risks of emergencies evolving substantially to include what other risks?

    Emergencies now must encompass areas such as cyberterrorism, product tampering, biological threats, and ecological terrorism.

8. On a federal level what are the three primary governmental agencies with regulatory responsibilities related to emergency response?

    For the public sector, the primary governmental agency is the Federal Emergency Management Agency and in the private sector, the two primary agencies are the Occupational Safety and Health Administration and the Environmental Protection Agency.

9. Identify three areas that must be addressed in a written emergency action plan according to OSHA standards.
    - Procedures for reporting fires and other emergencies
    - Emergency escape procedures and emergency escape route assignments
    - Procedures to be followed by employees who remain to operate critical plant operations before they evacuate
    - Procedures to account for all employees after emergency evacuation has been completed
    - Procedures to be followed by employees performing rescue or medical duties
    - Identification of the names or regular job titles of persons and departments who can be contacted for further information or explanation on duties under plan

- Maintain an alarm system which has a distinctive signal for each purpose and the system shall comply with 29 C.F.R. 1910.165
- Designate and train a sufficient number of persons to assist in the safe and orderly emergency evacuation of employees
- Review the emergency action plan with each employee covered by the plan when the plan is developed or the employee is assigned initially to a job, when the employee's responsibilities under the plan change, or when the plan is changed.

10. What is the primary purpose of the OSHA Chemical Process Safety Management standard?

    The primary purpose of this standard is to eliminate or minimize the consequences of catastrophic releases of toxic, reactive, flammable, or explosive
    chemicals.

11. What are three examples of duties that a fire brigade may perform during a fire emergency?
    - Sound the alarm and aid in employee evacuation
    - Shut off machinery and utilities and assure fire suppression systems are working properly and fire doors are closed
    - Move motor vehicles away from plant
    - Direct fire fighters to scene of fire
    - Stand by at sprinkler valves
    - Extinguish the fire and maintain a fire watch after the fire is extinguished
    - Assist with salvage operations and put fire protection equipment back into service

12. One of the purposes of the EPA's Emergency Planning and Community Right-to-Know Act is to establish requirements for federal, state, and local governmental agencies, and many business facilities regarding emergency response to environmental emergencies.

    Increase public knowledge of and access to information on the presence of toxic chemicals in communities, releases of toxic chemicals into the environment, and waste management activities involving toxic chemicals.

13. What is one of the purposes of the Integrated Contingency Plan that is published by the National Response Team?
    - Provide a mechanism for consolidating multiple facility response plans into one plan that can be used during an emergency

- Improve coordination of planning and response activities within the facility and with public and commercial responders
- Minimize duplication of effort and unnecessary paperwork burdens, and simplify plan development and maintenance

14. Another important decision in planning a response strategy is determining the extent to which the organization is willing to commit time and money to developing and implementing an emergency response plan. What are some examples of costs associated with the development and implementation of an emergency response team?

    Examples of costs include personal protective equipment, fire and other emergency equipment and supplies, medical costs, as well as training costs.

15. Using the NRT Integrated Contingency Plan, the core plan is intended to do what?

    Identify the essential steps necessary to initiate, conduct, and terminate an emergency response action: recognition, notification, and initial response, including assessment, mobilization, and implementation.

16. One of the OSHA requirements for First Aid is that in the absence of an infirmary, clinic, or hospital in near proximity to the workplace that is used for the treatment of all injured employees, a person or persons shall be adequately trained to render first aid. According to OSHA interpretations, what is used to evaluate "near proximity"?

    Near Proximity: In areas where accidents resulting in suffocation, severe bleeding, or other life-threatening or permanently disabling injury or illness can be expected, a 3 to 4 minute response time, from time of injury to time of administering first aid, is required. In other circumstances, i.e., where a life-threatening or permanently disabling injury is an unlikely, a longer response time such as 15 minutes is acceptable.

17. Identify at least two of the suggested measures to address the media as part of the overall emergency response plan?
    - Designate a safe area that is away from the flow of emergency traffic for all media vehicles and for all media communications
    - Maintain security in the designated media area and prohibit media representatives from access to the emergency area
    - Identify a specific member of management to be the spokesperson for the company and allow no other employees to talk with the media.

The person selected should have experience in public relations and in dealing with the media.
- Send the media to appropriate areas to acquire video footage when necessary
- Provide informational packets with company information to the media
- Review all information by legal counsel prior to its presentation and keep questions from the media to a minimum

18. What is the primary reason for investigating loss incidents involving fires?

The primary reason is to determine the causes so that preventive measures can be taken to prevent future occurrences.

Index

Page numbers in italics indicate figures and tables.

access control, 317
acetylene: in Class I electrical hazardous locations, 49; as fire hazard, 75–76
active fire protection systems, in building construction, 104
ADA. *See* Americans with Disabilities Act (ADA)
ADAAG. *See* Americans with Disabilities Accessibility Guidelines (ADAAG)
administrative controls, in fire protection management, 225–27
AED. *See* automated external defibrillators (AED)
aerated powder coating operations, 157–58
air: hydrogen and, 70; thermal conductivity of, 28
air-conditioning. *See* heating, ventilation, and air-conditioning (HVAC)
alarm and detection systems, 171–87; annunciation of, 180–81, 279; building codes for, 174; in building design, 329–30; classifications for, 172; in emergency action plans, 265, 277–80; in emergency-response plans, 229, 246; for flame, 178–79; for heat, 175–76, *176*; hot work permit programs and, 257; initiating devices for, *173*, 173–81; inspection, testing, and maintenance of, 185–87; installation of, 174; notification devices for, 181–82, *182*; power supplies for, 172–73; for radiant-energy, 178–79; reporting systems for, 182–85; for smoke, 174–78, *177*; testing of, 278–80
aluminum: in Class II electrical hazardous locations, 50; as fuel, 21; thermal conductivity of, 28
ambulatory health-care occupancy, life safety and, 137
American National Standards Institute (ANSI): on alarm and detection systems, 280; on hydrogen piping, 71; National Electrical Safety Code of, 46; on oxygen, 77
American Petroleum Institute (API): on liquid storage tanks, 64; on liquid underground tanks, 68
American Society for Testing and Materials (ASTM), on fire resistance, 110–11

395

American Society of Mechanical Engineers (ASME): on hydrogen containers, 70; on liquefied hydrogen, 73; on liquid storage tanks, 64; on liquid underground tanks, 68; on oxygen, 77
Americans with Disabilities Accessibility Guidelines (ADAAG), 277
Americans with Disabilities Act (ADA), 292, 296
ammonium nitrate, 92–94
annunciation, of alarm and detection systems, 180–81, 279
ANSI. *See* American National Standards Institute (ANSI)
antifreeze, in sprinkler systems, 209
API. *See* American Petroleum Institute (API)
arcing, 31; Class I electrical equipment requirements and, 52; dust explosions and, 96; electrical fires from, 46
area of coverage, of sprinkler heads, 214
arson: as building design threat, 313; wildfires from, 4
ASME. *See* American Society of Mechanical Engineers (ASME)
assembly occupancy: exit and, 140; life safety and, 136
ASTM. *See* American Society for Testing and Materials (ASTM)
atmospheric tanks, for flammable/combustible liquids, 63
attended vehicles, for explosives transporting, 89
automated external defibrillators (AED), 272
automatic detection systems, 3

back draft, 26
backflow preventers, for sprinkler systems, 208–9
baffles, 116

beams: ammonium nitrate and, 93; in building construction, 105, 108; in building design, 322; structural steel in, 33; of wood, 108
black powder, 85
blasting agents, 84–91; use of, 90–91
BLEVE. *See* boiling-liquid expanding-vapor explosion (BLEVE)
BLS. *See* Bureau of Labor Statistics (BLS)
BOCA. *See* Building Officials and Code Administration (BOCA)
boilers: explosions from, 97–100; as ignition sources, 9; maintenance of, 99–100
boiling-liquid expanding-vapor explosion (BLEVE), 25, 94–96, *95*
boiling point, 32
bollards, in building design, 318, 331
bonding wire: for dip-tank operations, 161, 166; for liquid transfer, 59, *60*, 68–69
BPAT. *See* Building Performance Assessment Team (BPAT)
brick: in building construction, 109; emissivity of, 30; in exterior walls, 326
British thermal unit (Btu), 26; boilers and, 100; fire load and, 120
building codes, 125–27; for alarm and detection systems, 174; for boilers, 100; fire doors in, 117–18; on hazard identification, 224; on high-rise buildings, 124; history of, 125; occupancy and commodity classifications of, 121–22; on sprinkler systems inspections, 216; Station Night Club fire and, 131
building construction, 103–27; fire load and, 120; fire protection in, 104, 114–20; fire resistance in, 104, 110–11, *113*; fire spread and, 115–17; gypsum board in, 110; interior finish in, 118–19; major types of, 111–14; masonry

and brick in, 109; occupancy and commodity classifications of, 120–23; reinforced concrete in, 110; steel in, 107–8; structural elements in, 104–6; terminology, 104; venting in, 118; wood in, 107–8, *108*
building contents, 119–20
building design: alarm and detection systems in, 329–30; bollards in, 318, 331; cladding in, 325–26; columns in, 320–22, 324; communication system in, 330–31; doors in, 319–20; electrical equipment in, 326–27; elevators in, 329; emergency generator in, 327–28; for emergency mitigation, 313–33; exterior frame in, 320–22; exterior walls in, 325–26; fire control center in, 329; floors in, 323–24; fuel in, 328; green space in, 331; HVAC in, 329; interior columns in, 324; interior limited access areas in, 315; interior public access in, 315; interior walls in, 324–25; means of egress in, 327; mechanical equipment in, 326–27; neighbors in, 319; off-site areas in, 316–19; parking in, 316–19; perimeter protection in, 316; planters in, 318, 331; point of entry in, 315–16; post-failure analysis for, 332; risk assessment in, 313–14; roof in, 322–23; site mitigation and, 314–19; smoke-control systems in, 330; sprinkler systems in, 220; standpipe hose systems in, 220; threats to, 313; ventilation in, 328–29; virtual reality software for, 331–32; walls in, 324–26; windows in, 319–20
building fire alarm systems, 172
building information, in emergency action plans, 273–74
building materials: characteristics of, 107–10. *See also specific types*
Building Officials and Code Administration (BOCA), 125

Building Performance Assessment Team (BPAT), 332
building structural damage, fire inspections for, 249
bulb sprinkler heads, 211–12
bulk oxygen systems, 77
Bureau of Labor Statistics (BLS), 8–9
burning rate, of oxidizers, 91
business continuity: manager, on emergency action plans, 267; plans, in emergency-response plans, 248–49
business occupancy, life safety and, 138

CAD. *See* computer-aided design (CAD)
carbon black, in Class II electrical hazardous locations, 50
carbon dioxide: from combustion, 24; as fire extinguisher agent, 193–94
carbon monoxide (CO): in back draft, 26; from combustion, 24
cardiopulmonary resuscitation (CPR), 245
carpet, ignition temperature of, 21
CAS. *See* Chemical Abstract Service (CAS)
cast iron, emissivity of, 30
CCTV. *See* closed-circuit television (CCTV)
ceilings: fire loading in, 120; fire spread in, 115; interior finish of, 118, 140; plastics in, 34; smoke detectors on, 177–78. *See also* sprinkler systems
ceiling temperature: alarm and detection systems for, 175, 176; for sprinkler heads, 212
Celsius temperature measurement, 26
central station fire alarm systems, 183–84
chain of command: in emergency action plans, 265; in evacuation, 298
charcoal: in black powder, 85; in Class II electrical hazardous locations, 50
Chemical Abstract Service (CAS), 39
chemical chain reaction, 21–22

chemical explosions, 83–84
chemical heat, 30
Chemical Process Safety Management, 232
chemical reactions: in chemical explosions, 84; as ignition source for flammable/combustible liquids, 69
Chemical Transportation Emergency Center (CHEMTREC), 42
chlorine, as oxidizer, 20
circuit overload, electrical fires from, 46
cladding, in building design, 325–26
Class A fire extinguishers, 199–200, *200*; in emergency action plans, 274
Class A fires: carbon dioxide for, 193; dry chemical fire extinguisher agents for, 195
Class B fire extinguishers, 199–200, *200*; in emergency action plans, 274
Class B fires: dry chemical fire extinguisher agents for, 195; flammable/combustible gases in, 190; flammable/combustible liquids in, 190
Class C fire extinguishers, 200, *200*; in emergency action plans, 274
Class C fires (electrical fires), 10; carbon dioxide for, 193; dry chemical fire extinguisher agents for, 195; human error in, 47
Class D fire extinguishers, 200, *200*
Class D fires, combustible metals in, 190
Class I commodities, 122
Class I electrical equipment requirements, 52
Class I electrical hazardous locations, 47–49; spray booths as, 157
Class I explosive magazine, 87
Class I flammable/combustible liquids, 55; in dip-tank operations, 164, 166
Class II commodities, 122
Class II electrical equipment requirements, 52–53

Class II electrical hazardous locations, 50; dust explosions and, 97
Class II explosive magazine, 87–88
Class II flammable/combustible liquids, 55; in dip-tank operations, 164
Class III commodities, 123
Class III electrical equipment requirements, 53
Class III electrical hazardous locations, 50–51
Class III flammable/combustible liquids, 55–56; in dip-tank operations, 164
Class III standpipe systems, 219–20
Class II standpipe systems, 219–20
Class I standpipe systems, 219–20
Class IV commodities, 123
Class K fire extinguishers, *200*
Class K fires, cooking media in, 190
Clean Air Act, 232
Clean Water Act, 238
clear-time effects, 321
closed-circuit television (CCTV), 317
CO. *See* carbon monoxide (CO)
coal, in Class II electrical hazardous locations, 50
coefficient of expansion, of steel, 107
coefficient of thermal conductivity, 28
columns: in building construction, 105; in building design, 320–22, 324; structural steel in, 33
combination systems, 172
combustible materials: hot work permit programs and, 256–57. *See also* flammable/combustible
combustible metals: in Class D fires, 190; fire extinguisher agents for, 18; as fire hazard, 33–34; in spray booths, 156; thermal conductivity of, 27; water in fire extinguishers and, 193
combustion, 22–30; physics of, 31–32; unique phenomena of, 24–26
combustion waves, in deflagration, 25

command center, in emergency action plans, 280
command post leader: on emergency management team, 268; in evacuation, 299–300
commodity classifications, of building construction, 120–23
communication system, in building design, 330–31
compartmentation, fire walls for, 116
Compressed Gas Association: on acetylene, 76; on oxygen, 77
computer-aided design (CAD), 331–32
concrete, thermal conductivity of, 28
conduction, heat transfer by, 27–28; for steel, 107
connectors, in building construction, 106
containment, of explosions, 25
Contingency Planning Requirements, of RCRA, 238
convection, heat transfer by, 29
cooking media: in Class K fires, 190; as fuel, 21
copper: specific heat of, 27; for sprinkler system piping, 208; thermal conductivity of, 28
corridors: for exit access, 140; fire spread through, 115
corrosive-resistance, of underground liquid tanks, 67
corrosive substances, as hazardous materials, 38
corrosive substances, explosives transporting and, 88
CPR. *See* cardiopulmonary resuscitation (CPR)
CPTED. *See* crime prevention through environmental design (CPTED)
CPVC, for sprinkler system piping, 208
Crane, Alanson, 198
Crescent City, Illinois, BLEVE at, 95–96
crime prevention through environmental design (CPTED), 318–19

critical temperature, of steel, 107
cross-sectional area, in conduction, 28
curtain walls, 106

day-care occupancy, life safety and, 136
dead load, in building construction, 104
dedicated function fire alarm systems, 172
deflagration, 25
deflagrations, flame front speed in, 84
deluge sprinkler system, 206; in emergency action plans, 274; valves in, 210–11
Department of Homeland Security, 14–15
detection, alarm, and communication systems, for life safety, 144
detection-and-notification systems, 2
detention and correctional occupancy, life safety and, 137
detonation, 25
detonations, flame front speed in, 84
diatomaceous earth, in dynamite, 85
diking, for liquid outside storage tanks, 65
dip-tank operations: drains for, 162; fire protection for, 166–67; hazardous processes in, 159–68, *160, 163*; inspection and testing of, 168; liquid level control in, 162; operation and maintenance of, 167–68; overflow prevention in, 161–62; piping systems for, 166; storage, handling, and distribution of flammable/combustible liquids for, 165; temperature in, 162–64; training for, 168; venting of, 164–65
direct pumping system, for sprinkler systems, 208
disabled persons, in evacuation, 292
distance: in conduction, 28; radiant heat transfer and, 30
doors: in building design, 319–20; in emergency action plans, 272; for exit,

142–43; fire spread through, 115. *See also* fire doors
drainage: in emergency action plans, 273; for liquid outside storage tanks, 65
drains, for dip-tank operations, 162
dry-barrel fire hydrants, 217–18
dry chemicals, as fire extinguisher agents, 195–96
dry-pipe sprinkler systems, 205; in emergency action plans, 274; valves in, 210; water flow alarms for, 211
dust: combustible, in Class II electrical hazardous locations, 50; explosions, 96–97; liquid housekeeping and, 70; storms, explosives and, 90
dustproof electrical equipment, in Class II electrical equipment requirements, 52–53
dynamite, 85

early suppression fast response (ESFR), sprinkler heads, 213, 214
educational occupancy, life safety and, 136
electrical equipment: in building design, 326–27; dust explosions and, 96; for electrostatic spray operations, 151–52; emergency-response plans for, 246; explosives transporting and, 89; hydrogen and, 71; liquefied hydrogen and, 73, 74; LP gas and, 80; NRTLs for, 53–54; oxygen and, 77; safe design of, 51–53; of spray booths, 153, 157; water in fire extinguishers and, 192
electrical fires. *See* Class C fires
electrical heat (resistance heating), 31
electric cords, electrical fires from, 46
electricity, as ignition source, 20, 45–54
electromagnetic waves, radiant heat transfer of, 29–30
electrostatic fluidized beds, 158–59
electrostatic spray operations, 150–52

elevators (shafts): alarm and detection systems initiating devices and, 174; in building design, 329; as firestop, 116; as means of egress, 292–93
Emergency-Action Plan Regulation, of OSHA, 238
emergency action plans: alarm and detection systems in, 265, 277–80; annual review of, 284; building information in, 273–74; command center in, 280; elevators in, 292–93; emergency management team in, 266–68; evacuation in, 263–311; fire brigades in, 268, 289–90; fire extinguishers in, 265, 274; floor captains in, 268; floor plans in, 271–73; insurance providers in, 267–68; life safety and, 144–46, 264; means of egress in, 265, 274; off-site data storage in, 280–81; post-event restoration in, 303–11; preplanning for, 276; revision of, 287; shutdown procedures in, 290–92, *291*; sprinkler systems in, 274–75; steps for creation of, 282–83; training in, 284–89; Urban Search and Rescue Grid in, 276; utilities in, 273, 281–82; vendor and contractor identification in, 269–71
emergency command center, in high-rise building fires, 124
emergency director, on emergency management team, 267
emergency generator, in building design, 327–28
emergency lighting, for exit, 141–42
emergency management team: in emergency action plans, 266–68; fire brigades on, 289–90; floor captains on, 268, 289–90
emergency medical care, in emergency-response plans, 244–45
Emergency Planning and Community Right-to-Know Act, 231, 235–36

INDEX 401

emergency planning and preparedness, 2
emergency response, 1–2
Emergency Response Guidebook, 36, 40
emergency-response plans: business recovery/continuity plans in, 248–49; cost and resource commitment to, 241; emergency medical care in, 244–45; EPA on, 235–36; FEMA on, 228; on fire brigades, 232–36; fire prevention in, 240; in fire program management, 226–49; formal agreements with nearby organizations in, 241–42; media control in, 247–48; NFPA on, 237–39; OSHA on, 229–34; on PPE, 233–34, 246–47; RCRA and, 236–37; shutdown procedures in, 246; training for, 245–46; in writing, 242–44
emissivity, 30
engineering controls, in fire protection management, 225–27
Environmental Protection Agency (EPA): on antifreeze in sprinkler systems, 209; on emergency-response plans, 235–36, 242–44, 248; GHS and, 56; NRT and, 238, 239; Oil Pollution Prevention Regulation of, 238; Risk Management Program Regulation of, 238; Risk Management Programs for Chemical Accident Release Prevention Standard of, 232; Standards for Owners and Operators of Hazardous Waste Treatment, Storage, and Disposal Facilities of, 236
EPA. See Environmental Protection Agency (EPA)
ESFR. See early suppression fast response (ESPR)
ethylene, in Class I electrical hazardous locations, 49
evacuation: accounting for employees and visitors in, 297–98; assistance devices for, 294–96; chain of command in, 298; command post leader in, 299–300; disabled persons in, 292; drill evaluation for, 302; elevators forb, 292–93; in emergency action plans, 263–311; in emergency-response plans, 229, 246; fire brigades in, 298; fire walls for, 116; floor captains in, 298; group leader in, 299; from high-rise building fires, 123; plans, for wildfires, 4; reentry after, 301–2; refuge areas in, 296; runners in, 299; staging areas in, 296–97; from Station Night Club fire, 130–31; from Triangle Shirtwaist fire, 130
exit, 140–43; access, 139–40; in emergency action plans, 272; signs, 141, *141. See also* means of egress
explosions, 25–26, 83–100; in aerated powder coating operations, 158; from boilers, 97–100; emergency-response plans for, 247; engineering controls for, 226; from flammable/combustible gases, 97; from hazardous processes, 149–50; from pressure vessels, 97–100; shock waves in, 83
explosion venting, for liquefied hydrogen, 74
explosive-proof electrical equipment, 52
explosives, 84–91; handling and storage of, 87–88; as hazardous materials, 36; labels and placards for, 85–87, *86, 87*; oxidizers and, 91–94; transporting, 88–90; use of, 90–91
explosive train, 86
exterior frame, in building design, 320–22
exterior walls, 106; in building design, 325–26
extra hazard fire load, 199; sprinkler system piping and, 208
extra hazard occupancies, 122

eye wash stations, in emergency action plans, 272

facilities maintenance manager, on emergency management team, 267
Fahrenheit temperature measurement, 26
Federal Emergency Management Agency (FEMA), 15; on business continuity plans, 248–49; on emergency action plans, 276; on emergency-response plans, 228, 248–49; fire inspections of, 252; on fire protection management, 226; MAT of, 332–33
feed mains, for sprinkler systems, 209
FEMA. *See* Federal Emergency Management Agency (FEMA)
fibers, Class III electrical equipment requirements for, 53
fire alarms. *See* alarm and detection systems
firearms, explosives and, 88
fire barrier wall, 106, 116
fire blanket, hot work permit programs and, 257
fire brigades: in emergency action plans, 268, 289–90; emergency-response plans on, 232–36; in evacuations, 298
fire casualties, fire inspections for, 249
fire control center, in building design, 329
fire dampers, 116
fire department: fire program management and, 224, 235, 251; sprinkler systems and, 215, *215*, 216; standpipe-hose systems and, 219–20
fire development, fire inspections for, 249
fire doors, 2; fire walls and, 117–18
fire escape, at Triangle Shirtwaist fire, 130
fire extinguishers/extinguishment, 2, 189–220; agents/methods of, *190*, 190–98; for dip-tank operations, 166; distribution and mounting of, 201–2; in emergency action plans, 265, 272, 274; in emergency-response plans, 230; explosives transporting and, 89; fire hydrants and, 217–20, *218*; fuel classifications for, 21; labeling of, 199; maintenance, inspection, and testing of, 202–3; portable, 198–99, *199*; rating system for, 199–200, *200*; for spray booths, 155; Station Night Club fire and, 132; training for, 203; types of, *199*; in workplace, 201. *See also* sprinkler systems
fire hazards, 45–79; acetylene as, 75–76; combustible metals as, 33–34; in emergency action plans, 265; in emergency-response plans, 230; emergency-response plans for, 231, 246; flammable/combustible gases as, 70–77; flammable/combustible liquids as, 54–70; gypsum board as, 34; hydrogen as, 70–74, *71*, *72*, *74*; liquefied hydrogen as, 72–74, *74*, *75*; LP gas as, 78–80, *80*; masonry/stone as, 33; oxygen as, 76–78; plastics as, 34; spray booths as, 152, 153; synthetic materials as, 34; textiles as, 35; wood as, 32–33
fire hydrants, 217–20, *218*
fire-ignition sequence, 249
fire inspections of, 252
fire investigations, in fire program management, 249–56, *253*–55
fire load: in building construction, 104, 110, 120; fire extinguishers and, 199
fire plume, 174; sprinkler heads and, 213
fire prevention, 1–3; for dip-tank operations, 161; in emergency action plans, 265; in emergency-response plans, 229–30, 240; in fire program management, 225–27; for oxidizers, 92
fire program management, 223–58; emergency-response plans in,

INDEX

226–49; fire investigations in, 249–56, 253–55; fire prevention in, 225–27; fire protection in, 225–27; hazard identification in, 224; hot work permit programs in, 256–25; measurement of effectiveness of, 227; risk quantification in, 225

fire protection, 1–3; administrative controls in, 225–27; in building construction, 104, 114–20; for dip-tank operations, 166–67; engineering controls in, 225–27; fire inspections for, 252; in fire program management, 225–27; maintenance of systems for, 250–51; in spray booths, 155–56

fire resistance: in building construction, 104, 110–11, *113*; exit and, 140; of gypsum board, 110; for hydrogen, 71

fire-retardant treatment, for wood, 108–9

fires: annual averages of, *10*; chemistry and physics of, 19–42, *20*, *22*, *41*; civilian injuries in, *12–13*; classification of, 189–90; economic impact of, 7–8; in high-rise buildings, 123–25; historically major, 6, *7*; igniting equipment in, *11–12*

fire service mains, sprinkler systems and, 208

fire spread: building construction and, 115–17; at Station Night Club fire, 130

firestops, 116

fire tetrahedron, 19–22, *20*, *22*

fire walls, 2, 106, 116–17; in building construction, 104; for dip-tank operations, 161; openings in, 117–18

first aid: in emergency action plans, 265, 272; OSHA on, 244; USDOT on, 40–42

flame: alarm and detection systems for, 175, 178–79. *See also* open flames

flame front speed, 84

flame spread: in building construction, 104; exit and, 140; of interior finishes, 119; of wood, 108, 109

Flame Spread Index (FSI), 119

flame supervisory unit, for boilers, 99

flammable/combustible gases: in Class B fires, 190; Class I electrical equipment requirements and, 52; in Class I electrical hazardous locations, 47–49; combustion of, 23; combustion physics of, 32; engineering controls for, 226; explosions from, 97; as fire hazards, 70–77; as fuel, 21; as hazardous materials, 36–37; heat of combustion of, 27; heat transfer by convection of, 29; thermal conductivity of, 27

flammable/combustible liquids: in Class B fires, 190; in Class I electrical hazardous locations, 47–49; combustion of, 23; combustion physics of, 31–32; containers and portable tanks for, 58, *58*; engineering controls for, 226; as fire hazards, 54–70; flash point of, 31, 32, 54–55, 57, 67, 68; GHS for, 56, 56–57; as hazardous materials, 37; hazardous processes with, 150–68; heat of combustion of, 27; heat transfer by convection of, 29; housekeeping for, 69–70; as ignition source, 69; inside storage rooms for, 61–63, *63*; NFPA on, 55–56; outside aboveground tanks for, 64–66; oxygen and, 77; sprinkler heads and, 214; storage and handling of, 57–69, 154–55; storage cabinets for, 61, *62*; storage tanks for, 63–64; tank vehicles and cars for, 68–69; transfer of, *58*, 58–60; underground tanks for, 66–68; ventilation requirements for, 57; workplace practices with, 69–70

flammable/combustible waste: in emergency-response plans, 230; as fire hazard, 70
flashover, 26
flash point, of flammable/combustible liquids, 31, 32, 54–55, 57, 67, 68
floor captains: in emergency action plans, 268, 289–90; in evacuations, 298
floor plans, in emergency action plans, 271–73
floors: in building design, 323–24; fire loading in, 120; fire spread in, 115; interior finish of, 118; plastics in, 34
fluorine, as oxidizer, 20
FM Global, 64
Fourier's law of heat conduction, 28
frictional heat: dust explosions and, 96; as ignition source for flammable/combustible liquids, 69; in spray booths, 154
FSI. *See* Flame Spread Index (FSI)
fuel (storage): in building design, 328; classification of, 21, *22*; in emergency action plans, 265; fire inspections for, 252; in fire tetrahedron, 19–22, *20*, *22*
fuel load: in high-rise building fires, 124; occupancy classification and, 120–21
fusible sprinkler heads, 212

gases. *See* flammable/combustible gases; toxic gases
general industrial occupancy, life safety and, 138
GHS. *See* Globally Harmonized System (GHS)
glass: thermal conductivity of, 28. *See also* windows
Globally Harmonized System (GHS), *56*, 56–57
global warming, 3
glycerin, in water for fire extinguishers, 192

grains, in Class II electrical hazardous locations, 50
gravity systems, for sprinkler systems, 208
green space, in building design, 331
ground failure, electrical fires from, 46
group leader: on emergency management team, 268; in evacuations, 299
gypsum board: in building construction, 110; in building construction interior finish, 118; as fire hazard, 34; as firestop, 116; ignition temperature of, 21

halogens: as fire extinguisher agent, 194–95; hydrogen and, 70
hazard identification: in emergency action plans, 264; in fire program management, 224
hazardous locations: NEC and, 47–51; spray booths as, 153
hazardous materials, *41*; classes of, 36–39; in emergency action plans, 272; information sources on, 39–42; water in fire extinguishers and, 193. *See also* explosives
Hazardous Materials Table (HMT), 40, *41*
hazardous processes: in dip-tank operations, 159–68, *160*, *163*; explosions from, 149–50; with flammable/combustible liquids, 150–68; in spray booths, 152–57
hazardous waste, emergency-response plans on, 231, 236–37
Hazardous Waste Operations and Emergency Response (HAZWOPR), of OSHA, 236, 238
hazards. *See* fire hazards
HAZWOPR. *See* Hazardous Waste Operations and Emergency Response (HAZWOPR)
HCN. *See* hydrogen cyanide (HCN)

health-care occupancy, life safety and, 136–37
heat: alarm and detection systems for, 175–76, *176*; alarm and detection systems initiating devices for, 175–76, *176*; from combustion, 23; in fire tetrahedron, 19–22, *20, 22*; sources of, 30–31; *versus* temperature, 26–27
heating, ventilation, and air-conditioning (HVAC): in building design, 329; dip-tank operations and, 159; fire dampers in, 116; as firestop, 116; as ignition source, 9, 20; liquefied hydrogen and, 74; for spray booths, 157
heat of combustion, 27
heat of decomposition, 30
heat of evaporation, of water, 191
heat of solution, 31
heat release rate (HRR), fire load and, 120
heat transfer, 27–28; in BLEVE, 95
heat transfer coefficient, 29
high hazard industrial occupancy, life safety and, 138
high-pressure gas switch, for boilers, 99
high-rise buildings, fires in, 123–25
HMT. *See* Hazardous Materials Table (HMT)
hot work permit programs, 256–57
household fire warning systems, 172
HRR. *See* heat release rate (HRR)
human resources, on emergency management team, 267
HVAC. *See* heating, ventilation, and air-conditioning (HVAC)
hydrofluoric acid, as oxidizer, 20
hydrogen: in Class I electrical hazardous locations, 49; as fire hazard, 70–74, *71, 72, 74*
hydrogen cyanide (HCN), 24
hydrogen peroxide: as hazardous materials, 37; as oxidizer, 20

IBC. *See* International Building Code (IBC)
ICBO. *See* International Conference of Building Officials (ICBO)
ICC. *See* International Code Council (ICC)
ICP. *See* Integrated Contingency Plan (ICP)
ICS. *See* Incident Command System (ICS)
ignition limiting timer, for boilers, 99
ignition source: alarm and detection systems initiating devices and, 175; in dip-tank operations, 159–61; for dust explosions, 96; electricity as, 20, 45–54; in emergency-response plans, 230; explosives and, 88, 90; explosives transporting and, 88; fire inspections for, 252; flammable/combustible liquids as, 69; in fuel tetrahedron, 20; LP gas and, 80; in spray booth, 152, 153
ignition temperature, 21; flashover and, 26; of wood, 108
Imperial Foods fire, 5
Incident Command System (ICS), 239
incident fires, emergency-response plans on, 230
industrial occupancy, life safety and, 138
infectious substances, as hazardous materials, 37–38
initiating devices, for alarm and detection systems, *173*, 173–81
insulation/insulators, 28; electrical fires from, 46; for electrostatic spray operations, 151; for steel, 108
insurance providers, in emergency action plans, 267–68
Integrated Contingency Plan (ICP), 238–39; as written emergency-response plan benchmark, 242–43
interior columns, in building design, 324
interior finish, in building construction, 118–19

interior limited access areas, in building design, 315
interior public access, in building design, 315
interior walls, in building design, 324–25
intermediate level sprinklers, 213
International Building Code (IBC), 121–22, 126–27; on exit access, 140; on life safety, 135
International Code Council (ICC), 126–27; on life safety, 135
International Conference of Building Officials (ICBO), 125
International Energy Conservation Code, 127
International Existing Building Code, 127
International Fire Code, 127
international fire experience, 8, *8*
International Fuel Gas Code, 127
International Green Construction Code, 127
International Mechanical Code, 127
International Plumbing Code, 127
ISO/TS, 225
isolation, of explosions, 25–26

joules, 26

Kelvin temperature measurement, 26
KNOX-BOX° Rapid Entry System, 274

LEL. *See* lower explosive limit (LEL)
LEPCs. *See* local emergency planning communities (LEPCs)
LFL. *See* lower flammable limit (LFL)
life safety, 129–46; codes and regulations for, 133–35; detection, alarm, and communication systems for, 144; for electrostatic spray operations, 151; emergency action plans and, 144–46, 264; fire inspections for, 252; human behavior during emergencies and, 132–33; occupancy classifications and, 135–39; smoke and, 144–45;

utilities and, 144–45; venting and, 144–45. *See also* evacuation; means of egress
light hazard fire load, 199; sprinkler system inspections for, 216; sprinkler system piping and, 208
light hazard occupancies, 121
lighting: for alarm and detection systems notification systems, 182; in dip-tank operations, 164; emergency, for exit, 141–42; in emergency-response plans, 241; for exterior protection, 318; in spray booth, 152; sprinkler heads and, 213
lightning: explosives and, 90; heat from, 31; as ignition source, 20; as ignition source for flammable/combustible liquids, 69; wildfires from, 4
line-type heat detectors, 176
liquefied hydrogen, 72–74, *74, 75*
liquefied petroleum gas. *See* LP gas (liquefied petroleum gas)
liquids. *See* flammable/combustible liquids
load-bearing walls, 106; parapets of, 116
local emergency planning communities (LEPCs), 235, 241, 243
logistics and procurement manager, on emergency management team, 267
loss of life. *See* life safety
lower explosive limit (LEL): for electrostatic spray operations, 152; of flammable/combustible liquids, 57
lower flammable limit (LFL), of flammable/combustible liquids, 57
low-pressure gas switch, for boilers, 99
low-pressure tanks, for flammable/combustible liquids, 63
low water cutoffs, for boilers, 99
LP gas (liquefied petroleum gas): BLEVE of, 95–96; as fire hazard, 78–80, *80*

magazines, for explosives handling and storage, 87–88

INDEX 407

magnesium: in Class II electrical hazardous locations, 50; as fuel, 21
main drain valve: sprinkler system inspections for, 217; in wet-pipe sprinkler systems, 210
masonry: in building construction, 109; as fire hazard, 33. *See also* brick
Massey, Curtis, 276
Massey Disaster Plan, 276
MAT. *See* Mitigation Assessment Team
material safety data sheet (MSDS): in emergency action plans, 265, 273–74; in emergency-response plans, 235
means of egress, 2; in building design, 327; capacity of, 143–44; elevators as, 292–93; in emergency action plans, 265, 274; general requirements for, 139–44; in high-rise building fires, 125; maintenance of, 146; venting for, 118. *See also* evacuation
mechanical equipment, in building design, 326–27
mechanical explosions, 83
mechanical heat, 31
mechanical rooms, in emergency action plans, 272
media control, in emergency-response plans, 247–48
media relations contact, on emergency management team, 267
mercantile occupancy, life safety and, 138
metals. *See specific metals*
mission statement, in emergency action plans, 265
Mitigation Assessment Team (MAT), 332–33
monoammonium phosphate, as fire extinguisher agent, 195
MSDS. *See* material safety data sheet (MSDS)
multiple occupancy, life safety and, 139
municipal fire alarm systems, 172
mutual aid, Station Night Club fire and, 132

National Electrical Code (NEC), of NFPA, 46; hazardous locations and, 47–51
National Electrical Safety Code, of American National Standards Institute (ANSI), 46
National Fire Incident Reporting System, 227
National Fire Protection Association (NFPA), 2–3, 14; on alarm and detection systems, 171–72, 279–80; on alarm and detection systems notification systems, 182; on alarm and detection systems power supplies, 173; on alarm and detection systems testing, inspection, and maintenance, 186–87; on building construction interior finish, 118–19; on building construction types, 111–12; on dip-tank operations, 159–61; on dip-tank operations drains, 162; documenting of fires by, 6; on electrostatic spray operations, 151; on emergency-response plans, 237–39; on exit access, 140; on fire classifications, 190–91; on fire hydrants, 218–19; fire inspections of, 252–56; on fire load, 199; on fire resistance, 111; on firestops, 116; on fire walls, 117; on flammable/combustible liquids, 55–56; on halogen fire extinguisher agents, 195; on hazard identification, 224; on hazardous materials, 39–40; on life safety, 130, 134–35; on liquid storage cabinets, 61; on liquid workplace practices, 69; on means of egress, 143; NEC of, 46; on oxidizers, 91; on risk quantification, 225; on smoke control, 145; on spray booth ventilation, 153–54; on sprinkler heads, 212, 214; on sprinkler systems, 131, 204, 208–9; on sprinkler systems inspections, 215–16; on standpipe hose systems, 220

National Institute for Occupational Safety and Health (NIOSH), 24
National Institute for the Certification of Engineering Technologies (NICET), 278
National Institute of Standards and Technology (NIST): on high-rise buildings, 123; on Station Night Club fire, 131
National Interagency Incident Management System (NIMS), 242
nationally recognized testing laboratories (NRTLs): for electrical equipment, 53–54; on liquid containers, 58
National Oceanic and Atmospheric Administration (NOAH), 3
National Response Team (NRT), 238, 239
NEC. See National Electrical Code (NEC)
neighbors, in building design, 319
NFPA. See National Fire Protection Association (NFPA)
NICET. See National Institute for the Certification of Engineering Technologies (NICET)
NIMS. See National Interagency Incident Management System (NIMS)
NIOSH. See National Institute for Occupational Safety and Health (NIOSH)
NIST. See National Institute of Standards and Technology (NIST)
nitric acid, as oxidizer, 20
nitroglycerin, in dynamite, 85
NOAH. See National Oceanic and Atmospheric Administration (NOAH)
noncombustible material, in building construction, 104
non-load bearing walls, 106
notification devices, for alarm and detection systems, 181–82, *182*

NRT. See National Response Team (NRT)
NRTLs. See nationally recognized testing laboratories (NRTLs)
nuclear heat, 31

oak, emissivity of, 30
occupancy classifications: of building construction, 120–23; exit and, 140–41; life safety and, 135–39
occupant load: of assembly occupancy, 136; means of egress and, 143–44
Occupational Safety and Health Administration (OSHA), 9, 13–14; on alarm and detection systems, 279–80; on ammonium nitrate, 92–94; on blasting agents, 85; on CO exposure, 24; on dip-tank operations, 159, 161; on electrical equipment safe design, 51; on electrical equipment testing, 53–54; on electrical hazardous locations, 47; on electrostatic spray operations, 151; Emergency-Action Plan Regulation of, 238; on emergency action plans, 145–46, 242, 264–65; on emergency medical care, 244–45; on emergency-response plans, 229–34; on explosives, 85, 90–91; on explosives transporting, 89; on fire brigades, 290; on fire extinguishers, 201; fire inspections of, 252; on first aid, 244; GHS and, 56–57; on hazard identification, 224; on hazardous locations, 51; HAZWOPR of, 236, 238; on hot work permit programs, 256–57; on hydrogen containers, 70; on life safety, 135; on liquefied hydrogen, 73; on liquid outside storage tanks, 65–66; on liquid storage and handling of, 57–69; on liquid workplace practices, 69; Phillips Petroleum Houston Chemical Complex explosion and, 150; Process Safety Management

Standards of, 238; on spray booth fire protection, 155–56; on spray booth ventilation, 154; on sprinkler heads, 214; on sprinkler systems inspections, 216
off-site areas, in building design, 316–19
off-site data storage, in emergency action plans, 280–81
Oil Pollution Prevention Regulation, of EPA, 238
oil-soaked rags, spontaneous ignition from, 21
Oklahoma City bombing, 316–17
open flames: dip-tank operations and, 159; explosives and, 88, 90; LP gas and, 80
opening pressure, for liquid outside storage tanks, 65
ordinary hazard fire load, 199; sprinkler system inspections for, 216
ordinary hazard occupancies, 121–22
organic peroxides, as hazardous materials, 37
ORM. See Other Regulated Materials (ORM)
OS&Y. See outside stem and yoke (OS&Y)
OSHA. See Occupational Safety and Health Administration (OSHA)
Other Regulated Materials (ORM), 39
outside stem and yoke (OS&Y), 209–10
oxidation, 23
oxidizers, 20; acetylene and, 76; combustion from, 22–30; explosives and, 91–94; explosives transporting and, 88; as hazardous materials, 37; liquefied hydrogen and, 74
oxygen: acetylene and, 76; in combustion, 24; dust explosions and, 96; as fire hazard, 76–78; in fire tetrahedron, 19–22, *20, 22*; hydrogen and, 70

panic bars, for exit doors, 142, 143

paper: in Class A fire, 190; as fuel, 21; liquefied hydrogen and, 74
parapets, 116
parking, in building design, 316–19
Parmalee, Henry S., 204
partial coverage, of alarm and detection systems initiating devices, 174
partition wall, 106
passive fire protection systems, 3; in building construction, 104
PB, for sprinkler system piping, 208
pendant sprinklers, 213
perchloric acid, as oxidizer, 91
perimeter protection, in building design, 316
personal protective equipment (PPE), emergency-response plans on, 233–34, 246–47
Phillips Petroleum Houston Chemical Complex explosion, 149–50
Pipeline and Hazardous Materials Safety Administration (PHMSA), 40
PIV. See post indicator valve (PIV)
planters, in building design, 318, 331
plaster, emissivity of, 30
plastics: in building construction interior finish, 118; in Class II electrical hazardous locations, 50; as fire hazard, 34
plywood, ignition temperature of, 21
point of entry, in building design, 315–16
polystyrene, ignition temperature of, 21
portable fire extinguishers, 2
post indicator valve (PIV), 209–10; sprinkler system inspections for, 216
potassium chlorate, as oxidizer, 91
potassium permanganate, as oxidizer, 91
power supplies: for alarm and detection systems, 172–73; in emergency action plans, 273
PPE. See personal protective equipment (PPE)
Pratt, Philip W., 204

preaction sprinkler systems, 205–6; valves in, 210–11
pressure relief devices, engineering controls for, 226
pressure vessels: explosions from, 97–100; for flammable/combustible liquids, 63
Process Safety Management Standards, of OSHA, 238
propane, in Class I electrical hazardous locations, 49
proprietary supervising station fire alarm systems, 184
propylene glycol, in water for fire extinguishers, 192
protected premises fire alarm systems, 172
public fire alarm systems, 183
pyrotechnics, Station Night Club fire and, 130–32

quick burning, of wood, 108

radiant-energy, alarm and detection systems for, 178–79
radiant heat, as ignition source for flammable/combustible liquids, 69
radiant heat transfer, 29–30
radioactive materials: as hazardous materials, 38; water in fire extinguishers and, 193
radio transmitters, explosives and, 90
rafters, in building construction, 105
railroad derailment, BLEVE from, 95–96
RCRA. See Resource Conservation and Recovery Act (RCRA)
refuge areas, in evacuation, 296
reinforced concrete, in building construction, 110
releasing fire alarm systems, 172
remote supervising station fire alarm systems, 185
reporting systems, for alarm and detection systems, 182–85

residential board and care occupancy, life safety and, 138
residential occupancy, life safety and, 137
resistance heating (electrical heat), 31
Resource Conservation and Recovery Act (RCRA), 236–37; Contingency Planning Requirements of, 238
risk assessment: in building design, 313–14; for emergency action plans, 282; for fire program management, 223, 225–27
Risk Management Program Regulation, of EPA, 238
Risk Management Programs for Chemical Accident Release Prevention Standard, of EPA, 232
risk management specialist, on emergency management team, 267
risk quantification, in fire program management, 225
roof: ammonium nitrate and, 93; in building construction, 114, 116; in building design, 322–23; drains, in emergency action plans, 273; slate, 125; of spray booths, 154; sprinkler systems and, 209
runners: on emergency management team, 268; in evacuation, 299
rupture disk device, for boilers, 99

Safety Data Sheets (SDS): of CHEMTREC, 42; of GHS, 56
safety factor, in building construction, 104
safety relief valve, for boilers, 99
safety science, 1
SARA. See Superfund Amendments and Reauthorization (SARA)
SCBA. See self-contained breathing apparatuses (SCBA)
SDS. See Safety Data Sheets (SDS)
self-contained breathing apparatuses (SCBA), 292

INDEX 411

September 11, 2001 terrorist attack, 7, 226, 275
SERC. *See* state emergency response commission (SERC)
sewer lines, in emergency action plans, 273
SFPE, on risk quantification, 225
shock waves: in detonation, 25; in explosions, 83
short circuits: electrical fires from, 46; explosives transporting and, 89
shutdown procedures: in emergency action plans, 290–92, *291*; in emergency-response plans, 246
shutters, fire walls and, 117
sidewall sprinklers, 213
site mitigation, building design and, 314–19
slate roof, 125
smoke: alarm and detection systems for, 174–78, *177*; in back draft, 26; baffles for, 116; from building construction interior finishes, 119; from building contents, 119; from combustion, 23–24; in high-rise building fires, 124; life safety and, 144–45; venting of, 118, 144–45
smoke-control systems, in building design, 330
smoke detectors, 171, 176–78, *177*; in emergency action plans, 272; with sprinkler systems, 275
smoking: dip-tank operations and, 161; explosives and, 88; as ignition source for flammable/combustible liquids, 69; liquefied hydrogen and, 74; oxygen and, 78; spray booths and, 156
sodium chlorite, as oxidizer, 91
sodium dichromate, as oxidizer, 91
sodium hypochlorite, as oxidizer, 91
sodium nitrate: in black powder, 85; in dynamite, 85; as oxidizer, 91
sodium persulfate, as oxidizer, 91

solids: combustion of, 23; combustion physics of, 31; as hazardous materials, 37; thermal conductivity of, 27
Southern Building Code Congress International, 125
sparking, 31; Class I electrical equipment requirements and, 52; dip-tank operations and, 159; electrical fires from, 46; explosives and, 88; explosives transporting and, 88; hot work permit programs and, 257; hot work permits for, 256; as ignition source for flammable/combustible liquids, 69; spray booths and, 156
special purpose industrial occupancy, life safety and, 138
specific gravity, 32
specific heat, 27; of water, 191
spontaneous heating, 30
spontaneous ignition, 21; as ignition source for flammable/combustible liquids, 69
spot-type heat detectors, 176
spray booth: drying, curing, and fusion apparatuses in, 156–57; ignition source in, 152, 153; operation and maintenance of, 156
spray booths: engineering controls for, 226; as fire hazard, 152; fire protection in, 155–56; flammable/combustible liquids storage and handling in, 154–55; hazardous processes in, 152–57; venting of, 153–54
sprinkler systems, 2, 203–17; for alarm and detection system initiating devices, 179–80; antifreeze in, 209; backflow preventers for, 208–9; in building construction, 104; in building design, 220; components of, 206, *207*; in dip-tank operations, 161–62; in emergency action plans,

265, 274–75; as engineering controls, 226; exit access and, 140; feed mains for, 209; fire department and, 215, *215*; fire service mains and, 208; heads for, 211–14, *212*; in high-rise building fires, 124; inspection of, 215–17; piping for, 209–10; smoke detectors with, 275; for spray booths, 155; Station Night Club fire and, 131; trip test for, 209; types of, 204–6; valves of, 209–11; water flow alarms for, 211; water supply and distribution for, 206–8
staging areas, in evacuation, 296–97
stainless steel, emissivity of, 30
stairways: alarm and detection systems initiating devices and, 174; in emergency action plans, 272; exit and, 140; fire spread through, 115
Standards for Owners and Operators of Hazardous Waste Treatment, Storage, and Disposal Facilities, of EPA, 236
stand-by power, in high-rise building fires, 124
standpipe hose systems, 2, 219–20; in building design, 220; sprinkler system inspections for, 216
state emergency response commission (SERC), 235
static electrical charge: in dip-tank operations, 161, 164; heat from, 31; liquid storage and, 59; liquid tank vehicles and cars and, 68
Station Night Club fire, 130–32
steel: in building construction, 107–8; for liquid storage tanks, 64; in reinforced concrete, 110; thermal conductivity of, 28
Steiner Tunnel Test, 119
stoichiometric combustion, 118
storage occupancy, life safety and, 139
storm drains, in emergency action plans, 273

sulfur, in black powder, 85
sulfuric acid, as oxidizer, 20
Superfund Amendments and Reauthorization (SARA), 231, 235–36
suppression, of explosions, 25
synthetic materials, as fire hazard, 34

temperature: alarm and detection systems initiating devices and, 174; in dip-tank operations, 162–64; HCN and, 24; *versus* heat, 26–27; radiant heat transfer and, 30; sprinkler heads and, 212, *212*. *See also* ceiling temperature; ignition temperature
temperature gradient, in conduction, 28
temperature limit control, for boilers, 99
textiles (ignitable fibers): Class III electrical equipment requirements for, 53; Class III electrical hazardous locations for, 50–51; as fire hazard, 35
titanium, as fuel, 21
TNT. *See* trinitrotoluene (TNT)
Toledo, Peoria, and Western Railroad Company, BLEVE of, 95–96
total complete coverage, for alarm and detection systems initiating devices, 174
toxic gases: from building contents, 119–20; emergency-response plans for, 247
traffic control, in emergency-response plans, 242
Triangle Shirtwaist factory fire, 5, 130
trinitrotoluene (TNT), 85
trip test, for sprinkler systems, 209, 217
trusses, in building construction, 106
Type I buildings, 112
Type II buildings, 112
Type III buildings, 112–13
Type IV buildings, 113
Type V buildings, 113

UEL. *See* upper explosive limit (UEL)

UFL. *See* upper flammable limit (UFL)
Underwriters' Laboratories (UL): on electrostatic spray operations, 151; on fire resistance, 110–11; on liquid storage tanks, 64; on liquid underground tanks, 68
uninterruptible power system, in emergency action plans, 273
United States Fire Administration (USFA), 14–15; on building contents, 119–20
upper explosive limit (UEL), of flammable/combustible liquids, 57
upper flammable limit (UFL), of flammable/combustible liquids, 57
Urban Search and Rescue Grid, in emergency action plans, 276
U.S. Department of Labor (USDOL): on electrical equipment safe design, 51–53; on emergency medical care, 244–45; on emergency-response plans, 230, 231; GHS and, 56–57; on hazardous locations, 51; on liquefied hydrogen, 73
U.S. Department of Transportation (USDOT): on explosives, 84–85; on explosives labels and placards, 85–86, *86*, *87*; on explosives transporting, 88–89; on first aid, 40–42; on hazardous materials, 36–42, *41*; on hydrogen containers, 70; on liquefied hydrogen, 73; on liquid containers, 58; on LP gas, 79; on oxygen, 77
USDOL. *See* U.S. Department of Labor (USDOL)
USDOT. *See* U.S. Department of Transportation (USDOT)
USFA. *See* United States Fire Administration (USFA)
U.S. Fire Administration, 226
utilities: in emergency action plans, 273, 281–82; explosives and, 90–91; life safety and, 144–45; in wildfires, 5

vapor density, 32
vapor pressure, 31
venting/ventilation: of aerated powder coating operations, 158; for boilers, 100; in building construction, 118; in building design, 328–29; of dip-tank operations, 164–65; for electrostatic spray operations, 152; engineering controls for, 226; of explosions, 25; for explosions, for liquefied hydrogen, 74; for hydrogen, 71; life safety and, 144–45; for liquefied hydrogen, 74; for liquid outside storage tanks, 65; for liquid underground tanks, 67; of spray booths, 153–54. *See also* heating, ventilation, and air-conditioning (HVAC)
virtual reality software, for building design, 331–32

walls: in building construction, 105–6; in building design, 324–26; fire loading in, 120; fire spread in, 115; interior finish of, 118, 140; plastics in, 34; of spray booths, 152; structural steel in, 33. *See also* fire walls
warning systems: for wildfires, 4, 5. *See also* alarm and detection systems
water: in emergency action plans, 273; as fire extinguisher agent, 191–93; specific heat of, 27. *See also* sprinkler systems
water flow alarms: sprinkler system inspections for, 217; for sprinkler systems, 211
welding and cutting: acetylene for, 75; dip-tank operations and, 159; dust explosions and, 96; hot work permit programs and, 257; hot work permits for, 256; as ignition source for flammable/combustible liquids, 69; liquefied hydrogen and, 74

wet-barrel fire hydrant, 217, *218*
wet-pipe sprinkler systems, 205, *207*; in emergency action plans, 274; heads for, 211–12; valves in, 210
wetting agents, in water for fire extinguishers, 192
wildfires, 3–6
windows: in building design, 319–20; in emergency action plans, 272; fire walls and, 117; at Triangle Shirtwaist fire, 130

wood: in building construction, 107–8, *108*; in building construction interior finish, 118; in Class A fire, 190; in Class II electrical hazardous locations, 50; as fire hazard, 32–33; as fuel, 21; heat of combustion of, 27; ignition temperature of, 21

yield stress, of steel, 107

zirconium, as fuel, 21

Made in the USA
Middletown, DE
28 August 2024

59893672R00235